Technical Application

Glencoe

Automotive *Excellence*

Volume 2

- **Engine Repair**
- **Heating & Air Conditioning**
- **Automatic Transmission & Transaxle**
- **Manual Drive Train & Axles**

Meets **NATEF** Standards

McGraw Hill **Glencoe**

New York, New York   Columbus, Ohio   Chicago, Illinois   Peoria, Illinois   Woodland Hills, California

Contributors

Ron Chappell
Santa Fe Community College
Gainesville, Florida

Erick Dodge
OCM BOCES
Syracuse, New York

Patrick Hart
New York Automotive and
  Diesel Institute
Jamaica, New York

Leo Van Delft
Tulsa Technology Center
Tulsa, Oklahoma

Ted Grekowicz
Michigan Technical Academy
  (Retired)
Romulus, Michigan

Terry Wicker
Franklin County High School
Carnesville, Georgia

Al Blethen
Shelton State Community
  College
Tuscaloosa, Alabama

Robert Porter
Center for Technical Studies of
  Montgomery County
Plymouth Meeting, Pennsylvania

Safety Notice

The reader is expressly advised to consider and use all safety precautions described in *Automotive Excellence Technical Applications Volume 2* or that might also be indicated by undertaking the activities described herein. In addition, common sense should be exercised to help avoid all potential hazards.

Publisher assumes no responsibility for the activities of the reader or for the subject matter experts who prepared *Automotive Excellence Technical Applications Volume 2*. Publisher makes no representation or warranties of any kind, including but not limited to, the warranties of fitness for particular purpose or merchantability, nor for any implied warranties related thereto, or otherwise. Publisher will not be liable for damages of any type, including any consequential, special or exemplary damages resulting, in whole or in part, from reader's use or reliance upon the information, instructions, warnings or other matter contained in *Automotive Excellence Technical Applications Volume 2*.

Brand Disclaimer

Publisher does not necessarily recommend or endorse any particular company or brand name product that may be discussed or pictured in *Automotive Excellence Technical Applications Volume 2*. Brand name products are used because they are readily available, likely to be known to the reader, and their use may aid in the understanding of the text. Publisher recognizes that other brand name or generic products may be substituted and work as well or better than those featured in *Automotive Excellence Technical Applications Volume 2*.

 **Glencoe**

The McGraw-Hill Companies

Send all inquiries to:
Glencoe/McGraw-Hill
3008 W. Willow Knolls Drive
Peoria, IL 61614

13-digit ISBN 978-0-07-874415-0
10-digit ISBN 0-07-874415-6

Printed in the United States of America
1 2 3 4 5 6 7 8 9 10   047   10 09 08 07 06

CONTENTS

AUTOMOTIVE TECHNICIAN'S HANDBOOK

Job Sheets

ENGINE REPAIR

Diagnostic Sheets

Job Sheets

CONTENTS

CONTENTS

CONTENTS

HEATING & AIR CONDITIONING

Diagnostic Sheets

Job Sheets

CONTENTS

CONTENTS

AUTOMATIC TRANSMISSION & TRANSAXLE

Diagnostic Sheets

Job Sheets

CONTENTS

CONTENTS

MANUAL DRIVE TRAIN & AXLES

Diagnostic Sheets

Job Sheets

CONTENTS

CONTENTS

READING VEHICLE IDENTIFICATION NUMBERS

NATEF Standard(s) for Engine Repair, Heating & Air Conditioning, Automatic Transmission & Transaxle, and Manual Drive Train & Axles:
A4 Locate and interpret vehicle and major component identification numbers (VIN, vehicle certification labels, and calibration decals).

PROCEDURES Refer to the vehicle service information for information on decoding vehicle identification numbers. Then decode the VIN on the vehicle provided by your instructor.

VIN (Vehicle Identification Number)

The VIN is a string of letters and digits. Each VIN is unique; only one vehicle has that particular combination of letters and numbers.

The location of the VIN on a vehicle varies. Usually, it is in the upper left corner of the dash panel on the driver's side. In this case, the VIN can be seen through the windshield from the outside of the vehicle. The VIN might also be on the driver's door jamb.

The VIN is the legal identifier of a vehicle. The VIN appears on vehicle titles and vehicle insurance cards. It is needed to complete a repair order. It is also needed to properly use a scan tool to read diagnostic trouble codes.

The format of the VIN has changed over the years. To learn what the letters and numbers mean, you must decode the VIN. Manufacturers use variations of the following common VIN format.

Number/Letter Position	Definition
1	Country of origin (country in which the vehicle was manufactured)
2	Manufacturer of the vehicle
3	Make (manufacturer's division)
4	Car line (vehicle type)
5	Series (specific vehicle model)
6	Body style
7	Type of restraint system
8	Engine type
9	Check digit (verifies authenticity of vehicle)
10	Model year
11	Plant location
12—17	Plant sequence number

(continued)

Samples of VIN characters and their descriptions are given below. You will need to refer to vehicle service information to decode the VIN on your vehicle.

Position	Definition	Character	Description
		Vehicle Identification Number (VIN) System	
1	Country of Origin	1	U.S.A.
2	Manufacturer	G	General Motors
3	Make	2	Pontiac
4	Car Line	W	Grand Prix
5	Series	J P	SE GT
6	Body Style	1 5	2 Door Coupe 4 Door Sedan
7	Restraint System	2	Active (Manual) Belts with Driver and Passenger Supplemental Inflatable Restraint
8	Engine Type	M K 1	6 Cylinder MFI High Output 3.1L (RPO Code L82) 6 Cylinder MFI High Output 3800 (RPO Code L36) 6 Cylinder MFI High Supercharged 3800 (RPO Code L67)
9	Check Digit	—	Check Digit
10	Model Year	W	1998
11	Plant Location	F	Fairfax II
12—17	Plant Sequence Number	—	Plant Sequence Number

_____ 1. Write down the VIN of your vehicle or an assigned vehicle.

VIN_____

_____ 2. To obtain information about the vehicle, you need to know what each letter and number in the VIN means. Refer to the vehicle service information for this information, and record it on the table on the next page.

Position	Character	Description
1		
2		
3		
4		
5		
6		
7		
8		
9		
10		
11		
12		
13		
14		
15		
16		
17		

Performance ✓ Checklist

Name _____ Date _____ Class _____

PERFORMANCE STANDARDS:
Level 4–Performs skill without supervision and adapts to problem situations.
Level 3–Performs skill satisfactorily without assistance or supervision.
Level 2–Performs skill satisfactorily, but requires assistance/supervision.
Level 1–Performs parts of skill satisfactorily, but requires considerable assistance/supervision.

Attempt (circle one): **1 2 3 4**

Comments:

_____

_____

_____

_____

PERFORMANCE LEVEL ACHIEVED: _____

_____ **1.** Safety rules and practices were followed at all times regarding this job.

_____ **2.** Tools and equipment were used properly and stored upon completion of this job.

_____ **3.** This completed job met the standards set and was done within the allotted time.

_____ **4.** No injury or damage to property occurred during this job.

_____ **5.** Upon completion of this job, the work area was cleaned correctly.

Instructor's Signature _____ Date_____

READING MAJOR COMPONENT IDENTIFICATION NUMBERS

NATEF Standard(s) for Engine Repair, Heating & Air Conditioning, Automatic Transmission & Transaxle, and Manual Drive Train & Axles:
A4 Locate and interpret vehicle and major component identification numbers (VIN, vehicle certification labels, and calibration decals).

PROCEDURES Refer to the vehicle service information for information on decoding major component identification numbers. Then read the vehicle certification label and interpret the body code plate (if present), and major component numbers on the vehicle provided by your instructor.

Vehicle Certification Label

The vehicle certification label is sometimes called the vehicle safety certification label. It is permanently located on the edge of the driver's door. It carries the following information:

- Name of manufacturer.
- Gross vehicle weight rating (GVWR).
- Gross axle weight rating, front and rear (GAWR).
- Certification statement.

- Vehicle class type.
- Vehicle identification number.
- Date of manufacture (month/year).

_____ 1. Refer to the vehicle service information for information on reading the vehicle certification label. Record this information on the lines below.

_____

_____

_____

_____

_____

Body Code Plate

Some manufacturers add a body code plate to their vehicles. This plate is located in the engine compartment. It carries letters and numbers that signify the following information:

- Paint procedure.
- Primary paint.
- Secondary paint.
- Interior trim code.
- Engine code.
- Vehicle order number.
- Vinyl roof code.

- Vehicle shell line.
- Car line.
- Price class.
- Body type.
- Transaxle codes.
- Market code.
- Vehicle identification number.

_____ 1. Refer to the vehicle service information for information on interpreting the body code plate. Record this information on the lines below.

_____

_____

_____

_____

_____

_____

Major Component Identification Numbers

_____ 1. Engines and transmissions are etched, stamped, or labeled with an identification number. Refer to the vehicle service information for information on locating and interpreting these numbers. Record this information on the lines below.

_____

_____

_____

_____

_____

_____

_____

Performance ✓ Checklist

Name _____ Date _____ Class _____

PERFORMANCE STANDARDS:
Level 4–Performs skill without supervision and adapts to problem situations.
Level 3–Performs skill satisfactorily without assistance or supervision.
Level 2–Performs skill satisfactorily, but requires assistance/supervision.
Level 1–Performs parts of skill satisfactorily, but requires considerable assistance/supervision.

Attempt (circle one): **1 2 3 4**

Comments:

_____

_____

_____

_____

PERFORMANCE LEVEL ACHIEVED: _____

_____ 1. Safety rules and practices were followed at all times regarding this job.
_____ 2. Tools and equipment were used properly and stored upon completion of this job.
_____ 3. This completed job met the standards set and was done within the allotted time.
_____ 4. No injury or damage to property occurred during this job.
_____ 5. Upon completion of this job, the work area was cleaned correctly.

Instructor's Signature _____ Date_____

COMPLETING A VEHICLE REPAIR ORDER

NATEF Standard(s) for Engine Repair, Heating & Air Conditioning, Automatic Transmission & Transaxle, and Manual Drive Train & Axles:
A1 Complete work order to include customer information, vehicle identifying information, customer concern, related service history, cause, and correction.

SAMPLE VEHICLE REPAIR ORDER   Vehicle Repair Order # _____

Date ____/____/____

Customer Name & Phone #: _____   Vehicle Make/Type: _____   VIN: _____   Mileage: _____

Service History: _____

Customer Concern: _____

Cause of Concern: _____

Suggested Repairs/Maintenance: _____

Services Performed: _____

	Parts			Labor	Time In:
Item	Description	Price	Diagnosis Time:		Time Complete:
1			Repair Time:		Total Hours:
2					
3			I hereby authorize the above repair work to be done using the necessary material, and hereby grant you and/or your employees permission to operate the vehicle herein described on streets, highways, or elsewhere for the purpose of testing and/or inspection. An express mechanic's lien is hereby acknowledged on above vehicle to secure the amount of repairs thereof.		
4					
5					
6			X _____		

PROCEDURES Refer to the vehicle service information for specifications and special procedures. Then prepare a vehicle repair order for the vehicle provided by your instructor.

_____ 1. **Write legibly.** Others will be reading what you have written.

_____ 2. **Make sure all information is accurate.** Inaccurate information will slow the repair process.

_____ 3. **Complete every part of the Vehicle Repair Order.** Every part must be completed.

_____ 4. **Number the Vehicle Repair Order.** This will help others track the repair.

_____ 5. **Date the Vehicle Repair Order.** This will help document the service history.

_____ 6. **Enter the Customer Name and Phone Number.** Make sure you have spelled the Customer Name correctly. Double-check the Phone Number.

_____ 7. **Enter the Vehicle Make/Type.** This information is essential.

_____ 8. **Enter the VIN (vehicle identification number).** This is a string of coded data that is unique to the vehicle. The location of the VIN depends on the manufacturer. It is usually found on the dashboard next to the windshield on the driver's side. The VIN is a rich source of information. It is needed to properly use a scan tool to read diagnostic trouble codes. Double-check the VIN to ensure accuracy.

_____ 9. **Enter the Mileage of the vehicle.** This information is part of the service history.

_____ 10. **Complete the Service History.** The service history is a history of all the service operations performed on a vehicle. The service history alerts the technician to previous problems with the vehicle. In the case of recurring problems, it helps the technician identify solutions that were ineffective.
   - A detailed service history is usually kept by the service facility where the vehicle is regularly serviced.
   - Information on service performed on the vehicle at other service centers is not available unless the customer makes it available. For this reason, ask the customer about service performed outside of the present service center.

_____ 11. **Identify the Customer Concern.** This should be a reasonably detailed and accurate description of the problem that the customer is having with the vehicle. The customer is usually the best source of information regarding the problem. This information can be used to perform the initial diagnosis.

_____ 12. **Identify the Cause of Concern.** This will identify the problem.

_____ 13. Ask the customer to read the text at the bottom of the Labor box. By signing on the line at the bottom of this box, the customer authorizes repair work on the vehicle according to the terms specified.

_____ 14. **Identify Suggested Repairs/Maintenance.** This will identify what needs to be done to correct the problem.

_____ 15. **Identify Services Performed.** This will identify the specific maintenance and repair procedures that were performed to correct the problem.

_____ 16. **Provide Parts information.** This includes a numbered list of items used to complete the repair. It includes a description of each part with the price of the part.

_____ 17. **Provide Labor information.** The Diagnosis Time and the Repair Time are totaled to give the Total Hours.

(_continued_)

Performance ✓ Checklist

Name _____ Date _____ Class _____

PERFORMANCE STANDARDS:
Level 4–Performs skill without supervision and adapts to problem situations.
Level 3–Performs skill satisfactorily without assistance or supervision.
Level 2–Performs skill satisfactorily, but requires assistance/supervision.
Level 1–Performs parts of skill satisfactorily, but requires considerable assistance/supervision.

Attempt (circle one): **1 2 3 4**

Comments:

_____

_____

_____

_____

PERFORMANCE LEVEL ACHIEVED: _____

_____ **1.** Safety rules and practices were followed at all times regarding this job.

_____ **2.** Tools and equipment were used properly and stored upon completion of this job.

_____ **3.** This completed job met the standards set and was done within the allotted time.

_____ **4.** No injury or damage to property occurred during this job.

_____ **5.** Upon completion of this job, the work area was cleaned correctly.

Instructor's Signature _____ Date _____

RESEARCHING VEHICLE
SERVICE INFORMATION

NATEF Standard(s) for Engine Repair, Heating & Air Conditioning, Automatic Transmission & Transaxle, and Manual Drive Train & Axles:

ER A3 Research applicable vehicle and service information, such as internal engine operation, vehicle service history, service precautions, and technical service bulletins.

HA A3 Research applicable vehicle and service information, such as heating and air conditioning system operation, vehicle service history, service precautions, and technical service bulletins.

AT A3 Research applicable vehicle and service information, such as transmission/transaxle system operation, fluid type, vehicle service history, service precautions, and technical service bulletins.

MD A3 Research applicable vehicle and service information, such as drive train system operation, fluid type, vehicle service history, service precautions, and technical service bulletins.

_____ 1. Check the available sources of vehicle service information. Vehicle service information consists of manufacturer service, maintenance, and repair information. This information is available in the following sources:

  • Vehicle service manuals, both print and online versions.

  • Dealer service bulletins.

  These references present the proper methods for performing diagnosis, service procedures, adjustments, maintenance, and repairs.

_____ 2. Check vehicle service manuals. These manuals are available as print copies and electronically. Increasingly, technicians are using online and electronic versions of vehicle service manuals. Note that the information in printed vehicle service manuals was current at the time of printing. Vehicle service manuals available online have the advantage of presenting more current information, with errors corrected.

_____ 3. Check dealer service bulletins. These bulletins contain service, maintenance, and repair information that became available after the publication of the vehicle service manual. These bulletins often cover special procedures.

(*continued*)

_____ 4. Vehicle service information is a useful reference. In referring to any source of vehicle service information, be aware of the following:

- Note that vehicle service manuals are divided into sections, each section dealing with a major automotive system. The section is usually divided into specifications, diagnosis, repair procedures, system operation, and special tools.
- Read all information carefully before you begin. Make sure you have identified the safety procedures, tools, and equipment needed to complete the procedure.
- Be attentive to the step-by-step format presented in vehicle maintenance and repair procedures.
- Note that illustrations often serve as the primary sources for repair procedures.
- Refer to exploded views to identify parts.
- Refer to exploded views to aid in order of assembly.
- Look for Hints, which provide additional information that will help you perform the repair more efficiently.

_____ 5. Read all Cautions and Warnings carefully to minimize the risk of personal injury.

_____ 6. Read all Notices carefully to minimize the risk of damage to vehicles, tools, and equipment.

_____ 7. Research the special tools needed for the procedure.

_____ 8. Use the special tools as recommended throughout the procedure. Failure to use special tools can result in an imperfect repair and, possibly, personal injury.

_____ 9. Research the vehicle service history. The vehicle owner is usually the best source for such information. If service on the vehicle has been performed at another facility, ask the customer to identify the repairs and maintenance procedures that were performed.

Performance ✓ Checklist

Name _____ Date _____ Class _____

PERFORMANCE STANDARDS:
Level 4–Performs skill without supervision and adapts to problem situations.
Level 3–Performs skill satisfactorily without assistance or supervision.
Level 2–Performs skill satisfactorily, but requires assistance/supervision.
Level 1–Performs parts of skill satisfactorily, but requires considerable assistance/supervision.

Attempt (circle one): **1 2 3 4**

Comments:

_____
_____
_____
_____

PERFORMANCE LEVEL ACHIEVED: _____

_____ 1. Safety rules and practices were followed at all times regarding this job.
_____ 2. Tools and equipment were used properly and stored upon completion of this job.
_____ 3. This completed job met the standards set and was done within the allotted time.
_____ 4. No injury or damage to property occurred during this job.
_____ 5. Upon completion of this job, the work area was cleaned correctly.

Instructor's Signature _____ Date_____

CONVERTING MEASUREMENTS

Safety First
- ☐ Wear safety glasses at all times.
- ☐ Follow all safety rules when using common hand tools.

Tools & Equipment:
- Scale
- Measuring cup
- Measuring rule
- Graduated measuring container
- Pyrometer

In your work as an automotive technician, you may need to convert measurements from the United States Customary (USC) system, or standard system, to the System of International Units (SI), or metric system. In automotive service and repair, the measurements most commonly converted are those relating to weight, length, liquid volume, temperature, and pressure.

The following problems will give you practice in converting measurements from the USC, or standard, system to the SI, or metric, system. The measuring device you use can be calibrated in either the USC, or standard, or the SI, or metric, measuring system. For each problem, record the measurement reading. Then convert the measurement you obtain to the other measurement system. For example, if you obtain the measurement in the USC, or standard, measurement system, convert the measurement to the SI, or metric system, and vice versa.

Locate a measurement conversion table to make your conversions.

Converting Measurements of Weight

_____ 1. Weigh a spark plug. The scale can be calibrated in the standard or metric measurement system.

Record the reading _____ Convert the reading _____

_____ 2. Weigh a thermostat. The scale can be calibrated in the standard or metric measurement system.

Record the reading _____ Convert the reading _____

_____ 3. Weigh a quart container of motor oil.

Record the reading _____ Convert the reading _____

Converting Measurements of Length

_____ 1. Measure a length of radiator hose.

Record the reading _____ Convert the reading _____

_____ 2. Measure a length of electrical wire.

Record the reading _____ Convert the reading _____

Converting Measurements of Liquid Volume

_____ 1. Pour a small amount of water into a measuring cup.

Record the reading _____ Convert the reading _____

(continued)

_____ 2. Using a container graduated for measurement, measure the amount of oil you add to an engine.

Record the reading _____   Convert the reading _____

_____ 3. Using a container graduated for measurement, measure the amount of coolant you add to a radiator.

Record the reading _____   Convert the reading _____

Converting Measurements of Temperature

_____ 1. Using a pyrometer, measure engine temperature.

Record the reading _____   Convert the reading _____

_____ 2. Inspect and test the thermostat. Measure the temperature at which the thermostat closes.

Record the reading _____   Convert the reading _____

_____ 3. Read the temperature on the room thermostat in your facility.

Record the reading _____   Convert the reading _____

Converting Measurements of Pressure

_____ 1. Using a tire gauge, take a reading of tire inflation pressure.

Record the reading _____   Convert the reading _____

_____ 2. Using a cooling system pressure tester, measure cooling system pressure.

Record the reading _____   Convert the reading _____

Performance ✓ Checklist

Name _____   Date _____   Class _____

PERFORMANCE STANDARDS:
Level 4–Performs skill without supervision and adapts to problem situations.
Level 3–Performs skill satisfactorily without assistance or supervision.
Level 2–Performs skill satisfactorily, but requires assistance/supervision.
Level 1–Performs parts of skill satisfactorily, but requires considerable assistance/supervision.

Attempt (circle one): **1 2 3 4**

Comments:

_____

_____

_____

PERFORMANCE LEVEL ACHIEVED: _____

_____ 1. Safety rules and practices were followed at all times regarding this job.

_____ 2. Tools and equipment were used properly and stored upon completion of this job.

_____ 3. This completed job met the standards set and was done within the allotted time.

_____ 4. No injury or damage to property occurred during this job.

_____ 5. Upon completion of this job, the work area was cleaned correctly.

Instructor's Signature _____   Date_____

PERFORMING FASTENER REPAIRS

NATEF Standard(s) for Engine Repair:

C3 Perform common fastener and thread repair to include: remove broken bolt, restore internal and external threads, and repair internal threads with thread insert.

Safety First
- ☐ Wear safety glasses at all times.
- ☐ Follow all safety rules when using common hand tools.

Tools & Equipment:

- Vehicle service information
- Appropriate tap and die set
- Die grinder
- Awl
- Variable speed reversible drill
- Stud remover
- Hammer
- Drill bit set (left-hand drill bits, if possible)
- Penetrating oil
- Punch-and-chisel set
- Screw extractor set

PROCEDURES Refer to the vehicle service information for specifications and special procedures. Then perform fastener repairs on the vehicle provided by your instructor.

A bolt that is broken below the surface can be drilled and removed with a screw extractor. The following is a general procedure:

_____ 1. Create a flat surface on the top of the broken bolt with a die grinder or a pin punch and hammer.

_____ 2. Use a center punch and hammer to make a dimple in the exact center of the broken bolt. This will locate the drill bit and prevent it from wandering from the center.

_____ 3. Determine the appropriate size screw extractor for the bolt and the proper size drill bit for that screw extractor. This information is usually provided in the extractor kit or marked on the side of the extractor.

_____ 4. Use a sharp drill bit and a slow speed (500 rpm or lower). Apply penetrating oil for lubrication and cooling and drill a hole through the broken bolt centerline. Be careful to drill a straight hole. If possible use a left-hand drill bit. This will put torque on the bolt in a left-hand direction. This may screw the broken bolt out.

_____ 5. Insert the screw extractor in the hole. Use the correct wrench or driver to turn it counterclockwise. It should wedge in the hole and begin to loosen the bolt. Some extractors must be driven into the hole with a hammer to lock them in place. Take care not to break the extractor. It is made of very hard steel and cannot be drilled out.

(continued)

_____ 6. If the extractor turns in the bolt and will not remove it, try drilling the bolt out. Use several drill bits to enlarge the hole in stages until the minor diameter of the bolt threads are reached. Then use a pick or awl to try to remove the remaining bolt threads. If this is successful, chase and reuse the original threads.

_____ 7. If the above is not successful, the next option is to drill the hole to accommodate a tap the size of the original bolt and tap new threads. Other options would be to drill the hole oversize and tap it for the next size bolt or install a thread insert.

The following is a general procedure for removing a bolt that is broken above the surface:

_____ 1. Apply penetrating oil.

_____ 2. Tap the bolt with a hammer to work oil into the threads.

_____ 3. Try to remove the bolt with locking pliers or a stud remover.

_____ 4. An alternate approach is to weld a flat washer to the bolt. Then weld a nut to the washer. A wire welder will allow better control of heat. Heat from the welding helps loosen the bolt. Then back out the bolt with a wrench.

_____ 5. If an extractor or a tap becomes broken in the fastener, take the part to a facility with a plasma cutter or an electrical discharge machine (EDM). This equipment will disintegrate the broken component until it crumbles. The threads and the hole in the original part should remain undamaged.

Performance ✓ Checklist

Name _____   Date _____   Class _____

PERFORMANCE STANDARDS:
Level 4–Performs skill without supervision and adapts to problem situations.
Level 3–Performs skill satisfactorily without assistance or supervision.
Level 2–Performs skill satisfactorily, but requires assistance/supervision.
Level 1–Performs parts of skill satisfactorily, but requires considerable assistance/supervision.

Attempt (circle one): **1 2 3 4**

Comments:

_____

_____

_____

_____

PERFORMANCE LEVEL ACHIEVED: _____

_____ 1. Safety rules and practices were followed at all times regarding this job.

_____ 2. Tools and equipment were used properly and stored upon completion of this job.

_____ 3. This completed job met the standards set and was done within the allotted time.

_____ 4. No injury or damage to property occurred during this job.

_____ 5. Upon completion of this job, the work area was cleaned correctly.

Instructor's Signature _____   Date _____

PERFORMING THREAD REPAIRS

NATEF Standard(s) for Engine Repair:

C3 Perform common fastener and thread repair to include: remove broken bolt, restore internal and external threads, and repair internal threads with thread insert.

Safety First
- ☐ Wear safety glasses at all times.
- ☐ Follow all safety rules when using common hand tools.

Tools & Equipment:
- Vehicle service information
- Variable speed reversible drill
- Air blow gun
- Hammer
- Tap-and-die set

- Proper size thread insert kit
- Penetrating oil
- Small chisel or punch
- Drill bit set
- Small magnet

PROCEDURES Refer to the vehicle service information for specifications and special procedures. Then perform thread repairs on the vehicle provided by your instructor.

_____ 1. External thread damage is usually a bolt, which is normally replaced with a new one. Minor external thread damage can be repaired with a thread chaser or a die.

_____ 2. Minor internal thread damage can be repaired with a tap.

_____ 3. When possible, major internal thread damage can be replaced by drilling the bore over size and threading it to the next bolt size with a tap.

_____ 4. Major internal thread damage can be repaired back to the original size with thread insert.

_____ 5. Thread insert kit will be needed that includes a proper drill bit, a special tap, an installation tool, and thread inserts that resemble a coil spring.

_____ 6. Use the drill bit to drill the damaged bore over size.

_____ 7. Tap the bore with the special thread insert tap and use compressed air to blow out all the metal chips.

_____ 8. Use the installation tool to screw in thread insert flush with the top of the bore.

_____ 9. Use a small chisel or punch to break off the installation tang at the bottom of the insert. This tang can be retrieved with a small magnet.

_____ 10. The internally threaded hole is now back to the original condition.

(continued)

Performance ✓ Checklist

Name _____ Date _____ Class _____

PERFORMANCE STANDARDS:
Level 4–Performs skill without supervision and adapts to problem situations.
Level 3–Performs skill satisfactorily without assistance or supervision.
Level 2–Performs skill satisfactorily, but requires assistance/supervision.
Level 1–Performs parts of skill satisfactorily, but requires considerable assistance/supervision.

Attempt (circle one): **1 2 3 4**

Comments:

_____

_____

_____

_____

PERFORMANCE LEVEL ACHIEVED: _____

_____ **1.** Safety rules and practices were followed at all times regarding this job.

_____ **2.** Tools and equipment were used properly and stored upon completion of this job.

_____ **3.** This completed job met the standards set and was done within the allotted time.

_____ **4.** No injury or damage to property occurred during this job.

_____ **5.** Upon completion of this job, the work area was cleaned correctly.

Instructor's Signature _____ Date_____

FOLLOWING GENERAL SAFETY GUIDELINES

Tools & Equipment:
- Ladder
- Safety stand
- Personal protective equipment
- Safety-blow nozzle
- Hand tool, electric power tool, and pneumatic tool

PROCEDURES Observe the safety demonstrations provided by your instructor or a safety professional. Read the safety materials provided by your instructor. Then complete the following tasks, as assigned by your instructor.

_____ 1. Explain how to contain and clean up spills and leaks on the floor.

_____ 2. Demonstrate the covering or guarding of all floor openings.

_____ 3. Identify all safety zones and aisles.

_____ 4. Demonstrate inspection of a ladder for safety.

_____ 5. Demonstrate the correct use of a stool or ladder.

_____ 6. Demonstrate connecting a vehicle to the exhaust system.

_____ 7. Locate the exits to be used in case of emergency.

_____ 8. Identify personal protective equipment and typical situations in which it should be used.

_____ 9. Demonstrate the proper use of a hand tool.

_____ 10. Demonstrate the proper use of an electric power tool.

_____ 11. Demonstrate the proper use of a pneumatic tool.

_____ 12. Identify and demonstrate the use of all machine guards and guards on pulleys and drive belts.

_____ 13. Describe the use of a safety stand.

_____ 14. Demonstrate lockout/tagout.

_____ 15. Identify the different types of electrical power available in the facility and the outlets for each.

_____ 16. Explain the use of a ground fault circuit interrupter.

_____ 17. Explain the purpose of a grounding prong.

_____ 18. Demonstrate the use of a safety-blow nozzle when cleaning with compressed air.

_____ 19. Describe the proper use of lifts.

_____ 20. Identify the safety hazards related to welding.

(continued)

_____ **21.** Demonstrate how to lift a heavy object in order to avoid back injury.

_____ **22.** Identify equipment available in the facility that can be used to lift dead weight.

_____ **23.** Explain the use of a safety jack.

_____ **24.** Explain how a machine or tool with a rotating or spinning motion could cause injury.

_____ **25.** Explain how to prevent arm and hand injuries from twisting or vibrating tools.

Performance ✓ Checklist

Name _____ Date _____ Class _____

PERFORMANCE STANDARDS:
Level 4–Performs skill without supervision and adapts to problem situations.
Level 3–Performs skill satisfactorily without assistance or supervision.
Level 2–Performs skill satisfactorily, but requires assistance/supervision.
Level 1–Performs parts of skill satisfactorily, but requires considerable assistance/supervision.

Attempt (circle one): **1 2 3 4**

Comments:

_____

_____

_____

_____

PERFORMANCE LEVEL ACHIEVED: _____

_____ **1.** Safety rules and practices were followed at all times regarding this job.

_____ **2.** Tools and equipment were used properly and stored upon completion of this job.

_____ **3.** This completed job met the standards set and was done within the allotted time.

_____ **4.** No injury or damage to property occurred during this job.

_____ **5.** Upon completion of this job, the work area was cleaned correctly.

Instructor's Signature _____ Date_____

DEVELOPING SAFE WORK HABITS

Unsafe acts are the main cause of accidents and injuries on the job. Get into the habit of working safely. You'll benefit yourself and your coworkers. It is important to keep your work area neat and clean. It is also important to have respect for tools and equipment. To help develop safe work habits, follow these rules.

_____ 1. Work quietly and give your full attention to the task at hand.

_____ 2. Never indulge in horseplay or other foolish actions. Someone could get seriously hurt.

_____ 3. Never point a compressed-air gun at another person. Air from a gun can enter the skin and possibly a blood vessel.

_____ 4. Keep your work area neat.

_____ 5. To ensure a good grip, wipe oil and grease off your hands and tools.

_____ 6. To prevent slips and falls, promptly clean up oil, grease, or other liquid spilled on the floor.

_____ 7. Never use compressed air to blow dirt from clothing.

_____ 8. Always wash your hands before eating to avoid ingesting hazardous substances.

_____ 9. Don't wear dangling sleeves or ties that can get caught in machinery and cause serious injury.

_____ 10. Do not wear sandals or open-toed shoes on the job. Wear steel-toed safety shoes or full leather shoes with non-skid rubber heels and soles.

_____ 11. To keep long hair out of machinery, tie it back or wear a cap.

_____ 12. Do not wear rings, bracelets, or watches when working around moving machinery or electrical equipment. Jewelry can catch in moving parts. A ring or bracelet can cause a short circuit. The metal in a ring or bracelet can become white-hot in an instant, causing severe burns.

_____ 13. Always use appropriate personal protective equipment (PPE) for your eyes, head, skin, feet, and ears.

_____ 14. Wear seat belts when driving your own or a customer's vehicle.

_____ 15. Always use the safety equipment required by the manufacturer on tire-changing machines.

_____ 16. Keep your hands away from the engine fan and accessory drive belts when an engine is running. Hands can get caught in the fan or between a belt and pulley. You could be badly cut or even lose fingers.

_____ 17. Do not stand directly in line with the engine fan when it is turning or the engine is running. A spinning fan might throw a blade. A flying fan blade may injure or kill anyone it strikes.

_____ 18. Be sure to position a floor jack properly. Always put safety stands in place before going under a vehicle.

_____ 19. Watch out for sparks flying from a grinding wheel or a welding job. Sparks can set hair or clothes on fire.

_____ 20. Keep jack handles out of aisles.

_____ 21. Stand the creeper against the wall or out of the way when not in use.

(continued)

_____ 22. When lifting an object, lift with your legs and not your back. Keep your back straight. If the object is too heavy, ask for help. Back injuries are painful, costly, and can cause permanent disability.

_____ 23. Always use the right tool for the job. The wrong tool could damage the part you are working on or could injure you.

_____ 24. Never put screwdrivers or other sharp objects in your pocket. You could cut or stab yourself.

_____ 25. Learn how to read the MSDS and labels for hazardous materials. Know how to read an MSDS for emergency information.

_____ 26. Wear appropriate PPE when working with hazardous materials.

_____ 27. Observe "No Smoking" areas. Never smoke near compressed gas cylinders, paint operations, flammable storage rooms, gasoline, or fuel stations.

_____ 28. Store flammable liquids, such as gasoline, solvents, and thinners, in approved safety cans.

_____ 29. Never spray water on a flammable liquid or gasoline fire. Because the burning liquid floats, water will only spread the fire.

_____ 30. Always wear eye protection when particles or liquid spray is flying about.

_____ 31. Always wear eye protection when using a grinding wheel and when welding.

_____ 32. Wear goggles to protect eyes when using solvents or other chemicals. If chemicals enter eyes, flush them with water at once. Review the MSDS to determine the correct first aid.

_____ 33. Avoid inhaling asbestos fibers. Use an approved respirator and a HEPA vacuum device or wet-wash method.

Performance ✓ Checklist

Name _____ Date _____ Class _____

PERFORMANCE STANDARDS:
Level 4–Performs skill without supervision and adapts to problem situations.
Level 3–Performs skill satisfactorily without assistance or supervision.
Level 2–Performs skill satisfactorily, but requires assistance/supervision.
Level 1–Performs parts of skill satisfactorily, but requires considerable assistance/supervision.

Attempt (circle one): **1 2 3 4**

Comments:

_____

_____

_____

_____

PERFORMANCE LEVEL ACHIEVED: _____

_____ 1. Safety rules and practices were followed at all times regarding this job.

_____ 2. Tools and equipment were used properly and stored upon completion of this job.

_____ 3. This completed job met the standards set and was done within the allotted time.

_____ 4. No injury or damage to property occurred during this job.

_____ 5. Upon completion of this job, the work area was cleaned correctly.

Instructor's Signature _____ Date_____

FOLLOWING FIRE SAFETY GUIDELINES

Tools & Equipment:
- Fire extinguishers
- Disposal containers for flammable or explosive materials
- Storage containers/cabinets for flammable or explosive materials

PROCEDURES Observe the safety demonstrations provided by your instructor or a safety professional. Read the safety materials provided by your instructor. Then complete the following tasks, as assigned by your instructor.

_____ 1. Describe the procedure for responding to a fire.

_____ 2. Demonstrate the stop, drop, and roll technique to follow if hair or clothing catches fire.

_____ 3. Locate fire extinguishers.

_____ 4. Define the four classes of fires and identify which fire extinguisher to use for each class of fire.

_____ 5. Demonstrate how to operate a Class A fire extinguisher, if provided by your instructor.

_____ 6. Demonstrate how to operate a Class B fire extinguisher, if provided by your instructor.

_____ 7. Demonstrate how to operate a Class C fire extinguisher, if provided by your instructor.

_____ 8. Demonstrate how to operate a Class D fire extinguisher, if provided by your instructor.

_____ 9. Locate the exits to be used in case of emergency.

_____ 10. Identify the location where individuals are to meet after exiting the building during an emergency.

_____ 11. Identify flammable and explosive materials.

_____ 12. Read and interpret labels on containers of flammable and explosive materials.

_____ 13. Explain precautions required when using flammable and explosive materials.

_____ 14. Locate storage containers/cabinets for flammable and explosive materials.

_____ 15. Locate disposal containers for flammable and explosive materials.

(continued)

Performance ✓ Checklist

Name _____ Date _____ Class _____

PERFORMANCE STANDARDS:
Level 4–Performs skill without supervision and adapts to problem situations.
Level 3–Performs skill satisfactorily without assistance or supervision.
Level 2–Performs skill satisfactorily, but requires assistance/supervision.
Level 1–Performs parts of skill satisfactorily, but requires considerable assistance/supervision.

Attempt (circle one): **1 2 3 4**

Comments:

_____

_____

_____

_____

PERFORMANCE LEVEL ACHIEVED: _____

_____ **1.** Safety rules and practices were followed at all times regarding this job.

_____ **2.** Tools and equipment were used properly and stored upon completion of this job.

_____ **3.** This completed job met the standards set and was done within the allotted time.

_____ **4.** No injury or damage to property occurred during this job.

_____ **5.** Upon completion of this job, the work area was cleaned correctly.

Instructor's Signature _____ Date_____

PRACTICING SAFETY WITH
HAZARDOUS MATERIALS

Tools & Equipment:
- Storage and dispensing containers for hazardous materials
- Battery and charging station
- Drain pans for oil and antifreeze
- MSDS sheets

- Tanks of compressed gases
- Disposal containers for hazardous materials

PROCEDURES Observe the safety demonstrations provided by your instructor or a safety professional. Read the safety materials provided by your instructor. Then complete the following tasks, as assigned by your instructor.

_____ 1. Identify all hazardous materials used in the facility.

_____ 2. Read and interpret labels on containers of hazardous materials.

_____ 3. Explain precautions required when using hazardous materials.

_____ 4. Locate any special storage containers/cabinets for hazardous materials.

_____ 5. Demonstrate dispensing a chemical from a special dispensing container.

_____ 6. Explain storage procedures for compressed gases.

_____ 7. Demonstrate safe battery charging.

_____ 8. Demonstrate the safe procedure for cleaning parts with solvents, if provided by your instructor.

_____ 9. Explain the procedures for handling hot metal wastes.

_____ 10. Demonstrate the procedures for handling scrap metal, filings, and chips.

_____ 11. Locate disposal containers for scrap metal, filings, and chips.

_____ 12. Locate disposal containers for liquid wastes.

_____ 13. Locate disposal containers for absorbent compounds.

_____ 14. Identify the different drain pans used for oil and antifreeze.

_____ 15. Locate the material safety data sheets (MSDS) for all hazardous materials.

_____ 16. Identify the categories of information on an MSDS and where each category is located.

(continued)

Performance ✓ Checklist

Name _____ Date _____ Class _____

PERFORMANCE STANDARDS:
Level 4–Performs skill without supervision and adapts to problem situations.
Level 3–Performs skill satisfactorily without assistance or supervision.
Level 2–Performs skill satisfactorily, but requires assistance/supervision.
Level 1–Performs parts of skill satisfactorily, but requires considerable assistance/supervision.

Attempt (circle one): **1 2 3 4**

Comments:

_____

_____

_____

_____

PERFORMANCE LEVEL ACHIEVED: _____

_____ **1.** Safety rules and practices were followed at all times regarding this job.

_____ **2.** Tools and equipment were used properly and stored upon completion of this job.

_____ **3.** This completed job met the standards set and was done within the allotted time.

_____ **4.** No injury or damage to property occurred during this job.

_____ **5.** Upon completion of this job, the work area was cleaned correctly.

Instructor's Signature _____ Date_____

READING EPA REGULATIONS

The Environmental Protection Agency (EPA) is an agency of the federal government that regulates the tracking, handling, and disposal of hazardous materials (HAZMAT). The EPA regulates several areas that affect the automotive repair industry. These areas include lead-acid automotive batteries, crude oil and gas waste, and solvents. Individual states regulate the disposal of automotive tires.

The primary source of information regarding current EPA regulations is the EPA Web site. This Web site supplies information about regulatory issues. It also contains links to supporting documents, including Reports to Congress. It also provides links to technical background documents.

Use the information on the EPA Web site to research current regulations on the disposal of automotive batteries, used motor oil, used oil filters, and solvents. Using the Web, research information on the regulations in your state regarding the disposal of scrap automotive tires.

_____ 1. **Automotive Batteries.** Automotive batteries contain lead, which is a heavy metal that can contaminate the environment when batteries are improperly disposed of. A typical lead-acid battery contains 60 to 80 percent recycled lead and plastic. Nearly 90 percent of all lead-acid batteries are recycled. Using the EPA Web site, identify the recycling procedures for lead-acid automotive batteries.

_____

_____

_____

_____

_____

_____

_____ 2. **Used Motor Oil.** Improper disposal of used motor oil results in environmental contamination. Using the EPA Web site, research approved methods for disposing of used motor oil.

_____

_____

_____

_____

_____ 3. **Used Oil Filters.** Improper disposal of used oil filters results in environmental contamination. Using EPA Web site, research approved methods for disposing of used oil filters.

_____

_____

_____

_____

(continued)

_____ 4. **Solvents.** In the automotive repair industry, solvents are used to clean automotive parts. Using the Web, research the state and federal regulations regarding the storage and use of the solvents in common use in your facility.

_____

_____

_____

_____

_____ 5. **Automotive Tires.** The hauling, processing, and storage of scrap tires is regulated. Markets now exist for about 80 percent of scrap tires. Each state makes its own scrap tire laws and regulations. These laws typically set the stage for rules for scrap tire storage, collection, processing, and use. States also establish programs to clean up old scrap tire stockpiles. They also establish the funding needed to accomplish that goal. In recent years, scrap tire legislation has been a priority in many states. Using the Web, research the regulations in your state regarding the disposal of scrap automotive tires.

_____

_____

_____

_____

_____

_____

Performance ✓ Checklist

Name _____ Date _____ Class _____

PERFORMANCE STANDARDS:
Level 4–Performs skill without supervision and adapts to problem situations.
Level 3–Performs skill satisfactorily without assistance or supervision.
Level 2–Performs skill satisfactorily, but requires assistance/supervision.
Level 1–Performs parts of skill satisfactorily, but requires considerable assistance/supervision.

Attempt (circle one): **1 2 3 4**

Comments:

_____

_____

_____

_____

PERFORMANCE LEVEL ACHIEVED: _____

_____ 1. Safety rules and practices were followed at all times regarding this job.

_____ 2. Tools and equipment were used properly and stored upon completion of this job.

_____ 3. This completed job met the standards set and was done within the allotted time.

_____ 4. No injury or damage to property occurred during this job.

_____ 5. Upon completion of this job, the work area was cleaned correctly.

Instructor's Signature _____ Date _____

READING SAFETY COLOR CODES

> **PROCEDURES** Safety color codes are in common use in the workplace. Read this summary of ways in which different colors are used to indicate cautions, warnings, and other information. On the lines following each summary, list those places in your workplace where the color is used in the ways described.

_____ 1. Red is used to indicate Danger. Red is commonly used to identify the following:
- Fire equipment.
- Exit signs.
- Panic buttons on machinery.
- Containers for flammable materials.
- Signs for hazardous materials.

List those areas in your workplace where **red** is used to indicate danger.

_____   _____

_____   _____

_____ 2. Orange is used to indicate Warning of machine hazards. It is often used to outline and draw attention to the following:
- Machine guards.
- Machine parts that can cut or crush.
- Pulleys, belts, and gears.
- Electrical boxes that have start and stop buttons, levers, or toggles.

List those areas in your workplace where **orange** is used to warn against machine hazards.

_____   _____

_____   _____

_____ 3. Yellow is used to indicate Caution. It is used to draw attention to the following:
- Critical machines parts such as wheels, levers, and controls.
- Used with black for Caution signs.
- Used with black diagonal stripes to indicate tripping or falling hazards, such as stairs.

List those areas in your workplace where **yellow** is used to signal the need for caution.

_____   _____

_____   _____

_____ 4. Blue is used to provide Information. It is commonly used on signs and bulletin boards.

List those areas in your workplace where **blue** is used to provide information.

_____   _____

_____   _____

(continued)

_____ 5. Green is used to indicate Safety. It is used along with white to show the location of:
  • First aid stations.
  • First aid kits.
  • Safety equipment.

List those areas in your workplace where **green** is used to identify the location of first aid stations and kits and safety equipment.

_____   _____

_____   _____

_____ 6. White is commonly used to indicate Boundaries. It is used alone or with black to direct traffic flow. Stripes or arrows show direction and identify barricades.

List those areas in your workplace where **white** is used alone or with black to show direction or indicate barricades.

_____   _____

_____   _____

_____ 7. Magenta or purple on yellow is used to indicate Radiation Caution. It is used in areas where radiation is present.

List those areas in your workplace where **magenta** or **purple on yellow** is used to indicate the presence of radiation.

_____   _____

_____   _____

Performance ✓ Checklist

Name _____ Date _____ Class _____

PERFORMANCE STANDARDS:
Level 4–Performs skill without supervision and adapts to problem situations.
Level 3–Performs skill satisfactorily without assistance or supervision.
Level 2–Performs skill satisfactorily, but requires assistance/supervision.
Level 1–Performs parts of skill satisfactorily, but requires considerable assistance/supervision.

Attempt (circle one): **1 2 3 4**

Comments:

_____

_____

_____

_____

PERFORMANCE LEVEL ACHIEVED: _____

_____ 1. Safety rules and practices were followed at all times regarding this job.
_____ 2. Tools and equipment were used properly and stored upon completion of this job.
_____ 3. This completed job met the standards set and was done within the allotted time.
_____ 4. No injury or damage to property occurred during this job.
_____ 5. Upon completion of this job, the work area was cleaned correctly.

Instructor's Signature _____ Date_____

DIAGNOSING THE DRIVE BELT, TENSIONER, AND PULLEY

NATEF Standard(s) for Engine Repair:
D4 Inspect, replace, and adjust drive belts, tensioners, and pulleys; check pulley and belt alignment.

DIRECTIONS: Fill in the blanks below by identifying (1) Safety First Practices that must be followed, (2) Tools and Equipment required, (3) three possible causes of drive belt, tensioner, and pulley problems, and (4) corrective actions that should be taken.

Safety First

- _____
- _____
- _____
- _____

Tools and Equipment Required:

- _____   • _____
- _____   • _____
- _____   • _____

QUICK ✔ Diagnostic for Cause(s)

List at least three common causes of drive belt, tensioner, and pulley problems:

1. _____
2. _____
3. _____
4. _____
5. _____

QUICK ✔ Diagnostic for Corrective Action(s)

List possible corrective action(s):

1. _____
2. _____
3. _____
4. _____
5. _____

Engine Repair

DIAGNOSING THE WATER PUMP

NATEF Standard(s) for Engine Repair:

D8 Inspect, test, remove, and replace water pump.

DIRECTIONS: Fill in the blanks below by identifying (1) Safety First Practices that must be followed, (2) Tools and Equipment required, (3) three possible causes of water pump problems, and (4) corrective actions that should be taken.

Safety First

- _____
- _____
- _____
- _____

Tools and Equipment Required:

- _____ • _____
- _____ • _____
- _____ • _____

QUICK ✔ Diagnostic for Cause(s)

List at least three common causes of water pump problems:

1. _____
2. _____
3. _____
4. _____
5. _____

QUICK ✔ Diagnostic for Corrective Action(s)

List possible corrective action(s):

1. _____
2. _____
3. _____
4. _____
5. _____

Name _____ Date _____ Class _____

DIAGNOSING OIL CONSUMPTION AND
ABNORMAL EXHAUST CONDITIONS

NATEF Standard(s) for Engine Repair:
A7 Diagnose the cause of excessive oil consumption, unusual engine exhaust color, odor, and sound; determine necessary action.

DIRECTIONS: Fill in the blanks below by identifying (1) Safety First Practices that must be followed, (2) Tools and Equipment required, (3) three possible causes of oil consumption and exhaust problems, and (4) corrective actions that should be taken.

Safety First

- _____
- _____
- _____
- _____

Tools and Equipment Required:

- _____   • _____
- _____   • _____
- _____   • _____

QUICK ✔ Diagnostic for Cause(s)

List at least three common causes of oil consumption and exhaust problems:

1. _____
2. _____
3. _____
4. _____

QUICK ✔ Diagnostic for Corrective Action(s)

List possible corrective action(s):

1. _____
2. _____
3. _____
4. _____

Engine Repair

Name _____   Date _____   Class _____

DIAGNOSING ENGINE NOISES AND VIBRATIONS

NATEF Standard(s) for Engine Repair:
A6 Diagnose engine noises and vibrations; determine necessary action.

DIRECTIONS: Fill in the blanks below by identifying (1) Safety First Practices that must be followed, (2) Tools and Equipment required, (3) three possible causes of engine noise and vibration problems, and (4) corrective actions that should be taken.

Safety First

- _____
- _____
- _____
- _____

Tools and Equipment Required:

- _____   • _____
- _____   • _____
- _____   • _____

QUICK ✔ Diagnostic for Cause(s)

List at least three common causes of engine noise and vibration problems:

1. _____
2. _____
3. _____
4. _____
5. _____

QUICK ✔ Diagnostic for Corrective Action(s)

List possible corrective action(s):

1. _____
2. _____
3. _____
4. _____
5. _____

COMPLETING A VEHICLE REPAIR ORDER FOR AN ENGINE REPAIR CONCERN

NATEF Standard(s) for Engine Repair:

A1 Complete work order to include customer information, vehicle identifying information, customer concern, related service history, cause, and correction.

SAMPLE VEHICLE REPAIR ORDER   Vehicle Repair Order # _____

Date ____/____/____

Customer Name & Phone #: _____   Vehicle Make/Type: _____   VIN: _____   Mileage: _____

Service History: _____

Customer Concern: _____

Cause of Concern: _____

Suggested Repairs/Maintenance: _____

Services Performed: _____

	Parts		
Item	Description		Price
1			
2			
3			
4			
5			
6			

Labor	Time In:
Diagnosis Time:	Time Complete:
Repair Time:	Total Hours:

I hereby authorize the above repair work to be done using the necessary material, and hereby grant you and/or your employees permission to operate the vehicle herein described on streets, highways, or elsewhere for the purpose of testing and/or inspection. An express mechanic's lien is hereby acknowledged on above vehicle to secure the amount of repairs thereof.

X _____

PROCEDURES Refer to the vehicle service information for specifications and special procedures. Then prepare a vehicle repair order for the vehicle provided by your instructor.

_____ 1. **Write legibly.** Others will be reading what you have written.

_____ 2. **Make sure all information is accurate.** Inaccurate information will slow the repair process.

_____ 3. **Complete every part of the Vehicle Repair Order.** Every part must be completed.

_____ 4. **Number the Vehicle Repair Order.** This will help others track the repair.

_____ 5. **Date the Vehicle Repair Order.** This will help document the service history.

_____ 6. **Enter the Customer Name and Phone Number.** Make sure you have spelled the Customer Name correctly. Double-check the Phone Number.

_____ 7. **Enter the Vehicle Make/Type.** This information is essential.

(continued)

_____ 8. **Enter the VIN (vehicle identification number).** This is a string of coded data that is unique to the vehicle. The location of the VIN depends on the manufacturer. It is usually found on the dashboard next to the windshield on the driver's side. The VIN is a rich source of information. It is needed to properly use a scan tool to read diagnostic trouble codes. Double-check the VIN to ensure accuracy.

_____ 9. **Enter the Mileage of the vehicle.** This information is part of the service history.

_____ 10. **Complete the Service History.** The service history is a history of all the service operations performed on a vehicle. The service history alerts the technician to previous problems with the vehicle. In the case of recurring problems, it helps the technician identify solutions that were ineffective.

  • A detailed service history is usually kept by the service facility where the vehicle is regularly serviced.

  • Information on service performed on the vehicle at other service centers is not available unless the customer makes it available. For this reason, ask the customer about service performed outside of the present service center.

_____ 11. **Identify the Customer Concern.** This should be a reasonably detailed and accurate description of the problem that the customer is having with the vehicle. The customer is usually the best source of information regarding the problem. This information can be used to perform the initial diagnosis. The customer has a passenger car. She says, "The car emits a large cloud of blue exhaust when it is first started." Enter her concern on the Customer Concern line.

_____ 12. **Identify the Cause of Concern.** This will identify the problem. In this case, there may be several possible causes. Enter the possible causes on the Cause of Concern line.

_____ 13. **Ask the customer to read the text at the bottom of the Labor box.** By signing on the line at the bottom of this box, the customer authorizes repair work on the vehicle according to the terms specified.

_____ 14. **Identify Suggested Repairs/Maintenance.** This will identify what needs to be done to correct the problem.

_____ 15. **Identify Services Performed.** This will identify the specific maintenance and repair procedures that were performed to correct the problem.

_____ 16. **Provide Parts information.** This includes a numbered list of items used to complete the repair. It includes a description of each part with the price of the part.

_____ 17. **Provide Labor information.** The Diagnosis Time and the Repair Time are totaled to give the Total Hours.

Performance ✓ Checklist

Name _____ Date _____ Class _____

PERFORMANCE STANDARDS:
Level 4–Performs skill without supervision and adapts to problem situations.
Level 3–Performs skill satisfactorily without assistance or supervision.
Level 2–Performs skill satisfactorily, but requires assistance/supervision.
Level 1–Performs parts of skill satisfactorily, but requires considerable assistance/supervision.

Attempt (circle one): **1 2 3 4**

Comments:

_____

_____

_____

_____

PERFORMANCE LEVEL ACHIEVED: _____

_____ **1.** Safety rules and practices were followed at all times regarding this job.
_____ **2.** Tools and equipment were used properly and stored upon completion of this job.
_____ **3.** This completed job met the standards set and was done within the allotted time.
_____ **4.** No injury or damage to property occurred during this job.
_____ **5.** Upon completion of this job, the work area was cleaned correctly.

Instructor's Signature _____ Date_____

Engine Repair

INSPECTING & REPLACING ENGINE COOLANT AND HEATER HOSES

NATEF Standard(s) for Engine Repair:
D5 Inspect and replace engine cooling and heater system hoses.

Safety First

☐ Wear safety glasses at all times.
☐ Follow all safety rules when using common hand tools.
☐ Always connect a vehicle's exhaust to a vent hose before you run an engine in a closed shop. Unvented exhaust fumes in a closed shop can cause death.
☐ Check Material Safety Data Sheets for chemical safety information.
☐ Disconnect electric fans from power source before working in the vicinity of the fan blades.
☐ Keep jewelry, loose clothing, and hair away from moving parts while an engine is running.
☐ Be careful of hot engine parts.
☐ Set the parking brake and place automatic transmission in PARK or manual transmission in NEUTRAL.

Tools & Equipment:

- Coolant drain pan
- Appropriate hose clamps
- Gasket cement
- Common hand tools
- Knife
- Vehicle service information

PROCEDURES Refer to the vehicle service information for specifications and special procedures. Then inspect and replace, as needed, any cooling or heating hoses in the engine provided by your instructor.

_____ 1. Write up a repair order.

_____ 2. Make sure you follow all procedures in the vehicle service information.

_____ 3. Inspect all the hoses for breaks, bulges, cuts, deteriorated hose material, and other damage.

_____ 4. Inspect all the hose clamps for looseness and breaks, then tighten or replace defective clamps.

_____ 5. Check the hoses for contact with other vehicle components by running your hand around them to feel for other components against them. *Note:* Hoses that rub on other components or are in contact with a hot engine part will wear faster by rubbing or burning.

_____ 6. Squeeze and bend the hoses to check for hoses that are hard, brittle, soft, spongy, or swollen.

_____ 7. Look up the manufacturer's recommended procedures for replacing defective or damaged hoses.

Source: _____ Page: _____

(continued)

Engine Repair

_____ 8. Replace any defective or damaged hoses according to the manufacturer's recommendations.

_____ 9. Be sure that the engine is cool and then remove the radiator cap. *Note:* Removing a radiator cap from a hot engine can be very dangerous. Also, be certain to disconnect the power to the coolant fan (when working in the vicinity of the fan blades) because it can come on at any time. Be sure to reconnect connections when the job is complete.

_____ 10. Drain the coolant into the coolant drain pan. *Note:* If the coolant is clean, save it to reuse. If the coolant needs to be discarded, be certain to dispose of it according to EPA regulations.

_____ 11. Loosen the clamps from any hose that needs to be replaced and remove the hose ends by twisting and then gently pulling them. *Note:* Have your instructor show you how to remove a stuck hose. It is sometimes necessary to carefully split the end of a hose with a razor knife.

_____ 12. Clean the hose fittings with an appropriate solvent or wire brush.

_____ 13. Install the new hoses, put the hose clamps back on, and tighten them. Position the clamps so they will be easy to remove in the future.

_____ 14. Fill the system with clean coolant. Leave the coolant level several inches below the radiator filler neck. Run the engine until it reaches normal operating temperature, and add coolant as the level drops. When coolant begins to circulate, indicating that the thermostat has opened, top off the radiator and replace the cap. Adjust the level of coolant in the recovery tank to the proper level and inspect the entire cooling system for leaks.

Performance ✓ Checklist

Name _____ Date _____ Class _____

PERFORMANCE STANDARDS:
Level 4–Performs skill without supervision and adapts to problem situations.
Level 3–Performs skill satisfactorily without assistance or supervision.
Level 2–Performs skill satisfactorily, but requires assistance/supervision.
Level 1–Performs parts of skill satisfactorily, but requires considerable assistance/supervision.

Attempt (circle one): **1 2 3 4**

Comments:

_____

_____

_____

_____

PERFORMANCE LEVEL ACHIEVED: _____

_____ 1. Safety rules and practices were followed at all times regarding this job.
_____ 2. Tools and equipment were used properly and stored upon completion of this job.
_____ 3. This completed job met the standards set and was done within the allotted time.
_____ 4. No injury or damage to property occurred during this job.
_____ 5. Upon completion of this job, the work area was cleaned correctly.

Instructor's Signature _____ Date_____

INSPECTING & REPLACING BELTS

NATEF Standard(s) for Engine Repair:
D4 Inspect, replace, and adjust drive belts, tensioners, and pulleys; check pulley and belt alignment.

Safety First

- ☐ Wear safety glasses at all times.
- ☐ Follow all safety rules when using common hand tools.
- ☐ Always connect a vehicle's exhaust to a vent hose before you run an engine in a closed shop. Unvented exhaust fumes in a closed shop can cause death.
- ☐ Check Material Safety Data Sheets for chemical safety information.
- ☐ Keep jewelry, loose clothing, and hair away from moving parts while an engine is running.
- ☐ Be careful of hot engine parts.
- ☐ Set the parking brake and place automatic transmission in PARK or manual transmission in NEUTRAL.

Engine Repair

Tools & Equipment:
- Straightedge
- Short ruler
- Alignment shims
- Belt tension gauge
- Common hand tools
- Vehicle service information

PROCEDURES Refer to the vehicle service information for specifications and special procedures. Then inspect and replace as necessary the belts on the engine provided by your instructor.

_____ 1. Write up a repair order.

_____ 2. Make sure you follow all procedures in the vehicle service information.

_____ 3. Look up the manufacturer's recommended belt tensions for new and used belts.

Source: _____ *Page:* _____

_____ 4. Inspect all the drive belts by looking for cracks, oil, glazing, fraying, and other damage. Replace any worn or damaged belts.

_____ 5. Clean all belt pulley surfaces with a clean rag and an appropriate degreaser.

_____ 6. Install the new belts with the directional arrow pointing in the direction of the belt travel. *Note:* Always replace both belts of a matched set and never force a belt onto a pulley with a screwdriver or pry bar.

_____ 7. After replacing any worn or damaged belts, check the alignment of all the belts. *Note:* See your instructor for the procedure to check belt alignment.

_____ 8. Use the alignment shims to adjust any pulleys that are not aligned.

(continued)

_____ 9. Check the belts for the correct tension by using the tension gauge. *Note:* The belt deflection test can be a preliminary test on all but v-ribbed belts by using the straightedge and short ruler. See your instructor to learn this procedure.

_____ 10. Adjust belt tensions according to the manufacturer's recommendations. *Note:* Be certain to adjust the belts according to whether they are new or used.

_____ 11. Tighten all the belt adjustment bolts, run the engine for 10 to 15 minutes, stop the engine, check all the belt tensions, and readjust the belts according to the manufacturer's specifications.

_____ 12. Before replacing a serpentine v-ribbed belt, draw a sketch of its pattern before removing it from the engine. *Note:* This step is not necessary if there is a drawing on the underside of the hood.

_____ 13. Remove the belt by using a suitable tool to flex the tensioner against spring tension.

_____ 14. Install the new belt on all the pulleys and check the alignment. *Note:* Be sure to fit the belt's v-grooves properly on each pulley and route the belt properly.

_____ 15. Use a suitable tool to flex the tensioner against spring tension.

_____ 16. To locate the belt around the last pulley, slowly release the tensioner. Verify that the belt is properly routed and did not slip out of place when the tensioner was released.

_____ 17. Check tensioner operation by pulling on the belt. The tensioner should flex and re-apply proper tension on the belt when the belt is released.

_____ 18. Run the engine and observe that the belt tracks properly and stays in position on all pulleys.

Performance ✓ Checklist

Name _____ Date _____ Class _____

PERFORMANCE STANDARDS:
Level 4–Performs skill without supervision and adapts to problem situations.
Level 3–Performs skill satisfactorily without assistance or supervision.
Level 2–Performs skill satisfactorily, but requires assistance/supervision.
Level 1–Performs parts of skill satisfactorily, but requires considerable assistance/supervision.

Attempt (circle one): **1 2 3 4**

Comments:

_____

_____

_____

_____

PERFORMANCE LEVEL ACHIEVED: _____

_____ 1. Safety rules and practices were followed at all times regarding this job.

_____ 2. Tools and equipment were used properly and stored upon completion of this job.

_____ 3. This completed job met the standards set and was done within the allotted time.

_____ 4. No injury or damage to property occurred during this job.

_____ 5. Upon completion of this job, the work area was cleaned correctly.

Instructor's Signature _____ Date_____

INSPECTING & REPLACING THERMOSTATS

NATEF Standard(s) for Engine Repair:
D6 Inspect, test, and replace thermostat and gasket.

Safety First
- ☐ Wear safety glasses at all times.
- ☐ Follow all safety rules when using common hand tools.
- ☐ Always connect a vehicle's exhaust to a vent hose before you run an engine in a closed shop. Unvented exhaust fumes in a closed shop can cause death.
- ☐ Check Material Safety Data Sheets for chemical safety information.
- ☐ Disconnect electric fans from the power source before working in the vicinity of the fan blades.
- ☐ Keep jewelry, loose clothing, and hair away from moving parts while an engine is running.
- ☐ Be careful of hot engine parts.
- ☐ Set the parking brake and place automatic transmission in PARK or manual transmission in NEUTRAL.

Tools & Equipment:
- Syringe
- Cooking pot
- Sealant or gasket
- Vehicle service information
- Hot plate
- Wire
- Torque wrench
- Thermometer
- Putty knife
- Common hand tools

PROCEDURES Refer to the vehicle service information for specifications and special procedures. Then inspect, test, and replace the thermostat in the engine provided by your instructor.

_____ 1. Write up a repair order.

_____ 2. Make sure you follow all procedures in the vehicle service information.

_____ 3. On a cool engine, remove the radiator cap. *Note:* Remove the radiator cap slowly.

_____ 4. Using the syringe, decrease the coolant level to beneath the thermostat housing (or partially drain the radiator). *Note:* Save the coolant if it is clean. If it needs to be discarded, be sure to dispose of it according to EPA regulations.

_____ 5. Remove the thermostat housing or cap. *Note:* You may need to tap the housing free with a rubber hammer.

_____ 6. Remove and rinse the thermostat.

_____ 7. Inspect the thermostat for damage such as cracks, breaks, corrosion, and damaged seals.

_____ 8. Replace any defective or damaged thermostats.

(continued)

Engine Repair

_____ 9. Look up the manufacturer's recommended thermostat temperature rating for the vehicle.

Source: _____ Page: _____

_____ 10. Replace the thermostat if its temperature rating (stamped on the engine side of the thermostat) does not match the manufacturer's recommended rating.

_____ 11. Refer to your instructor for performance testing an undamaged thermostat with water, the hot plate, and the thermometer.

_____ 12. Look up the manufacturer's recommended thermostat selection.

Source: _____ Page: _____

_____ 13. Install the correct thermostat and then reinstall the thermostat housing with a new gasket. *Note:* A high-temperature silicon sealant can also be used. See your instructor for the procedure.

_____ 14. Torque the bolts to specification, reconnect the radiator hose, and replace the coolant that you removed.

_____ 15. Start the engine and run it until it reaches operating temperature and the thermostat opens.

_____ 16. Check for leaks and turn off the engine.

_____ 17. Repair any leaks.

Performance ✓ Checklist

Name _____ Date _____ Class _____

PERFORMANCE STANDARDS:
Level 4–Performs skill without supervision and adapts to problem situations.
Level 3–Performs skill satisfactorily without assistance or supervision.
Level 2–Performs skill satisfactorily, but requires assistance/supervision.
Level 1–Performs parts of skill satisfactorily, but requires considerable assistance/supervision.

Attempt (circle one): **1 2 3 4**

Comments:

_____

_____

_____

_____

PERFORMANCE LEVEL ACHIEVED: _____

_____ 1. Safety rules and practices were followed at all times regarding this job.

_____ 2. Tools and equipment were used properly and stored upon completion of this job.

_____ 3. This completed job met the standards set and was done within the allotted time.

_____ 4. No injury or damage to property occurred during this job.

_____ 5. Upon completion of this job, the work area was cleaned correctly.

Instructor's Signature _____ Date_____

PERFORMING SYSTEM PRESSURE TESTS

NATEF Standard(s) for Engine Repair:
D3 Perform cooling system pressure tests; check coolant condition; inspect and test radiator, pressure cap, coolant recovery tank, and hoses; determine necessary action.

Safety First
- ☐ Wear safety glasses at all times.
- ☐ Follow all safety rules when using common hand tools.
- ☐ Always connect a vehicle's exhaust to a vent hose before you run an engine in a closed shop. Unvented exhaust fumes in a closed shop can cause death.
- ☐ Always follow all safety rules for removing radiator caps.
- ☐ Keep jewelry, loose clothing, and hair away from moving parts while an engine is running.
- ☐ Be careful of hot engine parts.
- ☐ Set the parking brake and place automatic transmission in PARK or manual transmission in NEUTRAL.

Engine Repair

Tools & Equipment:
- Cooling system analyzer
- Common hand tools
- Coolant thermometer
- Vehicle service information
- Voltmeter

PROCEDURES Refer to the vehicle service information for specifications and special procedures. Then test the system pressures in the engine provided by your instructor.

_____ 1. Write up a repair order.

_____ 2. Make sure you follow all procedures in the vehicle service information.

_____ 3. Look up the recommended system pressure and temperature specifications in the vehicle service information.

Source: _____ Page:_____

_____ 4. Test the radiator cap with the cooling system pressure tester by following the directions with the kit. *Note:* Refer to the kit's directions to perform all the pressure tests.

_____ 5. Pump the tester slowly while noting when the pressure reading on the gauge stops. *Note:* This is where the relief valve on the cap opens.

_____ 6. Release the tester and repeat steps #4 and #5. If the radiator cap's pressure reading is below the minimum specified–or 2 to 3 psi above the maximum–replace the cap.

(*continued*)

_____ 7. Test the engine's cooling system with the cooling system tester kit by attaching the tester to the radiator filler neck and pumping the tester until the pressure on the gauge reaches the pressure specified by the manufacturer's recommendations.

_____ 8. Keep pressure on the system for at least two minutes while watching the gauge for any pressure drop. *Note:* If the pressure holds steady, go to step #10.

_____ 9. If the pressure drops, inspect the system for external leaks. Repair any leaks you find.

_____ 10. Check for milky looking oil on the end of the oil dipstick. This is a sign of an internal coolant leak. *Note:* If coolant is in the oil, then the cylinder head, gasket, or block is damaged. See your instructor if you find contaminated oil.

_____ 11. Test the internal combustion components by first removing the radiator cap and allowing the engine to run until the engine reaches normal operating temperature, then turn off the engine. *Note:* Be careful to follow all safety rules when removing a radiator cap from a warm engine. Remove it slowly. Never attempt to remove a radiator cap from a hot engine.

_____ 12. Attach the pressure tester to the radiator according to the manufacturer's recommended procedure. If there is a leak, isolate the source with the power balance test. *Note:* Do not exceed the cap pressure rating.

_____ 13. Check the coolant temperature by first allowing the engine to cool and then checking and correcting the coolant level if necessary.

_____ 14. Run the engine until the thermostat opens and check the coolant temperature by placing the coolant thermometer in the neck of the radiator. *Note:* If the temperature is above or below the recommended temperature, the thermostat may be faulty.

_____ 15. With the engine off, conduct an electrolysis test by grounding the positive probe of the voltmeter to the radiator and inserting the negative probe into the coolant.

_____ 16. If the reading is over 0.5 volts, flush the system and replace the coolant.

Performance ✓ Checklist

Name _____ Date _____ Class _____

PERFORMANCE STANDARDS:
Level 4–Performs skill without supervision and adapts to problem situations.
Level 3–Performs skill satisfactorily without assistance or supervision.
Level 2–Performs skill satisfactorily, but requires assistance/supervision.
Level 1–Performs parts of skill satisfactorily, but requires considerable assistance/supervision.

Attempt (circle one): **1 2 3 4**

Comments:

_____

_____

_____

_____

PERFORMANCE LEVEL ACHIEVED: _____

_____ 1. Safety rules and practices were followed at all times regarding this job.

_____ 2. Tools and equipment were used properly and stored upon completion of this job.

_____ 3. This completed job met the standards set and was done within the allotted time.

_____ 4. No injury or damage to property occurred during this job.

_____ 5. Upon completion of this job, the work area was cleaned correctly.

Instructor's Signature _____ Date _____

INSPECTING & REPLACING WATER PUMPS

NATEF Standard(s) for Engine Repair:
D8 Inspect, test, remove, and replace water pump.

Safety First

- ☐ Wear safety glasses at all times.
- ☐ Follow all safety rules when using common hand tools.
- ☐ Disconnect electric fans from power source before working in vicinity of fan blades.
- ☐ Always connect a vehicle's exhaust to a vent hose before you run an engine in a closed shop. Unvented exhaust fumes in a closed shop can cause death.
- ☐ Check Material Safety Data Sheets for chemical safety information.
- ☐ Keep jewelry, loose clothing, and hair away from moving parts while an engine is running.
- ☐ Be careful of hot engine parts.
- ☐ Set parking brake and put automatic transmission in PARK or manual transmission in NEUTRAL.

Tools & Equipment:

- Syringe
- Putty knife
- Common hand tools
- Bench vise
- Sealant or gasket
- Vehicle service information
- Coolant drain pan
- Torque wrench

PROCEDURES Refer to the vehicle service information for specifications and special procedures. Then check and replace the water pump in the engine provided by your instructor.

_____ 1. Write up a repair order.

_____ 2. Make sure you follow all procedures in the vehicle service information.

_____ 3. Visually inspect the water pump for leaks in a cold engine.

_____ 4. Start the engine, listen for rumbling or grinding coming from the water pump, then turn off the engine. *Note:* A noisy pump indicates bad bearings or a bent pulley or shaft.

_____ 5. If there are no leaks visible, do a pressure test for leaks. Look for signs of leakage at the vent hole under the water pump.

_____ 6. Check for worn bearings by wiggling the fan or pump pulley up and down. *Note:* If the pump shaft is loose, the pump must be replaced (this condition can often cause the shaft seals to leak as well).

_____ 7. Remove the radiator cap and decrease the coolant level with the syringe until the coolant surface is easy to see. Run the engine until it reaches operating temperature.

_____ 8. While the engine is running at operating temperature, see if the water circulates. If the thermostat is open, the pump impeller is slipping on its shaft.

(continued)

_____ 9. Replace the water pump if it fails any of the preceding tests.

_____ 10. Look up the manufacturer's recommended procedure for replacing the water pump.

    Source: _____ *Page:*_____

_____ 11. Drain coolant into coolant drain pan. ***Note:*** If the coolant is clean, save it to reuse. If the coolant needs to be discarded, be certain to dispose of it according to EPA regulations.

_____ 12. Remove the water pump drive belt and any other belts that are in the way. Swivel belt-driven units away from the water pump. If an engine-driven fan is attached to the water pump, it must be removed.

_____ 13. Remove the water pump, then stuff rags in the pump port holes to prevent dirt or gasket material from entering the engine.

_____ 14. Transfer the fan studs and the pulley (if necessary) to the new water pump. ***Note:*** Refer to your instructor for hard-to-remove parts.

_____ 15. Mount the new pump, with the appropriate sealant and gaskets on its ports, on the engine and start the bolts by hand.

_____ 16. Use a torque wrench to tighten the bolts according to the manufacturer's specifications.

_____ 17. Replace all the parts that were removed during the disassembly process and check belt alignment and tension.

_____ 18. Replace the coolant, check for leaks, and repair as necessary.

_____ 19. Run the engine to operating temperature, check for coolant circulation, and turn off the engine to check it for leaks.

_____ 20. Repair any leaks.

_____ 21. Top off radiator, replace cap, and adjust level of coolant in the coolant recovery bottle.

_____ 22. Verify that there are no leaks under normal operating pressure.

Performance ✓ Checklist

Name _____ Date _____ Class _____

PERFORMANCE STANDARDS:
Level 4–Performs skill without supervision and adapts to problem situations.
Level 3–Performs skill satisfactorily without assistance or supervision.
Level 2–Performs skill satisfactorily, but requires assistance/supervision.
Level 1–Performs parts of skill satisfactorily, but requires considerable assistance/supervision.

Attempt (circle one): **1 2 3 4**

Comments:

_____

_____

_____

_____

PERFORMANCE LEVEL ACHIEVED: _____

_____ 1. Safety rules and practices were followed at all times regarding this job.

_____ 2. Tools and equipment were used properly and stored upon completion of this job.

_____ 3. This completed job met the standards set and was done within the allotted time.

_____ 4. No injury or damage to property occurred during this job.

_____ 5. Upon completion of this job, the work area was cleaned correctly.

Instructor's Signature _____ Date_____

INSPECTING & TESTING COOLING FANS

NATEF Standard(s) for Engine Repair:
D10 Inspect and test fan(s) (electrical or mechanical), fan clutch, fan shroud, and air dams.

☐ Wear safety glasses at all times.

☐ Follow all safety rules when using common hand tools.

☐ Always connect a vehicle's exhaust to a vent hose before you run an engine in a closed shop. Unvented exhaust fumes in a closed shop can cause death.

☐ Disconnect electric fans from the power source before working in the vicinity of the fan blades.

☐ Keep jewelry, loose clothing, and hair away from moving parts while an engine is running.

☐ Be careful of hot engine parts.

☐ Set the parking brake and place automatic transmission in PARK or manual transmission in NEUTRAL.

Engine Repair

Tools & Equipment:

- Voltmeter
- Common hand tools
- Test light
- Vehicle service information
- Jumper wires

PROCEDURES Refer to the vehicle service information for specifications and special procedures. Then inspect and test the cooling fan in the engine provided by your instructor.

_____ 1. Write up a repair order.

_____ 2. Make sure you follow all procedures in the vehicle service information.

_____ 3. Check the fan shroud for cracks and missing fasteners and replace as necessary.

_____ 4. Inspect the fan for loose or damaged blades and tighten or replace as necessary. ***Note:*** Never try to straighten or weld badly bent or cracked blades.

_____ 5. Look up the manufacturer's recommended procedure for removing and replacing the electric fan.

Source: _____ *Page:* _____

_____ 6. Inspect and test the fan using the manufacturer's recommended procedure. ***Note:*** Use the following guidelines and your instructor's advice to assist you.

_____ 7. Inspect the fan for loose or broken wires and for defective motor mounts.

(continued)

_____ 8. Repair any damage and run the engine until the fan comes on. You may need to block the airflow to the radiator with a piece of cardboard or drive the vehicle to make the fan start. *Note:* Keep jewelry, loose clothing, and hair away from moving parts while an engine is running.

_____ 9. Once the fan is running, check the airflow through the radiator by holding a piece of paper against the radiator (you will have to remove any cardboard). *Note:* If the fan is pulling enough air, the paper will stay on the radiator by itself. Be certain the radiator is clean for this procedure.

_____ 10. If the fan does not turn on, check the fuse and coolant temperature switch. *Note:* Refer to your instructor to check the temperature switch with the voltmeter or test light.

_____ 11. If necessary, replace the fuse or temperature switch and recheck fan operation after you have made any electrical repairs. *Note:* Follow the manufacturer's recommendations whenever you replace the fan temperature switch.

_____ 12. Look up the manufacturer's recommended procedure for testing, removing, and replacing a thermo-clutch fan.

   Source: _____   *Page:* _____

_____ 13. Test, repair, and replace the thermo-clutch fan according to the manufacturer's instructions.

Performance ✓ Checklist

Name _____   Date _____   Class _____

PERFORMANCE STANDARDS:
Level 4–Performs skill without supervision and adapts to problem situations.
Level 3–Performs skill satisfactorily without assistance or supervision.
Level 2–Performs skill satisfactorily, but requires assistance/supervision.
Level 1–Performs parts of skill satisfactorily, but requires considerable assistance/supervision.

Attempt (circle one): **1 2 3 4**

Comments:

_____

_____

_____

_____

PERFORMANCE LEVEL ACHIEVED: _____

_____ 1. Safety rules and practices were followed at all times regarding this job.
_____ 2. Tools and equipment were used properly and stored upon completion of this job.
_____ 3. This completed job met the standards set and was done within the allotted time.
_____ 4. No injury or damage to property occurred during this job.
_____ 5. Upon completion of this job, the work area was cleaned correctly.

Instructor's Signature _____   Date _____

TESTING, FLUSHING & HANDLING
ENGINE COOLANT

NATEF Standard(s) for Engine Repair:

D7 Test coolant; drain and recover coolant; flush and refill cooling system with recommended coolant; bleed air as required.

Safety First

- ☐ Wear safety glasses at all times.
- ☐ Follow all safety rules when using common hand tools.
- ☐ Always connect a vehicle's exhaust to a vent hose before you run an engine in a closed shop. Unvented exhaust fumes in a closed shop can cause death.
- ☐ Check Material Safety Data Sheets for chemical safety information.
- ☐ Disconnect electric fans from the power source before working in the vicinity of the fan blades.
- ☐ Keep jewelry, loose clothing, and hair away from moving parts while an engine is running.
- ☐ Be careful of hot engine parts.
- ☐ Set the parking brake and place automatic transmission in PARK or manual transmission in NEUTRAL.

Tools & Equipment:

- Coolant hydrometer
- Common hand tools
- Coolant refractometer
- Vehicle service information
- Coolant drain pan

PROCEDURES Refer to the vehicle service information for specifications and special procedures. Then test, drain, and replace the coolant in the engine provided by your instructor.

_____ 1. Write up a repair order.

_____ 2. Make sure you follow all procedures in the vehicle service information.

_____ 3. On a cool engine, remove the radiator cap. *Note:* Remove the radiator cap slowly and then run your finger around the inside of the neck to check for rust, oil, and scale. If you find engine oil, it indicates an engine leak. Transmission oil in the coolant indicates an oil cooler leak. Light brown coolant is rusty and must be replaced.

_____ 4. Run the engine until it reaches normal operating temperature and then turn it off.

_____ 5. Determine the coolant's age by checking maintenance records and check the protective qualities of the coolant by using the hydrometer or refractometer according to the manufacturer's instructions. *Note:* If the coolant fails any tests, you will have to replace it. See your instructor for special instructions to flush and drain the coolant system.

(continued)

Engine Repair

_____ 6. Replace coolant by first draining the old coolant into the coolant drain pan. *Note:* Be sure to dispose of it according to EPA regulations.

_____ 7. Look up the manufacturer's recommended coolant type and volume for the vehicle.

   Source: _____ *Page:* _____

_____ 8. If the system does not need to be flushed, close the engine drains, refill the system with the appropriate antifreeze, and add enough water to bring the level to "full." This should result in about a 50/50 solution. *Note:* Some technicians prefer to mix a 50/50 solution before putting it in the vehicle.

_____ 9. Fill the recovery reservoir to the hot or maximum level, leave the radiator cap off, and turn on the passenger compartment heater.

_____ 10. Run the engine until it reaches operating temperature.

_____ 11. Some manufacturers require bleeding of air from the highest point in the cooling system. Have your instructor supervise this operation. Follow the manufacturer's instructions. *Note:* The coolant will be very hot. Do not allow it to touch your skin.

_____ 12. Stop the engine, add more coolant if necessary, and install the radiator cap.

_____ 13. Run the engine until it reaches operating temperature and turn it off.

_____ 14. Check for leaks. *Note:* Check for leaks, especially around the thermostat housing, all the hoses, and the drains.

_____ 15. Repair any leaks.

Performance ✓ Checklist

Name _____ Date _____ Class _____

PERFORMANCE STANDARDS:
Level 4–Performs skill without supervision and adapts to problem situations.
Level 3–Performs skill satisfactorily without assistance or supervision.
Level 2–Performs skill satisfactorily, but requires assistance/supervision.
Level 1–Performs parts of skill satisfactorily, but requires considerable assistance/supervision.

Attempt (circle one): **1 2 3 4**

Comments:

_____

_____

_____

_____

PERFORMANCE LEVEL ACHIEVED: _____

_____ 1. Safety rules and practices were followed at all times regarding this job.

_____ 2. Tools and equipment were used properly and stored upon completion of this job.

_____ 3. This completed job met the standards set and was done within the allotted time.

_____ 4. No injury or damage to property occurred during this job.

_____ 5. Upon completion of this job, the work area was cleaned correctly.

Instructor's Signature _____ Date _____

REPLACING THE RADIATOR

NATEF Standard(s) for Engine Repair:
D9 Remove and replace radiator.

☐ Wear safety glasses at all times.

☐ Follow all safety rules when using common hand tools.

☐ Always connect a vehicle's exhaust to a vent hose before you run an engine in a closed shop. Unvented exhaust fumes in a closed shop can cause death.

☐ Check Material Safety Data Sheets for chemical safety information.

☐ Disconnect electric fans from the power source before working in the vicinity of the fan blades.

☐ Keep jewelry, loose clothing, and hair away from moving parts while an engine is running.

☐ Be careful of hot engine parts.

☐ Set the parking brake and place automatic transmission in PARK or manual transmission in NEUTRAL.

Tools & Equipment:

- Coolant hydrometer
- Bottlebrush
- Wire brush
- Coolant drain pan
- Cooling system analyzer
- Common hand tools
- Transmission drain pan
- Universal plugs
- Vehicle service information

PROCEDURES Refer to the vehicle service information for specifications and special procedures. Then remove, repair, and replace the radiator in the engine provided by your instructor.

_____ 1. Write up a repair order.

_____ 2. Make sure you follow all procedures in the vehicle service information.

_____ 3. Inspect and clean the outside of the radiator for bugs, leaves, gravel, etc. **Note:** Use a water hose to wash the radiator core.

_____ 4. Check the fan shroud for cracks and missing fasteners and replace as necessary.

_____ 5. Check the radiator cap seal for cracks, tears, dryness, and pressure rating. Clean it gently and replace damaged caps as necessary. **Note:** Use the cooling system analyzer to test the pressure rating of the cap.

_____ 6. Check the filler neck for damage. Clean its surfaces with the wire brush. **Note:** Be careful not to contaminate the coolant if you are going to reuse it.

_____ 7. Drain the coolant into the coolant pan and save it to reuse. **Note:** If you discard the coolant, be sure to follow EPA regulations.

(continued)

_____ 8. Look up the manufacturer's recommended procedure for removing and replacing the radiator.

   Source: _____ *Page:* _____

_____ 9. Check the shroud, A/C condenser, fan, temperature switches and connections, coolant recovery tank, radiator hoses, and overflow tube for damage and wear.

_____ 10. Clean the recovery tank using the bottlebrush. *Note:* Be sure you can read the marks on the tank.

_____ 11. Reinstall the radiator according to the manufacturer's recommendations. *Note:* If the radiator needs to be repaired, send it to a radiator repair shop.

_____ 12. Reinstall the fan and shroud. *Note:* Be sure that any damage has been repaired.

_____ 13. Reinstall the recovery tank.

_____ 14. Close all the drains and connect all the hoses and lines.

_____ 15. Replace the coolant and check for leaks. *Note:* If you removed a transmission cooler during the manufacturer's radiator removal procedure, replace the transmission fluid according to the manufacturer's instructions before you start the engine.

_____ 16. Run the engine until it reaches operating temperature and add coolant and transmission fluid as needed.

_____ 17. Turn off the engine and check for leaks.

_____ 18. Repair any leaks.

Performance ✓ Checklist

Name _____ Date _____ Class _____

PERFORMANCE STANDARDS:
Level 4–Performs skill without supervision and adapts to problem situations.
Level 3–Performs skill satisfactorily without assistance or supervision.
Level 2–Performs skill satisfactorily, but requires assistance/supervision.
Level 1–Performs parts of skill satisfactorily, but requires considerable assistance/supervision.

Attempt (circle one): **1 2 3 4**

Comments:

_____

_____

_____

_____

PERFORMANCE LEVEL ACHIEVED: _____

_____ 1. Safety rules and practices were followed at all times regarding this job.
_____ 2. Tools and equipment were used properly and stored upon completion of this job.
_____ 3. This completed job met the standards set and was done within the allotted time.
_____ 4. No injury or damage to property occurred during this job.
_____ 5. Upon completion of this job, the work area was cleaned correctly.

Instructor's Signature _____ Date_____

INSPECTING & REPLACING OIL COOLERS

NATEF Standard(s) for Engine Repair:
D11 Inspect auxiliary oil coolers; determine necessary action.

Safety First

☐ Wear safety glasses at all times.
☐ Follow all safety rules when using common hand tools.
☐ Always connect a vehicle's exhaust to a vent hose before you run an engine in a closed shop. Unvented exhaust fumes in a closed shop can cause death.
☐ Disconnect electric fans from the power source before working in the vicinity of the fan blades.
☐ Keep jewelry, loose clothing, and hair away from moving parts while an engine is running.
☐ Be careful of hot engine parts.
☐ Follow all safety rules when using compressed air.
☐ Check Material Safety Data Sheets for chemical safety information.
☐ Set the parking brake and place automatic transmission in PARK or manual transmission in NEUTRAL.

Tools & Equipment:
- Air compressor
- Line wrench
- Vehicle service information
- Air hose adapter
- Cleaning solvent
- Oil line plug
- Common hand tools

PROCEDURES Refer to the vehicle service information for specifications and special procedures. Then inspect and test the oil cooler in the engine provided by your instructor.

_____ 1. Write up a repair order.

_____ 2. Make sure you follow all procedures in the vehicle service information.

_____ 3. Inspect the oil cooler for leaks. **CAUTION:** The oil in the cooler is under pressure and oil may spray several inches from a leak.

_____ 4. If the cooler is within reach, clean the cooler by blowing compressed air or water through the fins from the opposite direction of the airflow. **CAUTION:** Be sure to protect your eyes and body when you use compressed air.

_____ 5. Look up the manufacturer's recommended procedure in the vehicle service information for removing, testing, cleaning, and installing the oil cooler. *Note:* Most oil line connections require two wrenches, one to loosen the fitting and a wrench to prevent the inlet or outlet from twisting.

Source: _____ *Page:* _____

(continued)

Engine Repair

_____ 6. Remove the cooler according to the manufacturer's recommendations and clean it with an appropriate solvent. *Note:* Carefully spray the solvent into one of the ports of the oil cooler. If the solvent does not flow easily through the cooler, it needs to be replaced or repaired. Send damaged or clogged coolers to a radiator repair shop.

_____ 7. Test the cooler for leaks by plugging one line and applying 10 psi of compressed air into the other line while submerging the cooler in water. *Note:* If it bubbles, it leaks and must be replaced. Send defective coolers to a radiator repair shop.

_____ 8. Reinstall a good oil cooler according to the manufacturer's recommendations.

_____ 9. Run the engine and check for leakage. Repair as necessary.

_____ 10. Re-adjust oil levels to correct for any lost oil.

Performance ✓ Checklist

Name _____ Date _____ Class _____

PERFORMANCE STANDARDS:
Level 4–Performs skill without supervision and adapts to problem situations.
Level 3–Performs skill satisfactorily without assistance or supervision.
Level 2–Performs skill satisfactorily, but requires assistance/supervision.
Level 1–Performs parts of skill satisfactorily, but requires considerable assistance/supervision.

Attempt (circle one): **1 2 3 4**

Comments:

_____

_____

_____

_____

PERFORMANCE LEVEL ACHIEVED: _____

_____ 1. Safety rules and practices were followed at all times regarding this job.
_____ 2. Tools and equipment were used properly and stored upon completion of this job.
_____ 3. This completed job met the standards set and was done within the allotted time.
_____ 4. No injury or damage to property occurred during this job.
_____ 5. Upon completion of this job, the work area was cleaned correctly.

Instructor's Signature _____ Date_____

INSPECTING, TESTING & REPLACING OIL PRESSURE AND TEMPERATURE SENDING UNITS

NATEF Standard(s) for Engine Repair:
D12 Inspect, test, and replace oil temperature and pressure switches and sensors.

Safety First

- [ ] Wear safety glasses at all times.
- [ ] Follow all safety rules when using common hand tools.
- [ ] Always connect a vehicle's exhaust to a vent hose before you run an engine in a closed shop. Unvented exhaust fumes in a closed shop can cause death.
- [ ] Disconnect electric fans from the power source before working in the vicinity of the fan blades.
- [ ] Keep jewelry, loose clothing, and hair away from moving parts while an engine is running.
- [ ] Be careful of hot engine parts.
- [ ] Follow all safety rules when using compressed air.
- [ ] Check Material Safety Data Sheets for chemical safety information.
- [ ] Set the parking brake and place automatic transmission in PARK or manual transmission in NEUTRAL.

Tools & Equipment:
- Variable rheostat or gauge tester
- Vehicle service information
- Multimeter
- Common hand tools

PROCEDURES Refer to the vehicle service information for specifications and special procedures. Then inspect, test, and replace the oil temperature and pressure sending units in the engine provided by your instructor.

_____ 1. Write up a repair order.

_____ 2. Make sure you follow all procedures in the vehicle service information.

_____ 3. Look up the manufacturer's recommended procedure in the vehicle service information for inspecting, testing, and replacing oil temperature and pressure sending units.

Source: _____ Page: _____

_____ 4. Inspect, test, and replace the oil temperature and pressure sending units according to the manufacturer's recommendations.

(continued)

Performance ✓ Checklist

Name _____ Date _____ Class _____

PERFORMANCE STANDARDS:
Level 4–Performs skill without supervision and adapts to problem situations.
Level 3–Performs skill satisfactorily without assistance or supervision.
Level 2–Performs skill satisfactorily, but requires assistance/supervision.
Level 1–Performs parts of skill satisfactorily, but requires considerable assistance/supervision.

Attempt (circle one): **1   2   3   4**

Comments:

_____

_____

_____

_____

PERFORMANCE LEVEL ACHIEVED: _____

_____ 1. Safety rules and practices were followed at all times regarding this job.

_____ 2. Tools and equipment were used properly and stored upon completion of this job.

_____ 3. This completed job met the standards set and was done within the allotted time.

_____ 4. No injury or damage to property occurred during this job.

_____ 5. Upon completion of this job, the work area was cleaned correctly.

Instructor's Signature _____ Date_____

CHANGING THE OIL

NATEF Standard(s) for Engine Repair:
D13 Perform oil and filter change.

Safety First

- ☐ Wear safety glasses at all times.
- ☐ Follow all safety rules when using common hand tools.
- ☐ Always connect a vehicle's exhaust to a vent hose before you run an engine in a closed shop. Unvented exhaust fumes in a closed shop can cause death.
- ☐ Keep jewelry, loose clothing, and hair away from moving parts while an engine is running.
- ☐ Be careful of hot engine parts.
- ☐ Follow all safety rules when using jacks and jack stands, ramps, or lifts.
- ☐ Set the parking brake and place automatic transmission in PARK or manual transmission in NEUTRAL.

Tools & Equipment:
- Filter wrench
- Shop light
- Common hand tools
- Oil drain pan
- Jacks and jack stands, ramps, or lift
- Rubber gloves
- Vehicle service information

PROCEDURES Refer to the vehicle service information for specifications and special procedures. Then change the oil in the engine provided by your instructor.

_____ 1. Write up a repair order.

_____ 2. Make sure you follow all procedures in the vehicle service information.

_____ 3. Run the engine until it reaches operating temperature. *Note:* Circulating the oil suspends contaminates so they will drain out with the oil.

_____ 4. Raise the vehicle now if you need to in order to reach the oil plug or filter.

_____ 5. Put the oil drain pan under the oil drain plug in the oil pan. **CAUTION:** If the oil plug is in the side of the oil pan, be prepared for the oil to "shoot" sideways when you remove the plug.

_____ 6. Loosen the oil plug several turns with a wrench until you can turn it by hand. **CAUTION:** Wear rubber gloves for this procedure to protect your hands from the hot oil.

_____ 7. Turn the plug out carefully by hand and allow the oil to drain for five or ten minutes. *Note:* Quickly pull the plug out of the drain.

_____ 8. Clean the drain plug with a clean cloth and check the washer for damage. Replace the washer if it is damaged.

_____ 9. Reinstall the drain plug and put the drain pan under the filter.

(continued)

_____ 10. Loosen the filter with the filter wrench and finish removing the filter by hand. Make sure that the filter O-ring comes off with the old filter.

_____ 11. Clean the filter sealing surface on the engine with a clean cloth. Coat the gasket on the new filter with a thin film of clean oil.

_____ 12. Install the new filter by hand, and tighten it ¾ to a full turn after it touches the engine.

_____ 13. Refill the crankcase with the correct amount and type of oil.

_____ 14. If the vehicle has a turbocharger, refer to your instructor before proceeding. Starting a turbocharged engine without oil will damage the turbo.

_____ 15. Run the engine and let it idle until the oil pressure light goes out or the gauge reads that there is enough pressure. *Note:* Refer to your instructor for more information on pressure loss.

_____ 16. Idle the engine for several minutes, and then turn it off.

_____ 17. Check for leaks around the drain plug and the oil filter. **CAUTION:** Do not overtighten the plug. Check the washer if the plug is leaking.

_____ 18. Wait five or ten minutes and then check the oil with the dipstick in the engine.

_____ 19. Make sure the oil level is in the SAFE range.

_____ 20. Add oil if it is low. **CAUTION:** Never overfill the crankcase.

_____ 21. If the vehicle is raised, lower it.

_____ 22. Dispose of used oil according to EPA regulations.

Performance ✓ Checklist

Name _____ Date _____ Class _____

PERFORMANCE STANDARDS:
Level 4–Performs skill without supervision and adapts to problem situations.
Level 3–Performs skill satisfactorily without assistance or supervision.
Level 2–Performs skill satisfactorily, but requires assistance/supervision.
Level 1–Performs parts of skill satisfactorily, but requires considerable assistance/supervision.

Attempt (circle one): **1 2 3 4**

Comments:

_____

_____

_____

_____

PERFORMANCE LEVEL ACHIEVED: _____

_____ 1. Safety rules and practices were followed at all times regarding this job.
_____ 2. Tools and equipment were used properly and stored upon completion of this job.
_____ 3. This completed job met the standards set and was done within the allotted time.
_____ 4. No injury or damage to property occurred during this job.
_____ 5. Upon completion of this job, the work area was cleaned correctly.

Instructor's Signature _____ Date_____

DIAGNOSING THE CAUSE OF EXCESSIVE OIL CONSUMPTION OR UNUSUAL EXHAUST COLORS, ODORS, AND SOUND

Engine Repair

NATEF Standard(s) for Engine Repair:

A7 Diagnose the cause of excessive oil consumption, unusual engine exhaust color, odor, and sound; determine necessary action.

Safety First

☐ Wear safety glasses at all times.

☐ Follow all safety rules when using common hand tools.

☐ Use exhaust vent if running engine in closed shop. Unvented exhaust fumes can cause death.

☐ Keep jewelry, loose clothing, and hair away from moving parts while an engine is running.

☐ Stay clear of cooling fan and hot surfaces.

☐ Follow all safety rules when using a lift or jack and jack stands.

☐ Make sure automatic transmission is in PARK or manual transmission is in NEUTRAL and set the parking brake.

Tools & Equipment:

- Vehicle service information
- Cylinder leakage tester
- Exhaust gas analyzer
- Common hand tools
- Cooling system pressure tester
- Lift or jack and jack stands
- Compression test gauge
- Scan tool
- Fender covers

PROCEDURES Refer to the vehicle service information for specifications and special procedures. Then diagnose the cause of excessive oil consumption or unusual exhaust color, odor, and sound in the vehicle provided by your instructor.

Excessive Oil Consumption

_____ 1. Write up a repair order.

_____ 2. Make sure you follow all procedures in the vehicle service information.

_____ 3. Check for proper PCV system operation. A defective PCV valve or a plugged system may cause excessive crankcase pressure, leading to excessive oil consumption.

_____ 4. Check for poor condition of piston rings and cylinder walls. A dry compression test with poor results followed by a much better wet compression test indicates poor rings or cylinder wall conditions, which will cause excessive oil consumption.

_____ 5. A cylinder leakage test that indicates excessive blow-by in the crankcase indicates poor condition of piston rings or cylinder walls, leading to excessive oil consumption.

_____ 6. Look for valve stem seals that are damaged, hardened with age, or missing, all of which contribute to excessive oil consumption.

(continued)

Unusual Engine Exhaust Color, Odor, and Sound

_____ 1. Write up a repair order.

_____ 2. Make sure you follow all procedures in the vehicle service information.

_____ 3. An excessive blue color of engine exhaust indicates high oil consumption. Perform proper test to determine cause of high oil consumption.

_____ 4. An excessive black color of engine exhaust indicates a rich fuel air mixture. Use an appropriate scanner or exhaust gas analyzer to determine the cause of an excessive rich mixture.

_____ 5. An excessive white color of engine exhaust indicates steam caused by coolant leaking into the combustion chamber. On a cold engine, install a pressure tester to the radiator. Start the engine and watch the pressure gauge. If pressure begins to build immediately, there is a combustion leak in the cooling system (likely through a blown cylinder head gasket or a crack in the cylinder head).

_____ 6. A burning oil smell in the vehicle exhaust usually indicates high engine oil consumption. A rotten egg smell of the vehicle exhaust, on vehicles equipped with catalytic converters, indicates an excessive rich fuel/air mixture.

_____ 7. Unusual noises in the vehicle exhaust system may indicate leaks in the system. A puffing sound at the tail pipe indicates an engine misfiring condition that may be the result of ignition problems or burned exhaust valves.

Performance ✓ Checklist

Name _____ Date _____ Class _____

PERFORMANCE STANDARDS:
Level 4–Performs skill without supervision and adapts to problem situations.
Level 3–Performs skill satisfactorily without assistance or supervision.
Level 2–Performs skill satisfactorily, but requires assistance/supervision.
Level 1–Performs parts of skill satisfactorily, but requires considerable assistance/supervision.

Attempt (circle one): 1 2 3 4

Comments:

_____

_____

_____

_____

PERFORMANCE LEVEL ACHIEVED: _____

_____ 1. Safety rules and practices were followed at all times regarding this job.

_____ 2. Tools and equipment were used properly and stored upon completion of this job.

_____ 3. This completed job met the standards set and was done within the allotted time.

_____ 4. No injury or damage to property occurred during this job.

_____ 5. Upon completion of this job, the work area was cleaned correctly.

Instructor's Signature _____ Date_____

TESTING CYLINDER COMPRESSION

NATEF Standard(s) for Engine Repair:
A2 Identify and interpret engine concern; determine necessary action.
A10 Perform cylinder cranking compression tests; determine necessary action.

Safety First

- ☐ Wear safety glasses at all times.
- ☐ Follow all safety rules when using common hand tools.
- ☐ Always connect a vehicle's exhaust to a vent hose before you run an engine in a closed shop. Unvented fumes in a closed shop can cause death.
- ☐ Keep jewelry, loose clothing, and hair away from moving parts while an engine is running.
- ☐ Be careful of hot engine parts.
- ☐ Protect eyes when cleaning around spark plugs with compressed air.
- ☐ Set the parking brake and place automatic transmission in PARK or manual transmission in NEUTRAL.

Tools & Equipment:
- Masking tape
- 30W oil
- Air compressor
- Compression tester
- Common hand tools
- Remote starter switch
- Vehicle service information

PROCEDURES Refer to the vehicle service information for specifications and special procedures. Then test the cylinder compression in the engine provided by your instructor.

_____ 1. Write up a repair order.

_____ 2. Make sure you follow all procedures in the vehicle service information.

_____ 3. Idle the engine until it reaches normal operating temperature.

_____ 4. Turn off the engine. Then use masking tape to tag and number each spark plug wire. Remove the plug wires.

_____ 5. Clean debris and dirt from around the spark plugs with compressed air. *Note:* Protect your eyes when you do this.

_____ 6. Remove and inspect the spark plugs.

_____ 7. Block open the throttle and choke plates. Then place the compression tester into a spark plug hole and connect the remote starter switch.

(continued)

_____ 8. Crank the engine for five compression cycles and note any changes in the gauge readings between compression cycles. *Note:* The battery must be fully charged for accurate compression readings. It may be wise to connect a battery charger to ensure identical cranking speed for all cylinders.

_____

_____ 9. Repeat the compression test on each of the remaining cylinders, record any change between compression cycle readings, and note the highest compression reading for each cylinder.

_____   _____

_____   _____

_____ 10. Turn off engine and compare results. *Note:* See the instructor for inconsistent readings.

_____ 11. Make a wet compression test if any of the following is present in the readings: A range of greater than 15% between cylinders, all readings are less than 100 psi, or the readings of 75% of the cylinders fall below the manufacturer's specifications.

_____ 12. If a wet compression test is necessary, it is done the same way as the dry test except one tablespoon of 30W oil is squirted into the cylinder through the spark plug hole first. *Note:* Crank the engine with the remote starter several times before installing the compression gauge to distribute the oil in the cylinders.

_____ 13. Record your readings.

_____   _____

_____   _____

_____ 14. Turn off engine, review findings, and discuss any inconsistencies with your instructor.

_____ 15. Replace all the spark plugs using care to follow the numbered tags on the wires.

Performance ✓ Checklist

Name _____ Date _____ Class _____

PERFORMANCE STANDARDS:
Level 4–Performs skill without supervision and adapts to problem situations.
Level 3–Performs skill satisfactorily without assistance or supervision.
Level 2–Performs skill satisfactorily, but requires assistance/supervision.
Level 1–Performs parts of skill satisfactorily, but requires considerable assistance/supervision.

Attempt (circle one): **1 2 3 4**

Comments:

_____

_____

_____

_____

PERFORMANCE LEVEL ACHIEVED: _____

_____ 1. Safety rules and practices were followed at all times regarding this job.

_____ 2. Tools and equipment were used properly and stored upon completion of this job.

_____ 3. This completed job met the standards set and was done within the allotted time.

_____ 4. No injury or damage to property occurred during this job.

_____ 5. Upon completion of this job, the work area was cleaned correctly.

Instructor's Signature _____ Date_____

TESTING CYLINDERS FOR LEAKS

NATEF Standard(s) for Engine Repair:

A11 Perform cylinder leakage tests; determine necessary action.

Safety First

- ☐ Wear safety glasses at all times.
- ☐ Follow all safety rules when using common hand tools.
- ☐ Always connect a vehicle's exhaust to a vent hose before you run an engine in a closed shop. Unvented fumes in a closed shop can cause death.
- ☐ Keep jewelry, loose clothing, and hair away from moving parts while an engine is running.
- ☐ Be careful of hot engine parts.
- ☐ Set the parking brake and place automatic transmission in PARK and manual transmission in NEUTRAL.
- ☐ Be careful of hot radiator parts and pressure.
- ☐ Be certain that the ignition switch is OFF during leakage tests to prevent electrical shock.

Tools & Equipment:

- Masking tape
- Short piece of heater hose
- Cylinder leakage tester
- Common hand tools
- Remote starter switch
- Vehicle service information

PROCEDURES Refer to the vehicle service information for specifications and special procedures. Then test the cylinders in the engine provided by your instructor.

_____ 1. Write up a repair order.

_____ 2. Make sure you follow all procedures in the vehicle service information.

_____ 3. Idle the engine until it reaches normal operating temperature.

_____ 4. Turn off the engine, and then tag and number each spark plug wire with masking tape. Remove the plug wires.

_____ 5. Clean the engine around the spark plugs with compressed air. **CAUTION:** Protect your eyes when you do this.

_____ 6. Loosen the radiator cap to prevent pressure from building up. **CAUTION:** Coolant in an engine that has been running is hot and under pressure.

_____ 7. Remove and inspect the spark plugs.

_____ 8. Block open the throttle and choke plates.

_____ 9. Carefully remove the radiator cap with a shop cloth.

(continued)

Engine Repair

_____ **10.** Zero the leakage gauge.

_____ **11.** With a wrench on the crankshaft pulley, rotate the engine until the piston on the cylinder being tested is at top dead center (TDC) on the compression stroke. Screw the air hose adapter of the cylinder leakage tester tightly into the spark plug hole.

_____ **12.** Put manual transmissions into first gear and block the wheels. *Note:* See vehicle service information or your instructor to keep an engine with an automatic transmission from turning.

_____ **13.** Turn on the compressed air and record the reading.

Note: If the cylinders move from TDC, "bump" them back into position with the remote starter. Be certain the ignition switch is off and return the transmission to NEUTRAL while doing this.

_____ **14.** Use the short piece of hose to listen for leaks. See the vehicles service information to diagnose the locations. *Note:* During this test, look or listen for signs of leakage from the crankcase, air intake, radiator, and exhaust system.

_____ **15.** Perform this test on all suspect cylinders (possibly cylinders that have failed a compression test).

_____ **16.** Remove the leakage tester and the remote starter switch, then replace the spark plugs and the spark plug wires.

Performance ✓ Checklist

Name _____ Date _____ Class _____

PERFORMANCE STANDARDS:
Level 4–Performs skill without supervision and adapts to problem situations.
Level 3–Performs skill satisfactorily without assistance or supervision.
Level 2–Performs skill satisfactorily, but requires assistance/supervision.
Level 1–Performs parts of skill satisfactorily, but requires considerable assistance/supervision.

Attempt (circle one): **1 2 3 4**

Comments:

_____

_____

_____

_____

PERFORMANCE LEVEL ACHIEVED: _____

_____ **1.** Safety rules and practices were followed at all times regarding this job.
_____ **2.** Tools and equipment were used properly and stored upon completion of this job.
_____ **3.** This completed job met the standards set and was done within the allotted time.
_____ **4.** No injury or damage to property occurred during this job.
_____ **5.** Upon completion of this job, the work area was cleaned correctly.

Instructor's Signature _____ Date_____

INSPECTING AN ENGINE ASSEMBLY FOR VACUUM LEAKS

NATEF Standard(s) for Engine Repair:

A8 Perform engine vacuum tests; determine necessary action.

Safety First

☐ Wear safety glasses at all times.

☐ Follow all safety rules when using common hand tools.

☐ Always connect a vehicle's exhaust to a vent hose before you run an engine in a closed shop. Unvented exhaust fumes in a closed shop can cause death.

☐ Keep jewelry, loose clothing, and hair away from moving parts while an engine is running.

☐ Be careful of hot engine parts.

☐ Set the parking brake and place automatic transmission in PARK or manual transmission in NEUTRAL.

Tools & Equipment:

• Vacuum gauge with fittings
• Vehicle service information

• Remote starter switch

• Common hand tools

PROCEDURES Refer to the vehicle service information for specifications and special procedures. Then inspect the engine assembly provided by your instructor.

_____ 1. Write up a repair order.

_____ 2. Make sure you follow all procedures in the vehicle service information.

_____ 3. Idle the engine until it reaches normal operating range.

_____ 4. Turn the engine off, wait for it to come to a complete stop, and connect a remote starter according to the vehicles service information.

_____ 5. Connect a vacuum gauge "T" between the vacuum line and the intake manifold port.
Note: Locations vary. Check the vehicle service information for proper location and reading.

Source: _____ Page: _____

_____ 6. Disable the ignition. Do not "over-crank" the engine, but crank the engine for ten seconds at a time until a good reading occurs. Record the highest vacuum gauge reading as the engine cranks.

_____

(continued)

_____ 7. Allow the starter to cool for five to ten minutes before repeating Step 6. Record the reading again. It should be a steady reading.

_____

_____ 8. Re-enable the ignition. Allow the starter to cool, then start and idle the engine. Record the idling vacuum reading. *Note:* See vehicle service information for correct idle reading.

_____

_____ 9. Hold the engine rpms steady at 2000 rpm or as specified by holding the accelerator steady with your foot. Record the engine "revving" vacuum.

_____

Note: You can watch rpm with a tachometer. See your instructor for any incorrect or uneven readings.

_____ 10. Stop the engine and remove the vacuum gauge. Replace the vacuum lines, if necessary.

Performance ✓ Checklist

Name _____ Date _____ Class _____

PERFORMANCE STANDARDS:
Level 4–Performs skill without supervision and adapts to problem situations.
Level 3–Performs skill satisfactorily without assistance or supervision.
Level 2–Performs skill satisfactorily, but requires assistance/supervision.
Level 1–Performs parts of skill satisfactorily, but requires considerable assistance/supervision.

Attempt (circle one): **1 2 3 4**

Comments:

_____

_____

_____

_____

PERFORMANCE LEVEL ACHIEVED: _____

_____ 1. Safety rules and practices were followed at all times regarding this job.
_____ 2. Tools and equipment were used properly and stored upon completion of this job.
_____ 3. This completed job met the standards set and was done within the allotted time.
_____ 4. No injury or damage to property occurred during this job.
_____ 5. Upon completion of this job, the work area was cleaned correctly.

Instructor's Signature _____ Date_____

TESTING CYLINDERS FOR POWER BALANCE

NATEF Standard(s) for Engine Repair:
A9 Perform cylinder power balance tests; determine necessary action.

Safety First

- ☐ Wear safety glasses at all times.
- ☐ Follow all safety rules when using common hand tools.
- ☐ Always connect a vehicle's exhaust to a vent hose before you run an engine in a closed shop. Unvented exhaust fumes in a closed shop can cause death.
- ☐ Keep jewelry, loose clothing, and hair away from moving parts while an engine is running.
- ☐ Be careful of hot engine parts.
- ☐ DO NOT short out any cylinder for more than ten seconds–this can damage the catalytic converter.
- ☐ Set the parking brake and place automatic transmission in PARK or manual transmission in NEUTRAL.

Tools & Equipment:
- Engine analyzer and manual
- Common hand tools
- Vehicle service information

PROCEDURES Refer to the vehicle service information for specifications and special procedures. Then test the cylinders in the engine provided by your instructor.

Note: Many engine analyzers will do this test automatically and make comparisons when programmed to the cylinder balance test.

_____ 1. Write up a repair order.

_____ 2. Make sure you follow all procedures in the vehicle service information.

_____ 3. Put on the parking brake.

_____ 4. Hook up the engine analyzer according to its manufacturer's instructions.

_____ 5. Idle the engine until it is within the normal operating temperature range.

_____ 6. Use the engine analyzer to short out each cylinder one at a time. *Note:* Do not short out any cylinder for more than ten seconds. Doing this can damage the catalytic converter.

_____ 7. Record the engine rpm as each cylinder is shorted out.

_____

_____

(continued)

Engine Repair

_____ **8.** Record the engine vacuum as each cylinder is shorted out.

_____

_____

_____ **9.** Stop the engine and remove the analyzer.

_____ **10.** Review your findings. Rpm should fall at least 5% as each cylinder is shorted out. The difference between the highest and lowest vacuum readings (range) should be less than 5%.

_____ **11.** Consider other causes for cylinder power differences, such as burned-out valves, worn pistons, a leaky head gasket, vacuum leaks, a maladjusted carburetor, faulty fuel injectors, poor ignition, or a computer problem.

Performance ✓ Checklist

Name _____ Date _____ Class _____

PERFORMANCE STANDARDS:
Level 4–Performs skill without supervision and adapts to problem situations.
Level 3–Performs skill satisfactorily without assistance or supervision.
Level 2–Performs skill satisfactorily, but requires assistance/supervision.
Level 1–Performs parts of skill satisfactorily, but requires considerable assistance/supervision.

Attempt (circle one): **1 2 3 4**

Comments:

_____

_____

_____

_____

PERFORMANCE LEVEL ACHIEVED: _____

_____ **1.** Safety rules and practices were followed at all times regarding this job.
_____ **2.** Tools and equipment were used properly and stored upon completion of this job.
_____ **3.** This completed job met the standards set and was done within the allotted time.
_____ **4.** No injury or damage to property occurred during this job.
_____ **5.** Upon completion of this job, the work area was cleaned correctly.

Instructor's Signature _____ Date_____

INSPECTING AN ENGINE ASSEMBLY FOR LEAKS

NATEF Standard(s) for Engine Repair:

A5 Inspect engine assembly for fuel, oil, coolant, and other leaks; determine necessary action.

Safety First

- ☐ Wear safety glasses at all times.
- ☐ Follow all safety rules when using common hand tools.
- ☐ Follow all safety rules when using a lift or jack and jack stands.
- ☐ Always connect a vehicle's exhaust to a vent hose before you run an engine in a closed shop. Unvented exhaust fumes in a closed shop can cause death.
- ☐ Keep jewelry, loose clothing, and hair away from moving parts while an engine is running.
- ☐ Be careful of hot engine parts.
- ☐ Set the parking brake and place automatic transmission in PARK or manual transmission in NEUTRAL.

Tools & Equipment:
- Lift or jack and jack stands
- Vehicle service information
- Shop light
- Common hand tools
- Replacement parts as needed

PROCEDURES Refer to the vehicle service information for specifications and special procedures. Then inspect the engine assembly provided by your instructor.

_____ 1. Write up a repair order.

_____ 2. Make sure you follow all procedures in the vehicle service information.

_____ 3. Use a shop light to inspect around the radiator cap for leaks. If you find indications of a leak, replace the radiator cap.

_____ 4. Check the filler neck tanks and core for distortion or seepage. If you find defects, the radiator will have to be repaired at a radiator repair shop.

_____ 5. Inspect around the clamps of the cooling system hoses. Tighten the clamps or replace the clamps and/or hoses if you find leaks.

_____ 6. Check each hose and squeeze each one in the middle. Replace any hose that feels weak or shows signs of cracking.

_____ 7. Inspect the carburetor and the area around it for fuel leaks. If there are leaks, tighten the carburetor screws. If they are already tight, ask your instructor to help you determine the cause of the leak.

_____ 8. Take note of any fuel odors that may indicate a leak. If such odors exist, ask your instructor to help you with a more in-depth inspection.

(continued)

_____   9. Inspect fuel lines and fittings for damage, wear, or dripping fuel. Tighten the fittings or replace lines and fittings that have problems.

_____ 10. Check the area around the fuel pump for evidence of fuel leaks. If you find any evidence of a leak, repair the lines or replace the pump as necessary.

_____ 11. Inspect the engine head gasket for oil seepage.

_____ 12. Check the top surfaces of the engine for oil seepage. *Note:* Seepage at this location is usually caused by a leaking gasket. Replace gaskets as necessary.

_____ 13. Start the engine and let it idle until the temperature gauge moves into the normal range.

_____ 14. Repeat Steps 5 through 12 while the engine is idling. Watch out for moving parts.

_____ 15. Turn off the engine, lower the hood, and raise the vehicle on a lift or jack up the vehicle and place it on jack stands.

_____ 16. Check the oil drain plug for oil leaks. Tighten the leaking plug or replace it as necessary.

_____ 17. Inspect the oil pan and gasket for evidence of oil leaks and/or road damage. If necessary, replace these components.

_____ 18. Inspect around the front and rear crankshaft seals for oil leaks. Replace as necessary.

_____ 19. Check the entire area underneath the vehicle (including structural members and the exhaust system) for oil, coolant, or fuel leaks. Re-inspect components and repair leaks as needed. Note: Liquids may travel along parts of the vehicle and appear some distance from the original leak.

Performance ✓ Checklist

Name _____ Date _____ Class _____

PERFORMANCE STANDARDS:
Level 4–Performs skill without supervision and adapts to problem situations.
Level 3–Performs skill satisfactorily without assistance or supervision.
Level 2–Performs skill satisfactorily, but requires assistance/supervision.
Level 1–Performs parts of skill satisfactorily, but requires considerable assistance/supervision.

Attempt (circle one): **1 2 3 4**

Comments:

_____

_____

_____

_____

PERFORMANCE LEVEL ACHIEVED: _____

_____   1. Safety rules and practices were followed at all times regarding this job.
_____   2. Tools and equipment were used properly and stored upon completion of this job.
_____   3. This completed job met the standards set and was done within the allotted time.
_____   4. No injury or damage to property occurred during this job.
_____   5. Upon completion of this job, the work area was cleaned correctly.

Instructor's Signature _____ Date_____

PERFORMING OIL PRESSURE TESTS

NATEF Standard(s) for Engine Repair:
D1 Perform oil pressure tests; determine necessary action.

Safety First
- ☐ Wear safety glasses at all times.
- ☐ Follow all safety rules when using common hand tools.
- ☐ Always connect a vehicle's exhaust to a vent hose before you run an engine in a closed shop. Unvented exhaust fumes in a closed shop can cause death.
- ☐ Keep jewelry, loose clothing, and hair away from moving parts while an engine is running.
- ☐ Be careful of hot engine parts.
- ☐ Check Material Safety Data Sheets for chemical safety information.
- ☐ Set the parking brake and place automatic transmission in PARK or manual transmission in NEUTRAL.

Tools & Equipment:
- Oil pressure test kit
- Pressure sensing unit socket
- Tachometer
- Common hand tools
- Vehicle service information

PROCEDURES Refer to the vehicle service information for specifications and special procedures. Then perform oil pressure tests on the engine provided by your instructor.

_____ 1. Write up a repair order.

_____ 2. Make sure you follow all procedures in the vehicle service information.

_____ 3. Check the engine oil level and condition with the dipstick. *Note:* If the oil is thin, it may be contaminated with fuel. If it is thick or dark, it needs to be changed. If the level is low, add oil until it is at its recommended level.

_____ 4. Look up the sensor's location in the vehicle service information.
   Source: _____ *Page:* _____

_____ 5. Locate the sensor and carefully remove it by using the special socket and a ratchet.
   CAUTION: Be very careful not to break the sensing unit.

_____ 6. Install an oil gauge in the sensor orifice. *Note:* Use an adapter if necessary.

_____ 7. Connect the appropriate tachometer to the engine.

_____ 8. Start the engine and allow it to idle until it has reached its normal operating temperature.

_____ 9. If the engine has a carburetor, release the fast idle mechanism by quickly depressing the accelerator pedal.

(continued)

Engine Repair

_____ **10.** Record the idle rpm on the tachometer.

_____

_____ **11.** Record the oil pressure reading from the oil pressure gauge.

_____

_____ **12.** Increase engine rpm to 2000 by depressing the accelerator pedal and record the oil pressure.

_____

_____ **13.** Turn off the engine.

_____ **14.** Look up the recommended oil pressures in the vehicle service information.

    Source: _____ *Page:* _____

_____ **15.** Compare your pressure readings to the manufacturer's specifications. ***Note:*** See your instructor for necessary repairs.

Performance ✓ Checklist

Name _____ Date _____ Class _____

PERFORMANCE STANDARDS:
Level 4–Performs skill without supervision and adapts to problem situations.
Level 3–Performs skill satisfactorily without assistance or supervision.
Level 2–Performs skill satisfactorily, but requires assistance/supervision.
Level 1–Performs parts of skill satisfactorily, but requires considerable assistance/supervision.

Attempt (circle one): **1 2 3 4**

Comments:

_____

_____

_____

_____

PERFORMANCE LEVEL ACHIEVED: _____

_____ **1.** Safety rules and practices were followed at all times regarding this job.

_____ **2.** Tools and equipment were used properly and stored upon completion of this job.

_____ **3.** This completed job met the standards set and was done within the allotted time.

_____ **4.** No injury or damage to property occurred during this job.

_____ **5.** Upon completion of this job, the work area was cleaned correctly.

Instructor's Signature _____ Date_____

DIAGNOSING ENGINE NOISES AND VIBRATIONS

NATEF Standard(s) for Engine Repair:

A6 Diagnose engine noises and vibrations; determine necessary action.

Safety First

- ☐ Wear safety glasses at all times.
- ☐ Follow all safety rules when using common hand tools.
- ☐ Use exhaust vent if running engine in closed shop. Unvented exhaust fumes can cause death.
- ☐ Keep jewelry, loose clothing, and hair away from moving parts while an engine is running.
- ☐ Stay clear of cooling fan and hot surfaces.
- ☐ Follow all safety rules when using a lift or jack and jack stands.
- ☐ Make sure automatic transmission is in PARK or manual transmission is in NEUTRAL and set the parking brake

Tools & Equipment:

- Vehicle service information
- Fender covers
- Engine diagnostic equipment
- Common hand tools
- Electronic vibration analyzer
- Automotive stethoscope
- Lift or jack and jack stands

PROCEDURES Refer to the vehicle service information for specifications and special procedures. Then diagnose the engine noises and vibrations in the vehicle provided by your instructor.

_____ 1. Write up a repair order.

_____ 2. Make sure you follow all procedures in the vehicle service information.

_____ 3. Visually check for defective or worn belts and pulleys. Identify defective belts and pulleys that need replacement.

_____ 4. Use an automotive electronic vibration analyzer or automotive stethoscope to test for defective engine-driven accessories. Repair or replace defective generator, power steering pump, etc. needing replacement.

_____ 5. Check for exhaust leakage. Identify components needing service or repair.

_____ 6. Listen for a high-pitched metallic knocking noise at crankshaft speed. If the noise is present at a fast and steady idle speed (where most pronounced), it can be greatly affected by grounding out the ignition for a particular cylinder. The noise would most likely be caused by a worn or damaged rod bearing on that cylinder. The connecting rod, crankshaft, and related bearings would likely require replacement. Proper operation of the engine lubrication system must also be verified.

(continued)

Engine Repair

_____ 7. Listen for a low-pitched knocking noise most pronounced when the engine is cold or operating under a load. This noise is likely caused by excessive main bearing clearance caused by wear or damaged drain bearings. The crankshaft would likely require machining or replacement. The bearings would also have to be replaced. Proper operation of the engine lubrication system must also be verified.

_____ 8. Listen for a muffled, hollow, knocking noise in an upper cylinder block. The noise is usually louder with a cold engine. This can be a piston knock or slap caused by a damaged piston or excessive clearance between the piston skirt and the cylinder wall. The presence of this noise may require replacement pistons and a possible cylinder reboring.

_____ 9. Listen for a high-pitched metallic double-knock noise in the cylinder area of the engine block. This noise may be caused by excessive clearance of a piston pin and could require replacement of a piston, piston pin, and connecting rod.

Performance ✓ Checklist

Name _____ Date _____ Class _____

PERFORMANCE STANDARDS:
Level 4–Performs skill without supervision and adapts to problem situations.
Level 3–Performs skill satisfactorily without assistance or supervision.
Level 2–Performs skill satisfactorily, but requires assistance/supervision.
Level 1–Performs parts of skill satisfactorily, but requires considerable assistance/supervision.

Attempt (circle one): **1 2 3 4**

Comments:

_____

_____

_____

_____

PERFORMANCE LEVEL ACHIEVED: _____

_____ 1. Safety rules and practices were followed at all times regarding this job.
_____ 2. Tools and equipment were used properly and stored upon completion of this job.
_____ 3. This completed job met the standards set and was done within the allotted time.
_____ 4. No injury or damage to property occurred during this job.
_____ 5. Upon completion of this job, the work area was cleaned correctly.

Instructor's Signature _____ Date_____

REMOVING A FRONT-WHEEL DRIVE ENGINE

Engine Repair

NATEF Standard(s) for Engine Repair:

A12 Remove and reinstall engine in a late model front-wheel drive or rear-wheel drive vehicle (OBDII or newer); reconnect all attaching components and restore the vehicle to running condition.

Safety First
- ☐ Wear safety glasses at all times.
- ☐ Follow all safety rules when using common hand and air tools.
- ☐ Follow all safety rules when using an engine stand.
- ☐ Follow all safety rules when using lift equipment.
- ☐ Wear impervious gloves when handling chemicals.
- ☐ Use chemicals in well-ventilated areas.
- ☐ Wear a respirator when using chemicals.
- ☐ Check Material Safety Data Sheets for chemical safety information.

Tools & Equipment:
- Appropriate engine stand
- Solvent and hose
- Lift equipment
- Scraper
- Impervious gloves
- Common hand and air tools
- Respirator
- Transaxle support fixture
- Vehicle service information

PROCEDURES Refer to the vehicle service information for specifications and special procedures. Then mark or tag linkages, hoses, belts, wires, and fasteners in the engine provided by your instructor.

_____ 1. Write up a repair order.

_____ 2. Make sure you follow all procedures in the vehicle service information for removing and cleaning the engine.

_____ 3. Make scribe marks around hinges and remove the hood if necessary.

_____ 4. Mark or tag linkages, hoses, belts, wires, and fasteners to be sure they are correctly reinstalled.

_____ 5. With the engine on a stand, prepare the engine for cleaning by protecting the necessary components according to the vehicle service information.

_____ 6. Scrape off any dirt and apply the cleaning solvent. Allow the cleaner to penetrate for five minutes.

_____ 7. Hose off remaining dirt. **Note:** Make sure that the cleaning chemicals used and the contaminated wastewater generated are not in violation of local codes for ground or sewer contamination. In some cases, it may be necessary to disassemble the engine and individually clean parts in approved parts-cleaning equipment.

(continued)

Performance ✓ Checklist

Name _____ Date _____ Class _____

PERFORMANCE STANDARDS:
Level 4–Performs skill without supervision and adapts to problem situations.
Level 3–Performs skill satisfactorily without assistance or supervision.
Level 2–Performs skill satisfactorily, but requires assistance/supervision.
Level 1–Performs parts of skill satisfactorily, but requires considerable assistance/supervision.

Attempt (circle one): **1 2 3 4**

Comments:

_____

_____

_____

_____

PERFORMANCE LEVEL ACHIEVED: _____

_____ **1.** Safety rules and practices were followed at all times regarding this job.

_____ **2.** Tools and equipment were used properly and stored upon completion of this job.

_____ **3.** This completed job met the standards set and was done within the allotted time.

_____ **4.** No injury or damage to property occurred during this job.

_____ **5.** Upon completion of this job, the work area was cleaned correctly.

Instructor's Signature _____ Date_____

REMOVING A REAR-WHEEL DRIVE ENGINE

NATEF Standard(s) for Engine Repair:

A12 Remove and reinstall engine in a late model front-wheel drive or rear-wheel drive vehicle (OBDII or newer); reconnect all attaching components and restore the vehicle to running condition.

Safety First

☐ Wear safety glasses at all times.
☐ Follow all safety rules when using common hand and air tools.
☐ Follow all safety rules when using an engine stand.
☐ Follow all safety rules when using lift equipment.
☐ Wear impervious gloves when handling chemicals.
☐ Use chemicals in well-ventilated areas.
☐ Wear a respirator when using chemicals.
☐ Check Material Safety Data Sheets for chemical safety information.

Engine Repair

Tools & Equipment:

- Appropriate engine stand
- Solvent and hose
- Common hand and air tools
- Scraper
- Impervious gloves
- Vehicle service information
- Respirator
- Lift equipment

PROCEDURES Refer to the vehicle service information for specifications and special procedures. Then mark or tag linkages, hoses, belts, wires, and fasteners in the engine provided by your instructor.

_____ 1. Write up a repair order.

_____ 2. Make sure you follow all procedures in the vehicle service information for removing and cleaning the engine.

_____ 3. Make scribe marks around hinges and remove the hood if necessary.

_____ 4. Mark or tag linkages, hoses, belts, wires, and fasteners to be sure they are correctly reinstalled.

_____ 5. With the engine on a stand, prepare the engine for cleaning by protecting the necessary components according to the vehicle service information.

_____ 6. Scrape off any dirt and apply the cleaning solvent. Allow the cleaner to penetrate for five minutes.

_____ 7. Hose off remaining dirt. **Note:** Make sure that the cleaning chemicals used and the contaminated wastewater generated are not in violation of local codes for ground or sewer contamination. In some cases, it may be necessary to disassemble the engine and individually clean parts in approved parts-cleaning equipment.

(*continued*)

Performance ✓ Checklist

Name _____ Date _____ Class _____

PERFORMANCE STANDARDS:
Level 4–Performs skill without supervision and adapts to problem situations.
Level 3–Performs skill satisfactorily without assistance or supervision.
Level 2–Performs skill satisfactorily, but requires assistance/supervision.
Level 1–Performs parts of skill satisfactorily, but requires considerable assistance/supervision.

Attempt (circle one): **1 2 3 4**

Comments:

_____

_____

_____

_____

PERFORMANCE LEVEL ACHIEVED: _____

_____ **1.** Safety rules and practices were followed at all times regarding this job.

_____ **2.** Tools and equipment were used properly and stored upon completion of this job.

_____ **3.** This completed job met the standards set and was done within the allotted time.

_____ **4.** No injury or damage to property occurred during this job.

_____ **5.** Upon completion of this job, the work area was cleaned correctly.

Instructor's Signature _____ Date_____

INSPECTING CYLINDERS FOR WEAR AND DAMAGE

NATEF Standard(s) for Engine Repair:
C4 Inspect and measure cylinder walls for damage, wear, and ridges; determine necessary action.

Safety First

- ☐ Wear safety glasses at all times.
- ☐ Follow all safety rules when using common hand tools.
- ☐ Follow all safety rules when using engine stands.

Engine Repair

Tools & Equipment:
- Telescoping gauge
- Outside micrometer
- Engine stands
- Inside micrometer
- Bore gauge
- Drop light
- Common hand tools
- Vehicle service information

PROCEDURES Refer to the vehicle service information for specifications and special procedures. Then inspect and measure cylinder walls for damage and wear in the engine provided by your instructor.

_____ 1. Write up a repair order.

_____ 2. Make sure you follow all procedures in the vehicle service information.

_____ 3. Look up the manufacturer's recommended procedure for inspecting and measuring cylinder walls.

   Source: _____ *Page:* _____

_____ 4. Inspect the inside of each cylinder by using a drop light.

_____ 5. Check each cylinder for scuffs, gouges, burns, discoloration, glazed spots, and other damage. **CAUTION:** Minor damage can sometimes be repaired by honing the inside of the cylinder wall (Job Sheet ER-47) or by replacing a cylinder sleeve.

_____ 6. Look up and record the manufacturer's specified tolerances for out-of-round and taper of cylinder walls.

   _____

_____ 7. Measure the cylinder walls (the area of ring travel) for taper and out-of-round with the telescoping gauge and outside micrometer, the bore gauge, or the inside micrometer (according to your instructor). ***Note:*** Take at least two measurements at both the top (just under the ridge area) and the bottom (where the piston stops) of the cylinder.

_____ 8. If any of the cylinder walls cannot be easily repaired by honing or replacing a sleeve, send the block to a machine shop to have all of the cylinders bored.

(continued)

Performance ✓ Checklist

Name _____   Date _____   Class _____

PERFORMANCE STANDARDS:
Level 4–Performs skill without supervision and adapts to problem situations.
Level 3–Performs skill satisfactorily without assistance or supervision.
Level 2–Performs skill satisfactorily, but requires assistance/supervision.
Level 1–Performs parts of skill satisfactorily, but requires considerable assistance/supervision.

Attempt (circle one): **1 2 3 4**

Comments:

_____

_____

_____

_____

PERFORMANCE LEVEL ACHIEVED: _____

_____ **1.** Safety rules and practices were followed at all times regarding this job.

_____ **2.** Tools and equipment were used properly and stored upon completion of this job.

_____ **3.** This completed job met the standards set and was done within the allotted time.

_____ **4.** No injury or damage to property occurred during this job.

_____ **5.** Upon completion of this job, the work area was cleaned correctly.

Instructor's Signature _____   Date_____

REMOVING CYLINDER WALL RIDGES AND DISASSEMBLING ENGINE BLOCKS

NATEF Standard(s) for Engine Repair:

C4 Inspect and measure cylinder walls for damage, wear, and ridges; determine necessary action.

Safety First
- ☐ Wear safety glasses at all times.
- ☐ Follow all safety rules when using common hand tools.
- ☐ Follow all safety rules when using engine stands.
- ☐ Check Material Safety Data Sheets for chemical safety information.

Tools & Equipment:
- Ridge reamer
- Common hand tools
- Engine stands
- Vehicle service information
- Scraper
- Cleaning solvent

PROCEDURES Refer to the vehicle service information for specifications and special procedures. Then remove the ridges from the cylinder walls in the engine provided by your instructor.

_____ 1. Write up a repair order.

_____ 2. Make sure you follow all procedures in the vehicle service information.

_____ 3. Look up the manufacturer's recommended procedure for disassembling an engine.

Source: _____ *Page:* _____

_____ 4. Make a note of all the damaged areas of the engine so you will repair them before reassembling the engine.

_____

_____ 5. Remove the pistons according to the manufacturer's recommendations. **CAUTION:** The pistons will be damaged if ridges in the cylinder walls have not been removed first.

_____ 6. Check the internal cylinder walls for ridges by running a fingernail over the wall. *Note:* If your fingernail catches on a ridge, the ridge must be removed.

_____ 7. Clean any carbon buildup from this area by gently removing it with the scraper.

_____ 8. Remove the ridge by following the ridge reamer manufacturer's instructions.

_____ 9. Remove the piston according to the engine manufacturer's recommendations in the vehicle service information. *Note:* Before assembly, the engine block must be cleaned.

_____ 10. Remove all the other removable parts on the engine before cleaning the block. *Note:* You can leave copper cam bearings in place.

(continued)

_____ **11.** Determine the correct procedure for cleaning the block. *Note:* Cast iron blocks can be cleaned in a hot tank, but aluminum blocks cannot. Appropriate solvents can be used, or a block can be cleaned by heating it in an oven.

_____ **12.** Flush the block thoroughly according to your instructor and remove all the caustic cleaning solution. *Note:* Cleaning the block removes all the rust and flakes from the outside of the engine, and all the oil galleries and the water jacket.

_____ **13.** Lubricate a cast iron block immediately after it has been cleaned to prevent it from rusting. *Note:* A cast iron block will begin to rust right after it has been cleaned.

_____ **14.** Reassemble the engine according to your instructor's directions.

Performance ✓ Checklist

Name _____ Date _____ Class _____

PERFORMANCE STANDARDS:
Level 4–Performs skill without supervision and adapts to problem situations.
Level 3–Performs skill satisfactorily without assistance or supervision.
Level 2–Performs skill satisfactorily, but requires assistance/supervision.
Level 1–Performs parts of skill satisfactorily, but requires considerable assistance/supervision.

Attempt (circle one): **1 2 3 4**

Comments:

_____

_____

_____

_____

PERFORMANCE LEVEL ACHIEVED: _____

_____ **1.** Safety rules and practices were followed at all times regarding this job.
_____ **2.** Tools and equipment were used properly and stored upon completion of this job.
_____ **3.** This completed job met the standards set and was done within the allotted time.
_____ **4.** No injury or damage to property occurred during this job.
_____ **5.** Upon completion of this job, the work area was cleaned correctly.

Instructor's Signature _____ Date_____

DISASSEMBLING & INSPECTING CYLINDER HEADS AND VALVE TRAINS

NATEF Standard(s) for Engine Repair:

B2 Visually inspect cylinder head(s) for cracks; check gasket surface areas for warpage and leakage; check passage condition.

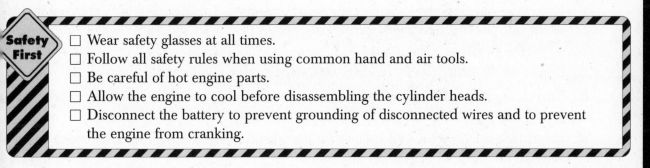

Safety First

☐ Wear safety glasses at all times.

☐ Follow all safety rules when using common hand and air tools.

☐ Be careful of hot engine parts.

☐ Allow the engine to cool before disassembling the cylinder heads.

☐ Disconnect the battery to prevent grounding of disconnected wires and to prevent the engine from cranking.

Engine Repair

Tools & Equipment:

- Common hand and air tools
- Straightedge
- Vehicle service information
- Other equipment as recommended by the manufacturer
- Feeler gauge
- Rotary wire brush

PROCEDURES   Refer to the vehicle service information for specifications and special procedures. Then disassemble and inspect the cylinder heads and valve trains in the engine provided by your instructor.

_____ 1. Write up a repair order.

_____ 2. Make sure you follow all procedures in the vehicle service informaton.

_____ 3. On a separate sheet of paper, draw a brief sketch of the cylinder head(s) for reassembling cylinder head(s) and valve trains.

_____ 4. Drain the engine coolant, clean the engine, and support the timing chain (if necessary) before beginning to take apart the cylinder head. Intake and exhaust manifolds may have to be removed from the head, or in some cases, they can be removed with the head.

_____ 5. As necessary, mark or label all attachments and fasteners before removing them.

_____ 6. As you remove the valve lifters, put them in labeled plastic bags.

_____ 7. Loosen the cylinder head bolts in the correct sequence.

_____ 8. Carefully and slowly pry the cylinder head away from the lower engine block at the edges. **CAUTION:** Do not pry against the ports. This can crack the head.

(continued)

_____ 9. Place the head on a head stand to prevent bending the valves. **CAUTION:** Do not place the head facedown on a bench–this can bend open valves on an overhead cam head if the cam is still in place.

_____ 10. Inspect the cylinder head for cracks, burned areas, and warpage. See your instructor about magnafluxing, dye penetrating, and gauging for finding damage. *Note:* Damage is often found at the exhaust seat or between chambers. To properly check for cracks, the valves will have to be removed and the area around the valve seats and guides will have to be cleaned with a rotary wire brush.

_____ 11. Inspect the valve train for loose nuts and bolts, broken valve springs, bent push rods, poorly seated valve locks and retainers, damaged or missing stem seals, and worn cam lobes, rocker arms, or followers.

_____ 12. Inspect the oil drain holes for blockage by inserting a wire into the drain holes.

_____ 13. After cleaning the gasket surface of the head, check for warpage using a straightedge and feeler gauge.

_____ 14. Refer to the vehicle service information for repairing damaged cylinder heads and valve trains. See your instructor for reassembly procedures.

Performance ✓ Checklist

Name _____   Date _____   Class _____

PERFORMANCE STANDARDS:
Level 4–Performs skill without supervision and adapts to problem situations.
Level 3–Performs skill satisfactorily without assistance or supervision.
Level 2–Performs skill satisfactorily, but requires assistance/supervision.
Level 1–Performs parts of skill satisfactorily, but requires considerable assistance/supervision.

Attempt (circle one): **1 2 3 4**

Comments:

_____

_____

_____

_____

PERFORMANCE LEVEL ACHIEVED: _____

_____ 1. Safety rules and practices were followed at all times regarding this job.
_____ 2. Tools and equipment were used properly and stored upon completion of this job.
_____ 3. This completed job met the standards set and was done within the allotted time.
_____ 4. No injury or damage to property occurred during this job.
_____ 5. Upon completion of this job, the work area was cleaned correctly.

Instructor's Signature _____   Date_____

INSPECTING & TESTING VALVE SPRINGS

NATEF Standard(s) for Engine Repair:

B3 Inspect valve springs for squareness and free height comparison; determine necessary action.

Safety First
- ☐ Wear safety glasses at all times.
- ☐ Follow all safety rules when using common hand tools.

Tools & Equipment:
- Caliper
- Valve spring compressor
- Beam-type foot-pound torque wrench
- Combination square
- Valve spring tester
- Common hand tools
- Vehicle service information

PROCEDURES Refer to the vehicle service information for specifications and special procedures. Then remove, test, and replace any defective valve springs in the engine provided by your instructor.

_____ 1. Write up a repair order.

_____ 2. Make sure you follow all procedures in the vehicle service information.

_____ 3. Remove the rocker arm or rocker follower, if there is one.

_____ 4. Using a valve spring compressor, remove the valve springs.

_____ 5. Check the spring for "squareness" by rotating it next to a combination square on a flat surface. Record the greatest gap. *Note:* If the gap is greater than 1/329 for each inch of length, it is not square and must be replaced.

_____ 6. Check and record the free height (uncompressed length) of the spring by measuring it with the calipers. *Note:* If the height varies by more than 10% of the manufacturer's specified length, replace the spring.

_____ 7. Determine the valve spring-loaded length specification and adjust the valve spring tester to this height.

_____ 8. Place the valve spring on the tester and use a beam-type foot-pound torque wrench to apply pressure to the spring.

_____ 9. As the tone sounds, note the indicator on the torque wrench and multiply by two. This is the tension applied to the spring at the specified height. If this is not within the manufacturer's specifications, replace the spring.

(continued)

Engine Repair

Performance ✓ Checklist

Name _____   Date _____   Class _____

PERFORMANCE STANDARDS:
Level 4–Performs skill without supervision and adapts to problem situations.
Level 3–Performs skill satisfactorily without assistance or supervision.
Level 2–Performs skill satisfactorily, but requires assistance/supervision.
Level 1–Performs parts of skill satisfactorily, but requires considerable assistance/supervision.

Attempt (circle one): **1 2 3 4**

Comments:

_____

_____

_____

_____

PERFORMANCE LEVEL ACHIEVED: _____

_____ 1. Safety rules and practices were followed at all times regarding this job.

_____ 2. Tools and equipment were used properly and stored upon completion of this job.

_____ 3. This completed job met the standards set and was done within the allotted time.

_____ 4. No injury or damage to property occurred during this job.

_____ 5. Upon completion of this job, the work area was cleaned correctly.

Instructor's Signature _____   Date_____

INSPECTING & RECONDITIONING VALVE GUIDES

Engine Repair

NATEF Standard(s) for Engine Repair:
B5 Inspect valve guides for wear; check valve stem-to-guide clearance; determine necessary action.

Safety First

☐ Wear safety glasses at all times.
☐ Follow all safety rules when using common hand tools.
☐ Follow all safety rules when using power tools.
☐ Check Material Safety Data Sheets for chemical safety information.

Tools & Equipment:

- Nylon guide brush
- Knurling lubricant
- Dial indicator
- Vehicle service information
- Ball gauge
- Knurling arbor
- Guide-cleaning compound
- Drill with a speed reducer
- Valve guide reamer
- Common hand tools

PROCEDURES Refer to the vehicle service information for specifications and special procedures. Then inspect the valve guide provided by your instructor.

_____ 1. Write up a repair order.

_____ 2. Make sure you follow all procedures in the vehicle service information.

_____ 3. Remove the valve from the guide according to the appropriate vehicle service information.

_____ 4. Inspect the valve guide for looseness and for burned, worn, or cracked areas.

_____ 5. Measure the guide by placing a small hole gauge into the guide and taking measurements at several places along the length of the guide. Measure the valve guide with a dial indicator.

_____ 6. Check the valve guide for too much clearance between the valve stem and the guide by inserting the valve into the guide, opening the valve to the normal open position, and wiggling the valve head while measuring this movement with a dial indicator. *Note:* See your instructor for how much movement is too much.

_____ 7. If a guide is unevenly or excessively worn according to the manufacturer's specifications, replace or recondition it. *Note:* Insert valve guides should be replaced in a machine shop.

_____ 8. Begin reconditioning a valve guide by removing any carbon deposits with a guide-cleaning compound. The valve guide can be reconditioned by knurling.

_____ 9. Lubricate the guide with a knurling lubricant and select the proper size of knurling arbor.

(continued)

_____ **10.** Position the arbor on top of the guide. **CAUTION:** If you are using a power drill to knurl, see your instructor for safety rules.

_____ **11.** Drive the arbor evenly through the guide. *Note:* The guide should now be smaller than the valve stem. If the stem fits in at all, the guide is too worn and must be replaced.

_____ **12.** After knurling, ream the guide to the proper diameter by inserting a reamer into it and slowly turning it clockwise to prevent the reamer from dulling.

_____ **13.** Clean the valve guide with a nylon brush or compressed air.

_____ **14.** If the bore of the valve guide becomes too big from being reconditioned, see your instructor. It is possible that the guide can be reamed oversize to accommodate an oversize valve stem.

Performance ✓ Checklist

Name _____ Date _____ Class _____

PERFORMANCE STANDARDS:
Level 4–Performs skill without supervision and adapts to problem situations.
Level 3–Performs skill satisfactorily without assistance or supervision.
Level 2–Performs skill satisfactorily, but requires assistance/supervision.
Level 1–Performs parts of skill satisfactorily, but requires considerable assistance/supervision.

Attempt (circle one): **1 2 3 4**

Comments:

_____

_____

_____

_____

PERFORMANCE LEVEL ACHIEVED: _____

_____ **1.** Safety rules and practices were followed at all times regarding this job.
_____ **2.** Tools and equipment were used properly and stored upon completion of this job.
_____ **3.** This completed job met the standards set and was done within the allotted time.
_____ **4.** No injury or damage to property occurred during this job.
_____ **5.** Upon completion of this job, the work area was cleaned correctly.

Instructor's Signature _____ Date_____

INSPECTING CAMSHAFT DRIVES

NATEF Standard(s) for Engine Repair:

B12 Inspect camshaft drives (including wear and backlash, sprocket and chain wear); determine necessary action.

Safety First

☐ Wear safety glasses at all times.
☐ Follow all safety rules when using common hand tools.
☐ Follow all safety rules when using air tools.
☐ Follow all safety rules when using a lift or a jack and jack stands.
☐ Always connect a vehicle's exhaust to a vent hose before you run an engine in a closed shop. Unvented exhaust fumes in a closed shop can cause death.
☐ Keep jewelry, loose clothing, and hair away from moving parts while an engine is running.
☐ Be careful of hot engine parts.
☐ Set the parking brake and place automatic transmission in PARK or manual transmission in NEUTRAL.

Tools & Equipment:
- Air tools
- Common hand tools
- Torque wrench
- Vehicle service information
- Timing light

PROCEDURES Refer to the vehicle service information for specifications and special procedures. Then inspect and replace any damaged or worn parts in the camshaft drive in the engine provided by your instructor.

_____ 1. Write up a repair order.

_____ 2. Make sure you follow all procedures in the vehicle service information.

_____ 3. Look up the procedure for disassembling the engine to inspect the camshaft drive.

   Source: _____ _Page:_ _____

_____ 4. Disassemble the engine to inspect the parts.

_____ 5. Check the sprockets and chains for looseness and wear, or check the gears for damaged teeth.

_____ 6. Replace worn or damaged parts as you reassemble the camshaft drive according to the manufacturer's specifications. **CAUTION:** Defective components of the camshaft drive can damage the valves and pistons. Follow the manufacturer's instructions for timing the sprockets or gears.

_____ 7. After engine assembly, verify and adjust coolant and oil levels.

_____ 8. If the ignition timing is adjustable, follow the manufacturer's instructions to properly adjust it.

(continued)

Engine Repair

Performance ✓ Checklist

Name _____ Date _____ Class _____

PERFORMANCE STANDARDS:
Level 4–Performs skill without supervision and adapts to problem situations.
Level 3–Performs skill satisfactorily without assistance or supervision.
Level 2–Performs skill satisfactorily, but requires assistance/supervision.
Level 1–Performs parts of skill satisfactorily, but requires considerable assistance/supervision.

Attempt (circle one): **1 2 3 4**

Comments:

_____

_____

_____

_____

PERFORMANCE LEVEL ACHIEVED: _____

_____ **1.** Safety rules and practices were followed at all times regarding this job.

_____ **2.** Tools and equipment were used properly and stored upon completion of this job.

_____ **3.** This completed job met the standards set and was done within the allotted time.

_____ **4.** No injury or damage to property occurred during this job.

_____ **5.** Upon completion of this job, the work area was cleaned correctly.

Instructor's Signature _____ Date_____

INSPECTING & MEASURING THE CAMSHAFT

NATEF Standard(s) for Engine Repair:

B14 Inspect camshaft for runout, journal wear, and lobe wear.

Safety First

☐ Wear safety glasses at all times.
☐ Follow all safety rules when using common hand tools.

Tools & Equipment:

- Outside micrometer
- Dial indicator
- V-blocks
- Common hand tools
- Vehicle service information

PROCEDURES Refer to the vehicle service information for specifications and special procedures. Then inspect and measure the camshaft provided by your instructor.

_____ 1. Write up a repair order.

_____ 2. Make sure you follow all procedures in the vehicle service information.

_____ 3. Inspect the surface of the camshaft for pitting, scoring, burning, or chipping. *Note:* Replace a pitted, scored, or chipped camshaft.

_____ 4. Use the micrometer to measure the outside diameter of the camshaft journal at four locations. Record your findings.

_____

_____ 5. Look up the manufacturer's specifications for allowable journal diameter. *Note:* Send the camshaft to be reground or replace it if the journals are not within the manufacturer's recommended specification. This is often no more than .0029 or .05 mm tolerance.

Source: _____ *Page:* _____

_____ 6. Measure each of the cam lobes and record your findings.

_____

_____ 7. Look up the manufacturer's specifications for allowable lobe size. *Note:* Send the camshaft to be reground or replace it if the lobes are not within the manufacturer's recommended specification. This is often no more than .0059 or .13 mm tolerance.

Source: _____ *Page:* _____

(continued)

Engine Repair

_____ 8. Measure and record the camshaft runout with a dial indicator by placing the camshaft on two V-blocks. ***Note:*** See your instructor for runout measurement procedures.

_____

_____ 9. Look up the manufacturer's specifications for allowable camshaft runout. ***Note:*** Send the camshaft to be straightened and reground or replace it if the camshaft runout is not within the manufacturer's recommended specification. This is often no more than .0059 or .13 mm tolerance.

Source: _____ *Page:* _____

_____ 10. Inspect the cam flange, cylinder head thrust surface, and engine block. Replace the camshaft if the flange or any contact area is worn or damaged.

Performance ✓ Checklist

Name _____ Date _____ Class _____

PERFORMANCE STANDARDS:
Level 4–Performs skill without supervision and adapts to problem situations.
Level 3–Performs skill satisfactorily without assistance or supervision.
Level 2–Performs skill satisfactorily, but requires assistance/supervision.
Level 1–Performs parts of skill satisfactorily, but requires considerable assistance/supervision.

Attempt (circle one): **1 2 3 4**

Comments:

_____

_____

_____

_____

PERFORMANCE LEVEL ACHIEVED: _____

_____ 1. Safety rules and practices were followed at all times regarding this job.
_____ 2. Tools and equipment were used properly and stored upon completion of this job.
_____ 3. This completed job met the standards set and was done within the allotted time.
_____ 4. No injury or damage to property occurred during this job.
_____ 5. Upon completion of this job, the work area was cleaned correctly.

Instructor's Signature _____ Date_____

INSPECTING & SERVICING PUSHRODS AND ROCKER ARMS

NATEF Standard(s) for Engine Repair:

B9 Inspect pushrods, rocker arms, rocker arm pivots and shafts for wear, bending, cracks, looseness, and blocked oil passages (orifices); determine necessary action.

Safety First

- ☐ Wear safety glasses at all times.
- ☐ Follow all safety rules when using common hand tools.
- ☐ Follow all safety rules when using power tools.
- ☐ Check Material Safety Data Sheets for chemical safety information.

Tools & Equipment:

- Micrometer
- Dial indicator
- Common hand tools
- V-blocks
- Thin probe
- Vehicle service information
- $1/16"$ welding rod
- Cleaning solvent

PROCEDURES Refer to the vehicle service information for specifications and special procedures. Then inspect and repair the pushrods and rocker arm assemblies provided by your instructor.

_____ 1. Write up a repair order.

_____ 2. Make sure you follow all procedures in the vehicle service information.

_____ 3. Clean the pushrods, rocker arms, rocker arm pivots, and shafts with the appropriate cleaning solvent provided by your instructor. Dry the parts gently with a clean cloth.

_____ 4. Clean any lubrication orifices in the pushrods by gently driving the 1/169 weld rod through them.

_____ 5. Inspect the pushrod ends for chips and wear.

_____ 6. Place the pushrod with its ends hanging over the edges on a corner of a bench.

_____ 7. Roll the pushrod and observe the ends. If it is bent, it will wobble.

_____ 8. Replace any bent, pitted, chipped, or worn pushrods.

_____ 9. Disassemble the rocker arm by removing the keeper from the end of the shaft. **Note:** Place the parts on a bench in the order they were removed. Be sure to note if they were removed from the right or left end of the shaft.

_____ 10. Inspect the rocker arm for pits and wear.

_____ 11. Inspect the rocker arm bushing (if used) inside the mounting hole for discoloration or wear. (Refer to the vehicle service information to replace worn rocker arm bushings.)

(continued)

_____ **12.** Inspect the rocker arm pivot and shaft for signs of wear or damage.

_____ **13.** Measure the diameter of the shaft with the micrometer and record it.

_____

_____ **14.** Compare the diameter measurement to the manufacturer's specifications in the vehicle service information. Replace the shaft if it is not within the manufacturer's specifications.

Source: _____ *Page:* _____

_____ **15.** Check the oil orifices in the shaft and remove any blockages with the thin probe.

_____ **16.** Put the rocker arm shaft across the V-blocks.

_____ **17.** Measure the run-out of the rocker arm shaft and record it.

_____ **18.** Compare it to the manufacturer's specifications. Refer to your instructor if it is not within specifications.

Source: _____ *Page:* _____

_____ **19.** Inspect the remaining rocker arm components and replace any damaged or worn parts. ***Note:*** Pedestal mounts rarely wear out. Do not replace them unless they have surface wear.

_____ **20.** Place the rocker arm shaft so the holes are all facing down and reassemble the rocker arm assembly. ***Note:*** Facing the holes downward provides lubrication to the shaft.

Performance ✓ Checklist

Name _____ Date _____ Class _____

PERFORMANCE STANDARDS:
Level 4–Performs skill without supervision and adapts to problem situations.
Level 3–Performs skill satisfactorily without assistance or supervision.
Level 2–Performs skill satisfactorily, but requires assistance/supervision.
Level 1–Performs parts of skill satisfactorily, but requires considerable assistance/supervision.

Attempt (circle one): **1 2 3 4**

Comments:

_____

_____

_____

_____

PERFORMANCE LEVEL ACHIEVED: _____

_____ **1.** Safety rules and practices were followed at all times regarding this job.

_____ **2.** Tools and equipment were used properly and stored upon completion of this job.

_____ **3.** This completed job met the standards set and was done within the allotted time.

_____ **4.** No injury or damage to property occurred during this job.

_____ **5.** Upon completion of this job, the work area was cleaned correctly.

Instructor's Signature _____ Date _____

INSPECTING & REPLACING LIFTERS

NATEF Standard(s) for Engine Repair:

B10 Inspect hydraulic or mechanical lifters; determine necessary action.

Engine Repair

Tools & Equipment:

- Numbered containers
- Small punch
- Common hand tools
- Snap-ring pliers
- Thin probe
- Vehicle service information
- Bristle brush
- Cleaning solvent

PROCEDURES Refer to the vehicle service information for specifications and special procedures. Then inspect and replace (if necessary) the lifter assemblies provided by your instructor.

_____ 1. Write up a repair order.

_____ 2. Make sure you follow all procedures in the vehicle service information.

_____ 3. Remove the lifters, mark their placement, and put them in the numbered containers. Refer to the vehicle service information as needed.

_____ 4. Inspect each lifter bottom for a convex shape. *Note:* To check the lifters, hold two of them bottom to bottom. You should be able to rock them against each other. If they don't rock, replace the lifter. Worn lifters can damage a good camshaft. Lifters and camshafts are usually replaced together.

_____ 5. Put a clean rag on a bench. If necessary, hydraulic lifters can be disassembled for inspection and cleaning.

_____ 6. Remove the snap ring on each lifter by holding the lifter with a small punch in the oil hole and then removing the snap ring with the snap-ring pliers. *Note:* Never inspect more than one lifter at a time or mix internal parts.

_____ 7. Put the internal parts on the rag and note the order in which you removed them.

_____ 8. Clean the lifter parts with an appropriate cleaning solvent.

_____ 9. Clean the oil orifices and open any other plugged holes with the thin probe.

(continued)

_____ **10.** Inspect the check valves, ball, and plunger. *Note:* If any of the parts show wear or damage, replace the lifter. Worn lifters cause oil leaks, reduce engine performance, and prevent oil from reaching the pushrod.

_____ **11.** Reassemble the lifter in the correct order and reinstall it according to the manufacturer's recommendations.

Performance ✓ Checklist

Name _____ Date _____ Class _____

PERFORMANCE STANDARDS:
Level 4–Performs skill without supervision and adapts to problem situations.
Level 3–Performs skill satisfactorily without assistance or supervision.
Level 2–Performs skill satisfactorily, but requires assistance/supervision.
Level 1–Performs parts of skill satisfactorily, but requires considerable assistance/supervision.

Attempt (circle one): **1 2 3 4**

Comments:

_____

_____

_____

_____

PERFORMANCE LEVEL ACHIEVED: _____

_____ **1.** Safety rules and practices were followed at all times regarding this job.
_____ **2.** Tools and equipment were used properly and stored upon completion of this job.
_____ **3.** This completed job met the standards set and was done within the allotted time.
_____ **4.** No injury or damage to property occurred during this job.
_____ **5.** Upon completion of this job, the work area was cleaned correctly.

Instructor's Signature _____ Date_____

INSPECTING ENGINE BLOCKS

NATEF Standard(s) for Engine Repair:

C2 Inspect engine block for visible cracks, passage condition, core and gallery plug condition, and surface warpage; determine necessary action.

Safety First
- ☐ Wear safety glasses at all times.
- ☐ Follow all safety rules when using common hand tools.
- ☐ Follow all safety rules when using an engine stand.

Engine Repair

Tools & Equipment:
- Steel rule
- Feeler gauge
- Vehicle service information
- File or whetstone
- Smooth file
- Brush
- Engine stand
- Common hand tools

PROCEDURES Refer to the vehicle service information for specifications and special procedures. Then inspect the engine block provided by your instructor.

_____ 1. Write up a repair order.

_____ 2. Make sure you follow all procedures in the vehicle service information.

_____ 3. Inspect the block by looking for cracks, rust, nicks, old repairs, and other damage on its surface. *Note:* Some cracks can be welded or repaired, but a cracked block usually must be replaced.

_____ 4. Remove the core hole plugs and oil gallery plugs. Inspect the core holes, and run a brush through the oil galleries and inspect them.

_____ 5. Clean the block's mating surfaces such as the deck, the main bearing caps, and the mounting pads for accessories with the file or the whetstone.

_____ 6. Check for warping by using the steel rule to "straight-edge" the block and its mating parts. Measure any warpage by slipping feeler gauges under the rule.

_____ 7. Look up the manufacturer's specifications for allowable warpage. *Note:* Any block that is warped beyond the specifications must be machined or replaced.

Source: _____ *Page:* _____

(continued)

Performance ✓ Checklist

Name _____ Date _____ Class _____

PERFORMANCE STANDARDS:
Level 4–Performs skill without supervision and adapts to problem situations.
Level 3–Performs skill satisfactorily without assistance or supervision.
Level 2–Performs skill satisfactorily, but requires assistance/supervision.
Level 1–Performs parts of skill satisfactorily, but requires considerable assistance/supervision.

Attempt (circle one): **1 2 3 4**

Comments:

_____

_____

_____

_____

PERFORMANCE LEVEL ACHIEVED: _____

_____ 1. Safety rules and practices were followed at all times regarding this job.

_____ 2. Tools and equipment were used properly and stored upon completion of this job.

_____ 3. This completed job met the standards set and was done within the allotted time.

_____ 4. No injury or damage to property occurred during this job.

_____ 5. Upon completion of this job, the work area was cleaned correctly.

Instructor's Signature _____ Date_____

INSPECTING & REPLACING CAMSHAFT BEARINGS

NATEF Standard(s) for Engine Repair:

B15 Inspect camshaft bearing surface for wear, damage, out-of-round, and alignment; determine necessary action.

Safety First
☐ Wear safety glasses at all times.
☐ Follow all safety rules when using common hand tools.

Tools & Equipment:
- Dial bore gauge
- Bearing driver
- Dial indicator
- Vehicle service information
- Plastigage®
- Torque wrench
- Flashlight
- V-blocks
- Straightedge
- Common hand tools

PROCEDURES Refer to the vehicle service information for specifications and special procedures. Then inspect and replace the camshaft bearings in the engine provided by your instructor.

Note: Replaceable cam bearings are usually replaced when a camshaft is removed and/or replaced. Inspecting the old bearings will offer clues that a camshaft is bent, that the head is warped, or that the block is not properly align bored.

_____ 1. Write up a repair order.

_____ 2. Make sure you follow all procedures in the vehicle service information.

_____ 3. Following the manufacturer's instructions, remove the camshaft from the engine and use a flashlight to inspect the bearings. Inspect them for pitting, scoring, and wear.

_____ 4. Many technicians send bearings to a machine shop to be replaced. If you install them yourself, be extremely careful not to mar the surfaces in any way and to keep the bearings clean. It is also very important to make sure that the oil holes in the bearings are in line with the oil holes in the engine. Bearings may be numbered, with number one located at the front of the engine.

_____ 5. For overhead cam engines with one-piece bearings, use the dial bore gauge to measure the diameter of the bearings. Record your findings. *Note:* Damaged bearings may indicate a bent camshaft or a warped head.

(continued)

_____ 6. Check cam runout with V-blocks and a dial indicator. Check journal bore alignment with a straightedge.

_____ 7. Replace one-piece bearings by driving the old bearings out, and the new bearings in with the bearing driver.

_____ 8. For overhead cam engines with two-piece bearings, label and remove each bearing cap, put Plastigage on the journal, replace the bearing cap, and torque it to the manufacturer's specifications.

_____ 9. Remove the bearing caps again and then measure the compressed Plastigage with the paper ruler included in the Plastigage package.

_____

_____ 10. Look up the manufacturer's specifications for allowable bearing wear.

Source: _____ *Page:* _____

_____ 11. When you replace a bearing, it is very important to make a note of the bearing's position in its cup before you remove it. Install the new bearing in the exact position in the cup and at the proper location number. *Note:* If location and position are not exact, the bearing bore will not be symmetrical.

_____ 12. Snap two-piece bearings into place, put the bearing caps back on in their original positions, and install the camshaft according to the manufacturer's instructions.

Performance ✓ Checklist

Name _____ Date _____ Class _____

PERFORMANCE STANDARDS:
Level 4–Performs skill without supervision and adapts to problem situations.
Level 3–Performs skill satisfactorily without assistance or supervision.
Level 2–Performs skill satisfactorily, but requires assistance/supervision.
Level 1–Performs parts of skill satisfactorily, but requires considerable assistance/supervision.

Attempt (circle one): **1 2 3 4**

Comments:

_____

_____

_____

_____

PERFORMANCE LEVEL ACHIEVED: _____

_____ 1. Safety rules and practices were followed at all times regarding this job.
_____ 2. Tools and equipment were used properly and stored upon completion of this job.
_____ 3. This completed job met the standards set and was done within the allotted time.
_____ 4. No injury or damage to property occurred during this job.
_____ 5. Upon completion of this job, the work area was cleaned correctly.

Instructor's Signature _____ Date_____

INSPECTING & MEASURING
CAMSHAFT BEARINGS

NATEF Standard(s) for Engine Repair:

C6 Inspect and measure camshaft bearings for wear, damage, out-of-round, and alignment; determine necessary action.

Safety First

☐ Wear safety glasses at all times.
☐ Follow all safety rules when using common hand tools.
☐ Follow all safety rules when using engine stands.

Tools & Equipment:
- Telescoping gauge
- Outside micrometer
- Small bore gauge
- Engine stands
- Common hand tools
- Vehicle service information

PROCEDURES Refer to the vehicle service information for specifications and special procedures. Then inspect and measure the camshaft bearings in the engine provided by your instructor.

_____ 1. Write up a repair order.

_____ 2. Make sure you follow all procedures in the vehicle service information.

_____ 3. Look up the finish and condition patterns for camshaft bearings.

   Source: _____ _Page:_ _____

_____ 4. Compare the finish and wear patterns of your camshaft bearings with those shown in your source.

_____ 5. If bearing finish matches any abnormal wear patterns, replace the old bearings with new ones. Be sure to discard the old bearings so they are not accidentally reused. **Note:** In normal practice, anytime a camshaft is removed for service and camshaft bearings are accessible, they are replaced to ensure sufficient oil pressure.

_____ 6. Camshaft bearings that pass the visual inspection can be measured to ensure that they are within specifications. Measure the inside diameter of each bearing by centering, tightening, and rotating the telescoping gauge inside the bearing to check it for roundness. **Note:** Center the telescoping gauge by making sure both of the ends are equal lengths inside the bearing. If available, a small bore gauge could be used to obtain these measurements.

_____ 7. Use an outside micrometer to measure the telescoping gauge to determine camshaft bore diameter and compare to specifications.

(continued)

_____ 8. If all the bearings are within specifications and round (the measurements inside each bearing are the same at different points), the bearings are still serviceable.

_____ 9. If one or two adjacent camshaft bearings are worn excessively or worn only on one side, the camshaft may be bent, or on an overhead camshaft engine, the cylinder head may be warped. The camshaft should be measured for runout, and the head should be checked for warpage.

Performance ✓ Checklist

Name _____ Date _____ Class _____

PERFORMANCE STANDARDS:
Level 4–Performs skill without supervision and adapts to problem situations.
Level 3–Performs skill satisfactorily without assistance or supervision.
Level 2–Performs skill satisfactorily, but requires assistance/supervision.
Level 1–Performs parts of skill satisfactorily, but requires considerable assistance/supervision.

Attempt (circle one): **1 2 3 4**

Comments:

_____

_____

_____

_____

PERFORMANCE LEVEL ACHIEVED: _____

_____ **1.** Safety rules and practices were followed at all times regarding this job.
_____ **2.** Tools and equipment were used properly and stored upon completion of this job.
_____ **3.** This completed job met the standards set and was done within the allotted time.
_____ **4.** No injury or damage to property occurred during this job.
_____ **5.** Upon completion of this job, the work area was cleaned correctly.

Instructor's Signature _____ Date_____

INSPECTING CRANKSHAFTS

NATEF Standard(s) for Engine Repair:

C7 Inspect crankshaft for end play, straightness, journal damage, keyway damage, thrust flange and sealing surface condition, and visual surface cracks; check oil passage condition; measure journal wear; check crankshaft sensor reluctor ring (where applicable); determine necessary action.

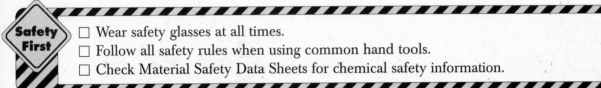

Safety First

☐ Wear safety glasses at all times.
☐ Follow all safety rules when using common hand tools.
☐ Check Material Safety Data Sheets for chemical safety information.

Tools & Equipment:
- Outside micrometer
- Hot tank or jet wash
- Common hand tools
- Vehicle service information

PROCEDURES Refer to the vehicle service information for specifications and special procedures. Then inspect and measure the crankshaft provided by your instructor.

_____ 1. Write up a repair order.

_____ 2. Make sure you follow all procedures in the vehicle service information.

_____ 3. Mark all bearing caps so you replace them in their correct locations.

_____ 4. Remove and clean the crankshaft in a hot tank or jet wash.

_____ 5. Check the crankshaft for cracks, scratches, pits, scoring, and other damage.

_____ 6. Measure the journals for out-of-round, taper, and undersize.

_____ 7. Inspect the outside surface of the journals for wear. Uneven wear patterns can indicate engine problems. If any damage is found, send the crankshaft for repair and further inspection by a machinist.

_____ 8. Reinspect any crankshaft that has been repaired for out-of-round, taper, and undersize; then install the correct bearings. *Note:* Ask your instructor for the correct bearing measurements.

_____ 9. Reinstall the crankshaft according to the vehicle service information. *Note:* If you store the crankshaft prior to reinstallation, store it standing on its end to prevent it from warping.

Source: _____ *Page:* _____

(continued)

Performance ✓ Checklist

Name _____ Date _____ Class _____

PERFORMANCE STANDARDS:
Level 4–Performs skill without supervision and adapts to problem situations.
Level 3–Performs skill satisfactorily without assistance or supervision.
Level 2–Performs skill satisfactorily, but requires assistance/supervision.
Level 1–Performs parts of skill satisfactorily, but requires considerable assistance/supervision.

Attempt (circle one): **1 2 3 4**

Comments:

_____

_____

_____

_____

PERFORMANCE LEVEL ACHIEVED: _____

_____ **1.** Safety rules and practices were followed at all times regarding this job.
_____ **2.** Tools and equipment were used properly and stored upon completion of this job.
_____ **3.** This completed job met the standards set and was done within the allotted time.
_____ **4.** No injury or damage to property occurred during this job.
_____ **5.** Upon completion of this job, the work area was cleaned correctly.

Instructor's Signature _____ Date _____

INSPECTING VIBRATION DAMPERS

NATEF Standard(s) for Engine Repair:

C14 Inspect, repair, or replace crankshaft vibration damper (harmonic balancer).

Safety First

- ☐ Wear safety glasses at all times.
- ☐ Follow all safety rules when using common hand tools.
- ☐ Follow all safety rules when using engine stands.
- ☐ Check Material Safety Data Sheets for chemical safety information.

Tools & Equipment:
- Engine stands
- Sleeve repair kit (if required)
- Common hand tools
- Vehicle service information
- Cleaning solvent

PROCEDURES Refer to the vehicle service information for specifications and special procedures. Then inspect and service the vibration damper in the engine provided by your instructor.

_____ 1. Write up a repair order.

_____ 2. Make sure you follow all procedures in the vehicle service information. Use the correct puller to remove the vibration damper.

_____ 3. Thoroughly clean the vibration damper. Be careful to use only cleaning solvents that are safe for rubber components. **CAUTION:** Some solvents will dissolve rubber.

_____ 4. Inspect all the parts by checking them for a tight fit. **CAUTION:** Never strike or pull on the outer metal ring of the vibration damper.

_____ 5. Look up the manufacturer's vibration damper part number for the engine provided by your instructor. *Note:* Other terms for a vibration damper are "harmonic dampener," "harmonic balancer," or "crank damper."

 Source: _____ *Page:* _____

_____ 6. Check the part number on the vibration damper provided by your instructor against the manufacturer's recommendation. *Note:* Installing the wrong damper can be worse than leaving the original damper in place.

_____ 7. Before reinstalling a vibration damper, clean and inspect the seal surface. Light scoring can be polished smooth with abrasive cloth.

_____ 8. Excessive scoring of the seal surface can be repaired by properly installing a sleeve kit. Follow the sleeve kit manufacturer's instructions. The front crankshaft seal also should be replaced.

(continued)

Engine Repair

Performance ✓ Checklist

Name _____ Date _____ Class _____

PERFORMANCE STANDARDS:
Level 4–Performs skill without supervision and adapts to problem situations.
Level 3–Performs skill satisfactorily without assistance or supervision.
Level 2–Performs skill satisfactorily, but requires assistance/supervision.
Level 1–Performs parts of skill satisfactorily, but requires considerable assistance/supervision.

Attempt (circle one): **1 2 3 4**

Comments:

_____

_____

_____

_____

PERFORMANCE LEVEL ACHIEVED: _____

_____ 1. Safety rules and practices were followed at all times regarding this job.

_____ 2. Tools and equipment were used properly and stored upon completion of this job.

_____ 3. This completed job met the standards set and was done within the allotted time.

_____ 4. No injury or damage to property occurred during this job.

_____ 5. Upon completion of this job, the work area was cleaned correctly.

Instructor's Signature _____ Date_____

INSPECTING ROD ALIGNMENT AND MAIN BEARING BORES

NATEF Standard(s) for Engine Repair:

C9 Identify piston and bearing wear patterns that indicate connecting rod alignment and main bearing bore problems; inspect rod alignment and bearing bore condition.

Safety First
☐ Wear safety glasses at all times.
☐ Follow all safety rules when using common hand tools.
☐ Follow all safety rules when using engine stands.

Tools & Equipment:

- Screwdriver
- Feeler gauge set
- Engine stands
- Common hand tools
- Vehicle service information
- Straightedge

PROCEDURES Refer to the vehicle service information for specifications and special procedures. Then inspect rod alignment and bearing bore condition in the engine provided by your instructor.

_____ 1. Write up a repair order.

_____ 2. Make sure you follow all procedures in the vehicle service information.

_____ 3. Look up the rod alignment and bearing bore condition wear indications in the vehicle service information.

Source: _____ Page: _____

_____ 4. On V-type engines, measure the side clearance by spreading the rods with the screwdriver, inserting the correct combination of feeler gauges between each set of rods, and recording the measurements. Repeat the measurement at two or three places around the circumference of the rods. *Note:* Change in side clearances indicates a bent rod.

_____

_____

_____ 5. Turn the crankshaft one full rotation to check the drag before removing the pistons. Torque should be equal throughout the complete turn. If it is not, possibly there is a bent or twisted rod. If the crankshaft turns smoothly after a particular rod is removed, it probably was the problem. If the crankshaft still does not turn smoothly after all the rods are removed, there is probably a bent crankshaft or an improper crankshaft line bore.

_____ 6. Remove the pistons according to the instructions in the vehicle service information.

(continued)

_____ 7. Inspect each piston for unusual wear patterns. *Note:* Unusual wear on a piston can indicate a twisted rod. Normal piston wear should be confined to a small area on the lower skirt. Any more wear than that may point to a bent rod, damaged crankshaft, or a damaged piston pin. A spiral or diagonal wear pattern on a piston skirt indicates a bent or twisted connecting rod. A binding piston wrist pin can also cause excessive piston skirt wear.

_____ 8. Inspect the rods and bearings for scratches, signs of uneven wear, and other damage.

_____ 9. Check the backs of the bearing inserts for signs of unusual wear that may mean the big end of the rod is distorted.

_____ 10. Send bent or twisted rods to a machine shop for repair or replace damaged rods with new ones.

_____ 11. Inspect the alignment of the main bearing bores in the block. With the main bearing caps, crankshaft, and upper main bearings removed, place a straightedge along the main bearing bores in the block along the centerline. The straightedge should indicate a misalignment if it does not contact all bores. If there is misalignment, the block will have to be line bored or replaced. If this problem is found, the machine shop should also check the crankshaft for warpage or out-of-round.

Performance ✓ Checklist

Name _____ Date _____ Class _____

PERFORMANCE STANDARDS:
Level 4–Performs skill without supervision and adapts to problem situations.
Level 3–Performs skill satisfactorily without assistance or supervision.
Level 2–Performs skill satisfactorily, but requires assistance/supervision.
Level 1–Performs parts of skill satisfactorily, but requires considerable assistance/supervision.

Attempt (circle one): **1 2 3 4**

Comments:

_____

_____

_____

_____

PERFORMANCE LEVEL ACHIEVED: _____

_____ 1. Safety rules and practices were followed at all times regarding this job.
_____ 2. Tools and equipment were used properly and stored upon completion of this job.
_____ 3. This completed job met the standards set and was done within the allotted time.
_____ 4. No injury or damage to property occurred during this job.
_____ 5. Upon completion of this job, the work area was cleaned correctly.

Instructor's Signature _____ Date _____

INSPECTING & MEASURING MAIN AND CONNECTING ROD BEARINGS

NATEF Standard(s) for Engine Repair:

C8 Inspect main and connecting rod bearings for damage and wear; determine necessary action.

Safety First

- ☐ Wear safety glasses at all times.
- ☐ Follow all safety rules when using common hand tools.
- ☐ Follow all safety rules when using engine stands.
- ☐ Check Material Safety Data Sheets for chemical safety information.

Tools & Equipment:
- Telescoping gauge
- Outside micrometer
- Engine stands
- Common hand tools
- Vehicle service information

PROCEDURES Refer to the vehicle service information for specifications and special procedures. Then inspect and measure the main and connecting rod bearings provided by your instructor.

_____ 1. Write up a repair order.

_____ 2. Make sure you follow all procedures in the vehicle service information.

_____ 3. Look up the finish and condition wear patterns for main and connecting rod bearings.

Source: _____ Page:_____

_____ 4. Inspect the main and connecting rod bearings and compare the finish and wear patterns to those shown in your source.

_____ 5. If any of the bearings have an abnormal wear pattern, replace the bearings and discard the old ones so they are not accidentally reused. In normal practice, all connecting rod and main bearings are replaced anytime a crankshaft is removed to ensure sufficient oil pressure.

_____ 6. Inside dimensions of bearings can be measured by reinstalling bearing caps and torquing them to specifications. Measure each of the bearings by inserting the telescoping gauge into the center of the bearing, centering the gauge, tightening it, and rotating it inside the bearing. *Note:* Center the telescoping gauge by making sure both of the ends are equal lengths inside the bearing. Remove the telescoping gauge and measure it with an outside micrometer to determine bearing inside diameter.

_____ 7. If the measurement on the telescoping gauge changes from point to point, the bearing or the bearing bore is out-of-round. If the bearing diameter measurements are equal inside the bearing, go to Step 10.

(continued)

Engine Repair

_____ **8.** If the bearing is out-of-round, remove the bearing and measure the bearing bore for roundness with the telescoping gauge.

_____ **9.** If the bearing bore is out-of-round, the block or connecting rod will have to be replaced or repaired.

_____ **10.** Look up the manufacturer's specifications for main and connecting rod bearing dimensions in the vehicle service information and compare them to your measurements.

Source: _____ *Page:* _____

_____ **11.** Normal manufacturer specifications are based on use of a serviceable crankshaft with standard diameter main and connecting rod journals. The crankshaft must be inspected and measured. Damaged crankshaft journals can be turned to a specific undersize diameter. This requires the use of undersized bearings (such as .010, .020, or .030 undersize).

Performance ✓ Checklist

Name _____ Date _____ Class _____

PERFORMANCE STANDARDS:
Level 4–Performs skill without supervision and adapts to problem situations.
Level 3–Performs skill satisfactorily without assistance or supervision.
Level 2–Performs skill satisfactorily, but requires assistance/supervision.
Level 1–Performs parts of skill satisfactorily, but requires considerable assistance/supervision.

Attempt (circle one): **1 2 3 4**

Comments:

_____

_____

_____

_____

PERFORMANCE LEVEL ACHIEVED: _____

_____ **1.** Safety rules and practices were followed at all times regarding this job.
_____ **2.** Tools and equipment were used properly and stored upon completion of this job.
_____ **3.** This completed job met the standards set and was done within the allotted time.
_____ **4.** No injury or damage to property occurred during this job.
_____ **5.** Upon completion of this job, the work area was cleaned correctly.

Instructor's Signature _____ Date_____

INSPECTING, MEASURING & REPLACING OIL PUMPS

NATEF Standard(s) for Engine Repair:
D2 Inspect oil pump gears or rotors, housing, pressure relief devices, and pump drive; perform necessary action.

Safety First
- ☐ Wear safety glasses at all times.
- ☐ Follow all safety rules when using common hand tools.
- ☐ Always connect a vehicle's exhaust to a vent hose before you run an engine in a closed shop. Unvented exhaust fumes in a closed shop can cause death.
- ☐ Keep jewelry, loose clothing, and hair away from moving parts while an engine is running.
- ☐ Be careful of hot engine parts.
- ☐ Check Material Safety Data Sheets for chemical safety information.
- ☐ Set the parking brake and place automatic transmission in PARK or manual transmission in NEUTRAL.

Tools & Equipment:
- Feeler gauges
- Ruler
- Cleaning solvent
- Vehicle service information
- Outside micrometer
- Inside micrometer
- Petroleum jelly
- Straightedge
- Tension gauge
- Common hand tools

PROCEDURES Refer to the vehicle service information for specifications and special procedures. Then remove, inspect, measure, and replace the oil pump components in the engine provided by your instructor.

_____ 1. Write up a repair order.

_____ 2. Make sure you follow all procedures in the vehicle service information.

_____ 3. Look up the recommended oil pump removal procedures in the vehicle service information. **Note:** Oil pumps usually demand little or no service. If the lubrication system has trouble or the engine is to be overhauled, then the pump should be removed and inspected. It is good practice to rebuild or replace the oil pump during engine overhaul.

Source: _____ Page: _____

_____ 4. Remove the oil pump and clean it and its parts thoroughly. **Note:** Be careful not to let the pump gears fall out.

_____ 5. Inspect all pump parts, including the pump housing, cover, and pressure relief device (including the bore) for cracks and other damage. **Note:** Always replace the pump if you find any damage.

(continued)

Engine Repair

_____ 6. Look up the recommended specifications for oil pump dimensions in the vehicle service information.

   Source: _____ *Page:*_____

_____ 7. Measure the endplay on both rotor and gear pumps by laying the straightedge across the face of the pump and measuring the distance between the gear or rotor and the straightedge with feeler gauges.

_____ 8. Measure the clearance between the inner and outer rotor and between the outer rotor and the housing with the feeler gauges.

_____ 9. Measure the lash (the space between the two sides of the meshing gears) and the gear side clearance on a gear pump with the feeler gauges.

_____ 10. Measure outer rotor thickness and wear on a rotor pump with the outside micrometer.

_____ 11. Measure the pump spring tension by either measuring the spring with the ruler or by measuring its tension with the tension gauge. *Note:* Some manufacturers recommend replacing worn pump springs. Others supply shims for adjusting spring tension or length.

_____ 12. If the pump is good, clean and oil all the parts and reassemble the pump.

_____ 13. Pack the pump with petroleum jelly before replacing the cover.

_____ 14. Make sure the pump turns smoothly. *Note:* See your instructor if it binds.

_____ 15. Clean the engine surfaces that will mate with the pump. Install new gaskets as necessary, and install the pump.

_____ 16. Reassemble the engine and add oil to the crankcase.

_____ 17. Run the engine and perform an oil pressure test. Check and repair all leaks.

Performance ✓ Checklist

Name _____ Date _____ Class _____

PERFORMANCE STANDARDS:
Level 4–Performs skill without supervision and adapts to problem situations.
Level 3–Performs skill satisfactorily without assistance or supervision.
Level 2–Performs skill satisfactorily, but requires assistance/supervision.
Level 1–Performs parts of skill satisfactorily, but requires considerable assistance/supervision.

Attempt (circle one): **1 2 3 4**

Comments:

_____

_____

_____

_____

PERFORMANCE LEVEL ACHIEVED: _____

_____ 1. Safety rules and practices were followed at all times regarding this job.

_____ 2. Tools and equipment were used properly and stored upon completion of this job.

_____ 3. This completed job met the standards set and was done within the allotted time.

_____ 4. No injury or damage to property occurred during this job.

_____ 5. Upon completion of this job, the work area was cleaned correctly.

Instructor's Signature _____ Date_____

REMOVING, INSPECTING & SERVICING AUXILIARY SHAFTS AND BEARINGS

NATEF Standard(s) for Engine Repair:
C13 Inspect auxiliary (balance, intermediate, idler, counterbalance or silencer) shaft(s); inspect shaft(s) and support bearings for damage and wear; determine necessary action; reinstall and time.

Safety First
☐ Wear safety glasses at all times.
☐ Follow all safety rules when using common hand tools.
☐ Follow all safety rules when using engine stands.
☐ Follow all safety rules when using compressed air.
☐ Check Material Safety Data Sheets for chemical safety information.

Tools & Equipment:
- Engine stands
- Brush
- Vehicle service information
- Torque wrench
- Air compressor
- Cleaning solvent
- Common hand tools

PROCEDURES Refer to the vehicle service information for specifications and special procedures. Then remove, inspect, and service the auxiliary shafts in the engine provided by your instructor.

_____ 1. Write up a repair order.

_____ 2. Make sure you follow all procedures in the vehicle service information.

_____ 3. Look up the manufacturer's recommendations for removing an auxiliary shaft or bearings.
    Source: _____ *Page:* _____

_____ 4. Using the appropriate tools, remove and replace bearings as necessary.

_____ 5. Clean the shafts with a brush and appropriate cleaning solvent, and then blow dry them with compressed air.

_____ 6. Inspect the shafts for burrs, hot spots, distortion, signs of abnormal wear, and overall condition. Replace any shafts that show signs of damage.

_____ 7. Using the proper assembly lubricant on shaft journals, follow the manufacturer's instructions for installation. Check for binding. Follow instructions for timing when installing drive belt(s) or chain(s).

_____ 8. Rotate the engine in the normal direction by hand for several rotations and verify that all timing marks will properly line up again. Also verify the position or adjustment of tensioners.

_____ 9. Replace all necessary covers and accessories.

(continued)

Performance ✓ Checklist

Name _____ Date _____ Class _____

PERFORMANCE STANDARDS:
Level 4–Performs skill without supervision and adapts to problem situations.
Level 3–Performs skill satisfactorily without assistance or supervision.
Level 2–Performs skill satisfactorily, but requires assistance/supervision.
Level 1–Performs parts of skill satisfactorily, but requires considerable assistance/supervision.

Attempt (circle one): **1 2 3 4**

Comments:

_____

_____

_____

_____

PERFORMANCE LEVEL ACHIEVED: _____

_____ **1.** Safety rules and practices were followed at all times regarding this job.

_____ **2.** Tools and equipment were used properly and stored upon completion of this job.

_____ **3.** This completed job met the standards set and was done within the allotted time.

_____ **4.** No injury or damage to property occurred during this job.

_____ **5.** Upon completion of this job, the work area was cleaned correctly.

Instructor's Signature _____ Date_____

INSPECTING & SERVICING VALVE SEATS

NATEF Standard(s) for Engine Repair:
B6 Inspect valves and valve seats; determine necessary action.

Safety First
- ☐ Wear safety glasses at all times.
- ☐ Follow all safety rules when using common hand tools.
- ☐ Follow all safety rules when using compressed air.
- ☐ Follow all safety rules concerning grinding stones. They can shatter and cause personal injury.
- ☐ Follow all safety rules when using air tools.

Engine Repair

Tools & Equipment:

- Grinding stones and mandrels
- Grinding pilots
- Powerhead
- Common hand tools
- Stone dressing tool
- Vehicle service information

PROCEDURES Refer to the vehicle service information for specifications and special procedures. Then inspect and service the valve seat provided by your instructor.

_____ 1. Write up a repair order.

_____ 2. Make sure you follow all procedures in the vehicle service information.

_____ 3. Remove the valves from the cylinder head and check the valve seats for severe pitting, burning, cracks, and recessed areas. *Note:* If you find any of these, send the valve seat to a machine shop to be repaired or replaced. Any required valve guide reconditioning must be done before you grind the seats.

_____ 4. To repair slightly worn valve seats, select the correct size of pilot that supports the grinding stone and that fits snugly inside the valve guide.

_____ 5. Determine the correct angle to be ground according to the vehicle service information. *Note:* The stone must be the correct angle and be slightly larger in diameter than the valve seat. The stone must be properly dressed before using it to grind a valve seat.

_____ 6. Choose a stone and fit it to the valve seat. Connect the stone and mandrel to a powerhead.

_____ 7. With the valve in the head, find the valve-to-seat contact point.

_____ 8. Remove the valve and place the grinding stone lightly against the point of contact.

_____ 9. Support and start the powerhead.

_____ 10. First, grind the face angle at 45°.

(continued)

_____ 11. Second, use the proper stone to grind the top angle at either 15° or 30° (refer to vehicle service information).

_____ 12. Third, use the proper stone to grind the throat angle at either 60° or 70°, depending on the valve seat. **Note:** These stones grind very fast. It is better to grind the throat angle by turning the stone by hand.

_____ 13. Manipulate the use of these three stones to create the proper width of the 45° seat and proper contact of the seat to the valve face.

_____ 14. Reinspect the valve seat for dark or pitted areas after you have ground it. **Note:** The seat should be smooth and shiny.

_____ 15. Regrind any rough or dark areas.

Performance ✓ Checklist

Name _____ Date _____ Class _____

PERFORMANCE STANDARDS:
Level 4–Performs skill without supervision and adapts to problem situations.
Level 3–Performs skill satisfactorily without assistance or supervision.
Level 2–Performs skill satisfactorily, but requires assistance/supervision.
Level 1–Performs parts of skill satisfactorily, but requires considerable assistance/supervision.

Attempt (circle one): **1 2 3 4**

Comments:

_____

_____

_____

_____

PERFORMANCE LEVEL ACHIEVED: _____

_____ 1. Safety rules and practices were followed at all times regarding this job.
_____ 2. Tools and equipment were used properly and stored upon completion of this job.
_____ 3. This completed job met the standards set and was done within the allotted time.
_____ 4. No injury or damage to property occurred during this job.
_____ 5. Upon completion of this job, the work area was cleaned correctly.

Instructor's Signature _____ Date _____

RESURFACING VALVES

NATEF Standard(s) for Engine Repair:

B6 Inspect valves and valve seats; determine necessary action.

Safety First
- ☐ Wear safety glasses at all times.
- ☐ Follow all safety rules when using common hand tools.
- ☐ Follow all safety rules when using power tools.
- ☐ Do not cut or grind sodium-filled valves. Sodium reacts violently with water.

Tools & Equipment:
- Valve grinding machine
- Rotary wire brush on bench grinder
- Common hand tools
- Vehicle service information

PROCEDURES Refer to the vehicle service information for specifications and special procedures. Then resurface the valves provided by your instructor.

_____ 1. Write up a repair order.

_____ 2. Make sure you follow all procedures in the vehicle service information.

_____ 3. Remove the valves from the head. Number the valves so you will return them to their original positions.

_____ 4. Inspect each valve and valve stem for blackening, wearing, or scorching and clean them with a rotary wire brush.

_____ 5. Inspect the keeper grooves and valve stem tips for wear and clean them with a rotary wire brush.

_____ 6. Inspect the valve heads for burning and wire brush the valves if they are only slightly scorched. **Note:** A valve will need to be ground to seat properly if the valve guides have been reconditioned or if the valve is worn, pitted, or burned.

_____ 7. Check the vehicle service information for specific information about resurfacing valves. **CAUTION:** Do not cut or grind sodium-filled valves. Sodium reacts violently with water. Follow safety procedures when disposing of any sodium-filled valves. If the sodium touches water, it can cause a fire.

_____ 8. Begin resurfacing a valve by adjusting the chuck head in the grinder to the manufacturer's specifications. Properly dress the grinding stone to be smooth and to grind at the proper angle. Insert the valve stem into the chuck close to the valve head. Rotate and chuck the valve again if it does not turn evenly in the grinder. **Note:** If this does not fix the problem, the valve is bent and must be replaced.

(continued)

Engine Repair

_____ **9.** Turn on the coolant and slowly move the valve face back and forth across the grinding stone. Remove less than .0005° per pass. *Note:* Keep the valve head in contact with the stone at all times during the grind and do not grind off any more than is absolutely necessary to remove the damage. Over-grinding can cause the face margin to be too thin and the valve will burn when it is used.

_____ **10.** Move the valve across the stone after it is clean until the sparking stops for an accurate face angle and finish. Discard the valve if there is less than 1/32" margin remaining.

_____ **11.** Properly re-face the valve stem. *Note:* It is important to remove the same amount of material from the stem as you removed from the face.

_____ **12.** Refer to the vehicle service information to replace the valves after the valve seats have been reconditioned.

Performance ✓ Checklist

Name _____ Date _____ Class _____

PERFORMANCE STANDARDS:
Level 4–Performs skill without supervision and adapts to problem situations.
Level 3–Performs skill satisfactorily without assistance or supervision.
Level 2–Performs skill satisfactorily, but requires assistance/supervision.
Level 1–Performs parts of skill satisfactorily, but requires considerable assistance/supervision.

Attempt (circle one): **1 2 3 4**

Comments:

_____

_____

_____

_____

PERFORMANCE LEVEL ACHIEVED: _____

_____ **1.** Safety rules and practices were followed at all times regarding this job.
_____ **2.** Tools and equipment were used properly and stored upon completion of this job.
_____ **3.** This completed job met the standards set and was done within the allotted time.
_____ **4.** No injury or damage to property occurred during this job.
_____ **5.** Upon completion of this job, the work area was cleaned correctly.

Instructor's Signature _____ Date_____

INSPECTING VALVE SPRING RETAINERS, ROTATORS, LOCKS, AND VALVE LOCK GROOVES

Engine Repair

NATEF Standard(s) for Engine Repair:

B4 Replace valve stem seals on an assembled engine; inspect valve spring retainers, locks, and valve grooves; determine necessary action.

Safety First
- ☐ Wear safety glasses at all times.
- ☐ Follow all safety rules when using common hand tools.

Tools & Equipment:
- Common hand tools
- Vehicle service information

PROCEDURES Refer to the vehicle service information for specifications and special procedures. Then inspect the valve spring retainers, rotators, locks, and valve lock grooves in the engine provided by your instructor.

_____ 1. Write up a repair order.

_____ 2. Make sure you follow all procedures in the vehicle service information.

_____ 3. Using the vehicle service information, locate a procedure(s) for removing and inspecting valve spring retainers, rotators, locks, and valve lock grooves. Record the source and page.

Source: _____   Page: _____

_____ 4. Inspect the valve spring retainers, valve locks, and rotators for cracks, warps, or other damage. Replace any damaged parts. *Note:* If you replace the valves, also replace the valve retainers.

_____ 5. Inspect the valve stem by closing the valve locks and moving the stem up and down. Ask your instructor for the acceptable amount of movement.

_____ 6. Replace any valve that has worn valve grooves.

(continued)

Performance ✓ Checklist

Name _____   Date _____   Class _____

PERFORMANCE STANDARDS:
Level 4–Performs skill without supervision and adapts to problem situations.
Level 3–Performs skill satisfactorily without assistance or supervision.
Level 2–Performs skill satisfactorily, but requires assistance/supervision.
Level 1–Performs parts of skill satisfactorily, but requires considerable assistance/supervision.

Attempt (circle one): **1 2 3 4**

Comments:

_____

_____

_____

_____

PERFORMANCE LEVEL ACHIEVED: _____

_____ **1.** Safety rules and practices were followed at all times regarding this job.

_____ **2.** Tools and equipment were used properly and stored upon completion of this job.

_____ **3.** This completed job met the standards set and was done within the allotted time.

_____ **4.** No injury or damage to property occurred during this job.

_____ **5.** Upon completion of this job, the work area was cleaned correctly.

Instructor's Signature _____   Date_____

REPLACING VALVE STEM SEALS

NATEF Standard(s) for Engine Repair:

B4 Replace valve stem seals on an assembled engine; inspect valve spring retainers, locks, and valve grooves; determine necessary action.

Safety First
- ☐ Wear safety glasses at all times.
- ☐ Follow all safety rules when using common hand tools.

Tools & Equipment:
- Valve spring compressor
- Vehicle service information
- Guide installation tool
- Common hand tools

PROCEDURES Refer to the vehicle service information for specifications and special procedures. Then replace valve stem seals on the valve stem assembly provided by your instructor.

Note: There are three different types of valve stem seals. They are the umbrella seal, the positive-type seal, and the O-ring seal. Note also that if head is assembled to engine, piston must be placed at top dead center on compression stroke and cylinder pressurized with compressed air. Using appropriate spring compressor, springs can be removed and seals replaced one cylinder at a time.

_____ 1. Write up a repair order.

_____ 2. Make sure you follow all procedures in the vehicle service information.

_____ 3. Use the valve spring compressor to remove the valve springs.

_____ 4. Using the vehicle service information for procedures, inspect the valve stem umbrella seal for breaks, brittleness, flattened areas, pinching, and obvious leaks.

_____ 5. Replace the umbrella seal. *Note:* When valves are disassembled, all the seals should be replaced. Damage other than normal wear, however, can help you diagnose other valve problems.

_____ 6. Inspect the positive valve stem seal for breaks, brittleness, flattened areas, pinching, and obvious leaks; then replace it.

_____ 7. Place a protective cap over the top of valve stem and force the seal over the guide with the seal installation tool.

_____ 8. Remove and inspect the O-ring stem seal. Look for breaks, brittleness, flattened areas, pinching, and obvious leaks. Do not replace it yet. *Note:* Umbrella seals are often used as replacement seals for O-rings because they are more efficient.

(continued)

Engine Repair

_____ 9. Install the valve spring assembly and retainer using the valve spring compressor.

_____ 10. Insert the O-ring and keepers into the valve stem grooves and remove the valve spring compressor.

Performance ✓ Checklist

Name _____ Date _____ Class _____

PERFORMANCE STANDARDS:
Level 4–Performs skill without supervision and adapts to problem situations.
Level 3–Performs skill satisfactorily without assistance or supervision.
Level 2–Performs skill satisfactorily, but requires assistance/supervision.
Level 1–Performs parts of skill satisfactorily, but requires considerable assistance/supervision.

Attempt (circle one): **1 2 3 4**

Comments:

_____

_____

_____

_____

PERFORMANCE LEVEL ACHIEVED: _____

_____ 1. Safety rules and practices were followed at all times regarding this job.
_____ 2. Tools and equipment were used properly and stored upon completion of this job.
_____ 3. This completed job met the standards set and was done within the allotted time.
_____ 4. No injury or damage to property occurred during this job.
_____ 5. Upon completion of this job, the work area was cleaned correctly.

Instructor's Signature _____ Date_____

REPLACING VALVE STEM SEALS ON AN ASSEMBLED ENGINE; INSPECTING VALVE SPRING RETAINERS, LOCKS, AND VALVE GROOVES

Engine Repair

NATEF Standard(s) for Engine Repair:

B4 Replace valve stem seals on an assembled engine; inspect valve spring retainers, locks, and valve grooves; determine necessary action.

Safety First
- ☐ Wear safety glasses at all times.
- ☐ Follow all safety rules when using common hand tools.
- ☐ Follow all safety rules when using valve spring compressors.

Tools & Equipment:
- Vehicle service information
- Common hand tools
- Seal installation tool
- Valve spring compressor (of the type that can be used on assembled engine)
- Spark plug hole-to-air hose adapter

PROCEDURES Refer to the vehicle service information for specifications and special procedures. Then, on the engine provided by your instructor, replace the valve stem seals and inspect the valve spring retainers, locks, and valve grooves.

_____ 1. Write up a repair order.

_____ 2. Make sure you follow all procedures in the vehicle service information.

_____ 3. Remove valve cover.

_____ 4. Replace valve stem seals on one cylinder at a time.

_____ 5. Remove spark plug from cylinder you choose to work on. Rotate crankshaft to place piston at top dead center on compression stroke.

_____ 6. Remove valve train parts as necessary (such as rocker arms, etc.) to gain access to valve springs.

_____ 7. Install air hose adapter to spark plug hole and pressurize cylinder to keep valves in closed position.

_____ 8. Using appropriate valve spring compressor, remove valve springs one at a time.

_____ 9. Remove old valve stem seal. Place seal installation tool (sleeve or guide) on end of valve stem. Lube and insert new umbrella or positive seal or stem.

_____ 10. Inspect valve spring retainers, locks, and stem grooves for damage.

_____ 11. Use valve spring compressor to reinstall spring and related parts.

(continued)

_____ **12.** When all cylinders have been serviced one at a time, replace all valve train parts and spark plugs.

_____ **13.** Adjust valves as necessary.

_____ **14.** Replace valve cover and any related parts.

Performance ✓ Checklist

Name _____ Date _____ Class _____

PERFORMANCE STANDARDS:
Level 4–Performs skill without supervision and adapts to problem situations.
Level 3–Performs skill satisfactorily without assistance or supervision.
Level 2–Performs skill satisfactorily, but requires assistance/supervision.
Level 1–Performs parts of skill satisfactorily, but requires considerable assistance/supervision.

Attempt (circle one): **1 2 3 4**

Comments:

_____

_____

_____

_____

PERFORMANCE LEVEL ACHIEVED: _____

_____ **1.** Safety rules and practices were followed at all times regarding this job.

_____ **2.** Tools and equipment were used properly and stored upon completion of this job.

_____ **3.** This completed job met the standards set and was done within the allotted time.

_____ **4.** No injury or damage to property occurred during this job.

_____ **5.** Upon completion of this job, the work area was cleaned correctly.

Instructor's Signature _____ Date_____

ADJUSTING VALVE TRAINS

NATEF Standard(s) for Engine Repair:

B11 Adjust valves (mechanical or hydraulic lifters).

Safety First

- ☐ Wear safety glasses at all times.
- ☐ Follow all safety rules when using common hand tools.
- ☐ Always connect a vehicle's exhaust to a vent hose before you run an engine in a closed shop. Unvented fumes in a closed shop can cause death.
- ☐ Keep jewelry, loose clothing, and hair away from moving parts while an engine is running.
- ☐ Be careful of hot engine parts.
- ☐ Set the parking brake and place automatic transmission in PARK or manual transmission in NEUTRAL.

Tools & Equipment:
- Feeler gauges
- Common hand tools
- Vehicle service information

PROCEDURES Refer to the vehicle service information for specifications and special procedures. Then adjust the valve train on the engine provided by your instructor.

Note: Some valve adjustments are done with the engine off and cold, some are done with the engine off and at normal operating temperature, and some are done with the engine running at normal operating temperature. Be sure to follow the manufacturer's instructions and specifications.

_____ 1. Write up a repair order.

_____ 2. Make sure you follow all procedures in the vehicle service information.

_____ 3. Adjust the valve train according to the manufacturer's specifications. **CAUTION:** It is important that a valve train be adjusted properly because loose or tight valves will damage an engine.

(continued)

Performance ✓ Checklist

Name _____ Date _____ Class _____

PERFORMANCE STANDARDS:
Level 4–Performs skill without supervision and adapts to problem situations.
Level 3–Performs skill satisfactorily without assistance or supervision.
Level 2–Performs skill satisfactorily, but requires assistance/supervision.
Level 1–Performs parts of skill satisfactorily, but requires considerable assistance/supervision.

Attempt (circle one): **1 2 3 4**

Comments:

_____

_____

_____

_____

PERFORMANCE LEVEL ACHIEVED: _____

_____ **1.** Safety rules and practices were followed at all times regarding this job.

_____ **2.** Tools and equipment were used properly and stored upon completion of this job.

_____ **3.** This completed job met the standards set and was done within the allotted time.

_____ **4.** No injury or damage to property occurred during this job.

_____ **5.** Upon completion of this job, the work area was cleaned correctly.

Instructor's Signature _____ Date_____

HONING & CLEANING CYLINDER WALLS

NATEF Standard(s) for Engine Repair:
C5 Deglaze and clean cylinder walls.

Safety First
- ☐ Wear safety glasses at all times.
- ☐ Follow all safety rules when using common hand tools.
- ☐ Follow all safety rules when using engine stands.
- ☐ Follow all safety rules when using power tools.
- ☐ Check Material Safety Data Sheets for chemical safety information.

Engine Repair

Tools & Equipment:
- Spring-loaded glaze breaker
- Ball-type glaze breaker
- Engine stands
- Honing tool set
- Variable speed electric drill
- Common hand tools
- Vehicle service information

PROCEDURES Refer to the vehicle service information for specifications and special procedures. Then hone and clean the cylinder walls in the engine provided by your instructor.

_____ 1. Write up a repair order.

_____ 2. Make sure you follow all procedures in the vehicle service information.

_____ 3. Put a small amount of honing oil on the inside of a cylinder. **CAUTION:** Always use honing oil. Never use kerosene during any part of the honing process. It will contaminate the cylinder.

_____ 4. Look up the manufacturer's recommended procedure for honing and deglazing cylinder walls. **_Note:_** See your instructor for how to use the different stones and honing tools.

Source: _____ Page:_____

_____ 5. Hone the inside of the cylinder by centering the honing tool inside the cylinder and running the honing tool at 450 rpm while gradually moving the device up and down within the area of ring travel at about one complete stroke per minute. **_Note:_** Do this until the wall is no longer glazed and has a crosshatch pattern at an angle of approximately 50° to 60° (flex hones are preferred because they leave the desired crosshatched pattern).

_____ 6. Clean the inside of the cylinder with soapy water. Rinse and dry.

_____ 7. Wipe the cylinder wall with a clean, oiled rag to remove any debris or grit left in the cylinder from the honing process. Repeat the cleaning process by using a clean cloth each time until the rag comes out clean.

(continued)

_____ 8. Inspect the cylinder for any other damage. *Note:* If there are any deep, vertical scratches, replace the cylinder sleeve, if possible, or send the engine block out to be bored and honed by a machinist.

_____ 9. Oil any ferrous parts to prevent them from oxidizing (rusting).

Performance ✓ Checklist

Name _____ Date _____ Class _____

PERFORMANCE STANDARDS:
Level 4–Performs skill without supervision and adapts to problem situations.
Level 3–Performs skill satisfactorily without assistance or supervision.
Level 2–Performs skill satisfactorily, but requires assistance/supervision.
Level 1–Performs parts of skill satisfactorily, but requires considerable assistance/supervision.

Attempt (circle one): **1** **2** **3** **4**

Comments:

_____

_____

_____

_____

PERFORMANCE LEVEL ACHIEVED: _____

_____ 1. Safety rules and practices were followed at all times regarding this job.
_____ 2. Tools and equipment were used properly and stored upon completion of this job.
_____ 3. This completed job met the standards set and was done within the allotted time.
_____ 4. No injury or damage to property occurred during this job.
_____ 5. Upon completion of this job, the work area was cleaned correctly.

Instructor's Signature _____ Date_____

INSPECTING, MEASURING & SERVICING PISTONS AND PINS

NATEF Standard(s) for Engine Repair:

C10 Inspect and measure pistons; determine necessary action.
C11 Remove and replace piston pin.

Safety First

☐ Wear safety glasses at all times.
☐ Follow all safety rules when using common hand tools.
☐ Follow all safety rules when using engine stands.
☐ Check all Material Safety Data Sheets for chemical safety information.

Engine Repair

Tools & Equipment:
- Ring expander
- Soft-faced vise jaws
- Engine stands
- Ring groove cleaner
- Outside micrometer
- Common hand tools
- Vehicle service information

PROCEDURES Refer to the vehicle service information for specifications and special procedures. Then inspect and service the pistons and pins in the engine provided by your instructor.

_____ 1. Write up a repair order.

_____ 2. Make sure you follow all procedures in the vehicle service information.

_____ 3. Thoroughly clean all components before inspecting and measuring them.

_____ 4. Place the piston rod in a vise with soft-faced vise jaws and remove the rings from the pistons with the ring expander. Be careful not to scratch the piston. **Note:** Do not reuse piston rings.

_____ 5. Carefully clean the ring grooves with the ring groove cleaner. **Note:** Use care not to remove any aluminum from the grooves.

_____ 6. Inspect the piston skirts for scuffs, cracks, and scores. **Note:** Some wear is normal, but most wear should be on the lower skirt.

_____ 7. Look up the manufacturer's specifications for piston skirt outside diameter.

   Source: _____ _Page:_____

_____ 8. Measure the piston skirt diameters with the outside micrometer. **Note:** The diameter at the bottom skirt diameter should be about 0.0127 mm wider than the top skirt diameter. Replace the piston if the difference is greater than the manufacturer's recommendations.

_____ 9. Check freedom of movement between the piston and the connecting rod. If binding is present, varnish may have to be removed from the piston pin. If there is excessive freedom or clearances, the piston, piston pin, or connecting rod may have to be replaced.

(_continued_)

Performance ✓ Checklist

Name _____ Date _____ Class _____

PERFORMANCE STANDARDS:
Level 4–Performs skill without supervision and adapts to problem situations.
Level 3–Performs skill satisfactorily without assistance or supervision.
Level 2–Performs skill satisfactorily, but requires assistance/supervision.
Level 1–Performs parts of skill satisfactorily, but requires considerable assistance/supervision.

Attempt (circle one): **1 2 3 4**

Comments:

_____

_____

_____

_____

PERFORMANCE LEVEL ACHIEVED: _____

_____ 1. Safety rules and practices were followed at all times regarding this job.
_____ 2. Tools and equipment were used properly and stored upon completion of this job.
_____ 3. This completed job met the standards set and was done within the allotted time.
_____ 4. No injury or damage to property occurred during this job.
_____ 5. Upon completion of this job, the work area was cleaned correctly.

Instructor's Signature _____ Date_____

INSPECTING & INSTALLING PISTON RINGS

NATEF Standard(s) for Engine Repair:
C12 Inspect, measure, and install piston rings.

Safety First

☐ Wear safety glasses at all times.
☐ Follow all safety rules when using common hand tools.
☐ Follow all safety rules when using engine stands.
☐ Check Material Safety Data Sheets for chemical safety information.

Engine Repair

Tools & Equipment:

- Ring expander
- Inside micrometer
- Common hand tools
- Vehicle service information
- Feeler gauge
- Soft-faced vise jaws
- Engine stands
- Outside micrometer
- Telescoping gauge
- Ring groove cleaner

PROCEDURES Refer to the vehicle service information for specifications and special procedures. Then inspect and install the piston rings in the engine provided by your instructor.

_____ 1. Write up a repair order.

_____ 2. Make sure you follow all procedures in the vehicle service information.

_____ 3. Thoroughly clean all components before inspecting and measuring them.

_____ 4. Place the piston rod in a vise with soft-faced vise jaws and remove the rings from the pistons with the ring expander. Be careful not to scratch the piston. *Note:* Never reuse piston rings.

_____ 5. Carefully clean the ring grooves with the ring groove cleaner. *Note:* Use care not to remove any aluminum from the grooves.

_____ 6. Look up the manufacturer's specifications for ring groove diameter.

   Source: _____ *Page:* _____

_____ 7. Check the ring grooves for wear and measure the groove diameter using the outside micrometer. *Note:* Replace any piston that does not meet the manufacturer's recommendations for groove diameter.

_____ 8. Inspect the piston pins for nicks, burrs, scratches, bends, and any uneven wear and replace any worn pins.

_____ 9. Check for binding pins and remove varnish if necessary.

_____ 10. Inspect the pin bushings by checking the piston rods for tightness. *Note:* If a pin is loose in the rod bushing, the bushing must be replaced.

(continued)

_____ 11. Look up the manufacturer's specifications for pin bushing inside diameter.

　　　　Source: _____ *Page:* _____

_____ 12. Measure and record the inside diameter of the pin bushing with either the telescoping gauge or the inside micrometer. *Note:* Replace any bushing that is not within the manufacturer's recommendations.

_____ 13. Many modern engines do not use rod bushings. The pin is pressed into the rod and floats within the pin bores in the pistons. If there is excessive play, the piston and pin must be replaced.

_____ 14. Place a compression ring in the appropriate bore. Invert a piston and squarely locate the ring in its approximate location. Then measure the end gap with a feeler gauge and compare to specifications. If the gap is excessive, the bore is excessively worn or you have the wrong rings. If the gap is too small, the ends of the ring can be filed.

_____ 15. Carefully install new piston rings by using the ring expanders. Be sure to follow the ring manufacturer's instructions carefully.

_____ 16. Use a feeler gauge to measure side clearance between the ring and ring groove. If this gap is excessive, the piston must be replaced.

_____ 17. Oil the rings and locate ring end gaps as instructed by the ring manufacturer.

_____ 18. Use a ring compressor to install the piston and rod assemblies.

Performance ✓ Checklist

Name _____ Date _____ Class _____

PERFORMANCE STANDARDS:
Level 4–Performs skill without supervision and adapts to problem situations.
Level 3–Performs skill satisfactorily without assistance or supervision.
Level 2–Performs skill satisfactorily, but requires assistance/supervision.
Level 1–Performs parts of skill satisfactorily, but requires considerable assistance/supervision.

Attempt (circle one): **1 2 3 4**

Comments:

_____

_____

_____

_____

PERFORMANCE LEVEL ACHIEVED: _____

_____ 1. Safety rules and practices were followed at all times regarding this job.

_____ 2. Tools and equipment were used properly and stored upon completion of this job.

_____ 3. This completed job met the standards set and was done within the allotted time.

_____ 4. No injury or damage to property occurred during this job.

_____ 5. Upon completion of this job, the work area was cleaned correctly.

Instructor's Signature _____ Date _____

INSPECTING & REPLACING PANS, COVERS, GASKETS, AND SEALS

NATEF Standard(s) for Engine Repair:
C15 Assemble engine block assembly.

Safety First

- ☐ Wear safety glasses at all times.
- ☐ Follow all safety rules when using common hand tools.
- ☐ Follow all safety rules when using air tools.
- ☐ Follow all safety rules when using a lift or jacks and jack stands.
- ☐ Always connect a vehicle's exhaust to a vent hose before you run an engine in a closed shop. Unvented exhaust fumes in a closed shop can cause death.
- ☐ Keep jewelry, loose clothing, and hair away from moving parts while an engine is running.
- ☐ Be careful of hot engine parts.
- ☐ Check Material Safety Data Sheets for chemical safety information.
- ☐ Set the parking brake and place automatic transmission in PARK or manual transmission in NEUTRAL.

Tools & Equipment:
- Putty knife
- Torque wrench
- Cleaning solvent
- Common hand tools
- Air tools
- Vehicle service information

PROCEDURES Refer to the vehicle service information for specifications and special procedures. Then inspect and replace the pans, covers, gaskets, and seals in the engine provided by your instructor.

_____ 1. Write up a repair order.

_____ 2. Make sure you follow all procedures in the vehicle service information.

_____ 3. Label the covers and remove them. *Note:* Labeling them will make it easier to replace them later.

_____ 4. Inspect all the pans and covers for dents, distortions, warping, and other visible damage such as holes and leaks. Replace any damaged parts. *Note:* Do not try to repair pans or covers. They protect very important engine parts from dirt and other elements.

_____ 5. Remove any stubborn gasket or seal material thoroughly with a putty knife. **CAUTION:** Be careful not to damage the underlying component with the putty knife.

(continued)

_____ 6. Replace the gaskets and seals according to the manufacturer's instructions. **Note:** Line the gasket holes up carefully with their corresponding holes in the engine. Use only sealant types that are approved by the manufacturer.

_____ 7. Replace all the covers and pans. **CAUTION:** Use care not to pinch the seals or to over-tighten the gaskets. See the manufacturer's recommended torque limits. Use appropriate seal drivers as needed to install the seals.

_____ 8. Verify and adjust all related fluid levels.

_____ 9. Run the engine and verify that there is no leakage at the gaskets and seals.

Performance ✓ Checklist

Name _____ Date _____ Class _____

PERFORMANCE STANDARDS:
Level 4–Performs skill without supervision and adapts to problem situations.
Level 3–Performs skill satisfactorily without assistance or supervision.
Level 2–Performs skill satisfactorily, but requires assistance/supervision.
Level 1–Performs parts of skill satisfactorily, but requires considerable assistance/supervision.

Attempt (circle one)**: 1 2 3 4**

Comments:

_____

_____

_____

_____

PERFORMANCE LEVEL ACHIEVED: _____

_____ 1. Safety rules and practices were followed at all times regarding this job.
_____ 2. Tools and equipment were used properly and stored upon completion of this job.
_____ 3. This completed job met the standards set and was done within the allotted time.
_____ 4. No injury or damage to property occurred during this job.
_____ 5. Upon completion of this job, the work area was cleaned correctly.

Instructor's Signature _____ Date_____

REASSEMBLING ENGINES

NATEF Standard(s) for Engine Repair:
C1 Disassemble engine block; clean and prepare components for inspection and reassembly.
C15 Assemble engine block assembly.

Safety First
- ☐ Wear safety glasses at all times.
- ☐ Follow all safety rules when using common hand tools.
- ☐ Follow all safety rules when using engine stands.
- ☐ Check Material Safety Data Sheets for chemical safety information.
- ☐ Follow all lifting rules when installing heavy components.

Engine Repair

Tools & Equipment:
- Engine stands
- Cleaning solution
- Vehicle service information
- Torque wrench
- Petroleum jelly
- Common hand tools
- Appropriate gaskets and sealants

PROCEDURES Refer to the vehicle service information for specifications and special procedures. Then reassemble the engine provided by your instructor.

_____ 1. Write up a repair order.

_____ 2. Make sure you follow all procedures in the vehicle service information and use recommended sealants.

_____ 3. Thoroughly clean and oil all engine components with the appropriate cleaning solutions and lubricants before reassembling. *Note:* The engine oil must dissolve the lubricant. Otherwise, the lubricant could clog the engine oil ports. Keep in mind that the slightest amount of dirt can also damage the components and cause the engine to fail early.

_____ 4. Install the appropriate camshaft bearings in the locations you marked during the disassembly process.

_____ 5. Using the proper assembly lubricant on camshaft lobes and journals, install the camshaft. Check for binding.

_____ 6. Install the camshaft retainer (if used) and verify camshaft endplay as necessary. See your instructor if it is not within specifications.

_____ 7. Carefully remove any burrs on the crankshaft and polish the journal surfaces.

_____ 8. Install the crankshaft from the top by rotating the engine on the stand so that the bottom of the engine is up. Carefully place the crankshaft in from this position. **CAUTION:** The crankshaft is heavy.

_____ 9. Install and align the timing components. Clean and lube the sprocket bores and shafts, and pull the sprockets in place together. Refrain from hammering on the sprockets.

(*continued*)

_____ 10. Install the timing cover, vibration damper, and water pump.

_____ 11. Clean the cylinder walls, pistons, and connecting rod assemblies. Remove burrs and put thread protectors on the connecting rod bolts.

_____ 12. Install the pistons and connecting rod assemblies. *Note:* Be certain that the pistons and connecting rods are mated according to the markings on them. Make sure all rod and main caps are in the correct location and torqued.

_____ 13. Clean the surface of the block for the oil pan. *Note:* Use a recommended gasket—not a sealant.

_____ 14. Clean the oil screen. *Note:* If the screen is damaged, replace it.

_____ 15. Pack the oil pump with petroleum jelly if required by the manufacturer, or pour oil into the pick-up screen. *Note:* This aids in priming later.

_____ 16. Install the oil pump, oil screen, and oil pan. *Note:* If the oil pan is damaged, replace it.

_____ 17. Clean all the mounting surfaces for the cylinder head(s), intake manifold, and exhaust manifold(s). *Note:* Make sure that the holes in the gaskets line up with the holes in the components.

_____ 18. Install the cylinder heads, valve train components, valve cover(s), intake manifold, and exhaust manifold(s).

_____ 19. Install oil plugs, core hole plugs, cam plugs, and motor mounts.

_____ 20. Install all engine accessory components that can be installed before installing the engine in the vehicle. **CAUTION:** Wait until after the engine is in place to install any engine accessory components that could be damaged while the engine is being installed in the vehicle.

_____ 21. To disassemble engine, reverse these steps. Follow vehicle service information closely.

Performance ✓ Checklist

Name _____ Date _____ Class _____

PERFORMANCE STANDARDS:
Level 4–Performs skill without supervision and adapts to problem situations.
Level 3–Performs skill satisfactorily without assistance or supervision.
Level 2–Performs skill satisfactorily, but requires assistance/supervision.
Level 1–Performs parts of skill satisfactorily, but requires considerable assistance/supervision.

Attempt (circle one): **1 2 3 4**

Comments:

_____

_____

_____

_____

PERFORMANCE LEVEL ACHIEVED: _____

_____ 1. Safety rules and practices were followed at all times regarding this job.

_____ 2. Tools and equipment were used properly and stored upon completion of this job.

_____ 3. This completed job met the standards set and was done within the allotted time.

_____ 4. No injury or damage to property occurred during this job.

_____ 5. Upon completion of this job, the work area was cleaned correctly.

Instructor's Signature _____ Date_____

INSPECTING & REPLACING TIMING BELTS, CAMDRIVE SPROCKETS, AND TENSIONERS

NATEF Standard(s) for Engine Repair:

B13 Inspect and replace timing belts (chains), overhead camdrive sprockets, and tensioners; check belt/chain tension; adjust as necessary.

Safety First

- ☐ Wear safety glasses at all times.
- ☐ Follow all safety rules when using common hand tools.
- ☐ Follow all safety rules when using air tools.
- ☐ Follow all safety rules when using a lift or a jack and jack stands.
- ☐ Always connect a vehicle's exhaust to a vent hose before you run an engine in a closed shop. Unvented exhaust fumes in a closed shop can cause death.
- ☐ Keep jewelry, loose clothing, and hair away from moving parts while an engine is running.
- ☐ Be careful of hot engine parts.
- ☐ Set the parking brake and place automatic transmission in PARK or manual transmission in NEUTRAL.

Tools & Equipment:

- Air tools
- Common hand tools
- Torque wrench
- Vehicle service information
- Timing light

PROCEDURES Refer to the vehicle service information for specifications and special procedures. Then inspect and replace any damaged or worn parts in the timing drive in the engine provided by your instructor.

_____ 1. Write up a repair order.

_____ 2. Make sure you follow all procedures in the vehicle service information.

_____ 3. Look up the procedure for disassembling the engine to inspect the timing components.

  Source: _____ *Page:* _____

_____ 4. Disassemble the engine to inspect the parts.

_____ 5. Check the belts, sprockets, and tensioners for looseness and wear.

_____ 6. Replace worn or damaged parts as you reassemble the timing drive according to the manufacturer's specifications. Observe all timing marks and properly set belt tension.

(continued)

_____ 7. Turn the engine several rotations by hand in the normal direction of rotation and re-check that all timing marks are still aligned.

_____ 8. After engine assembly, verify and adjust coolant and oil levels.

_____ 9. If the ignition timing is adjustable, follow the manufacturer's instructions to properly adjust it.

Performance ✓ Checklist

Name _____ Date _____ Class _____

PERFORMANCE STANDARDS:
Level 4–Performs skill without supervision and adapts to problem situations.
Level 3–Performs skill satisfactorily without assistance or supervision.
Level 2–Performs skill satisfactorily, but requires assistance/supervision.
Level 1–Performs parts of skill satisfactorily, but requires considerable assistance/supervision.

Attempt (circle one): **1 2 3 4**

Comments:

_____

_____

_____

_____

PERFORMANCE LEVEL ACHIEVED: _____

_____ 1. Safety rules and practices were followed at all times regarding this job.
_____ 2. Tools and equipment were used properly and stored upon completion of this job.
_____ 3. This completed job met the standards set and was done within the allotted time.
_____ 4. No injury or damage to property occurred during this job.
_____ 5. Upon completion of this job, the work area was cleaned correctly.

Instructor's Signature _____ Date_____

REMOVING & REINSTALLING CYLINDER HEADS AND GASKETS

Engine Repair

NATEF Standard(s) for Engine Repair:

B1 Remove and reinstall cylinder heads and gaskets; tighten according to manufacturer's specifications and procedures.

Safety First

☐ Wear safety glasses at all times.

☐ Follow all safety rules when using common hand tools.

☐ Follow all safety rules when removing the cylinder head. Cylinder heads can be very heavy, and dropping one may result in injury or damage to the vehicle.

Tools & Equipment:
- Vehicle service information
- Sealant
- Common hand tools
- Torque wrench
- Engine covers

PROCEDURES Refer to the vehicle service information for specifications and special procedures. Then remove and reinstall the cylinder head and gaskets in the vehicle provided by your instructor.

_____ 1. Write up a repair order.

_____ 2. Make sure you follow all procedures in the vehicle service informaiton.

_____ 3. Using the appropriate tools, remove the valve cover.

_____ 4. Depending on the type of engine, overhead cam engine (OHC) or overhead valve engine (OHV), remove the camshaft or rocker arm assembly according to the manufacturer's specifications. Keep all components in order for reinstallation.

_____ 5. Remove the remaining valve train components from the cylinder head.

_____ 6. Following procedure, remove the intake and exhaust manifold from the cylinder head.

_____ 7. Remove the cylinder head attaching bolts from the cylinder head and keep any reusable bolts in the same order for reinstallation.

_____ 8. Remove the cylinder head from the engine.

_____ 9. Remove the cylinder head gasket and inspect both the gasket and the cylinder head for signs of deterioration.

_____ 10. Make sure the engine cylinder head and block decks are clean and dry. The block deck usually has several dowel pins sticking above the deck surface that align the head gasket and head on the deck. Many cylinder head gaskets are installed without the use of a sealant. Sealant requirements can be found in the vehicle service information.

(continued)

_____ **11.** Note that many overhead camshaft (OHC) engines are classified as interference engines. Before installing an assembled cylinder head on these engines, the camshaft(s) and crankshaft must be positioned to the proper timing marks. This prevents interference between the valves and pistons.

_____ **12.** Place the cylinder head gasket on the block deck. Make sure it is the correct gasket and that the top of the gasket faces up. Some V-type engines use a different gasket on each bank. Incorrect gasket usage or positioning can block coolant passages and cause the engine to overheat.

_____ **13.** Carefully place the head over the dowel pins and onto the gasket surface. Make sure the head is fully seated on the gasket surface.

_____ **14.** Install the cylinder head bolts. Some head bolts enter the water jacket of the engine. They may require special sealing procedures.

_____ **15.** Tighten the head bolts to the proper torque value, using the procedures and specifications in the service information. As with other engine assembly bolts, some manufacturers recommend tightening cylinder head bolts using the torque/angle method.

Performance ✓ Checklist

Name _____ Date _____ Class _____

PERFORMANCE STANDARDS:
Level 4–Performs skill without supervision and adapts to problem situations.
Level 3–Performs skill satisfactorily without assistance or supervision.
Level 2–Performs skill satisfactorily, but requires assistance/supervision.
Level 1–Performs parts of skill satisfactorily, but requires considerable assistance/supervision.

Attempt (circle one): **1 2 3 4**

Comments:

_____

_____

_____

_____

PERFORMANCE LEVEL ACHIEVED: _____

_____ **1.** Safety rules and practices were followed at all times regarding this job.
_____ **2.** Tools and equipment were used properly and stored upon completion of this job.
_____ **3.** This completed job met the standards set and was done within the allotted time.
_____ **4.** No injury or damage to property occurred during this job.
_____ **5.** Upon completion of this job, the work area was cleaned correctly.

Instructor's Signature _____ Date_____

MEASURING & ADJUSTING CAMSHAFT TIMING

NATEF Standard(s) for Engine Repair:
B16 Establish camshaft(s) timing and cam sensor indexing according to manufacturer's specifications and procedure.

Safety First
- ☐ Wear safety glasses at all times.
- ☐ Follow all safety rules when using common hand tools.

Tools & Equipment:
- Protractor or degree wheel
- Vehicle service information
- Common hand tools
- Torque wrench

PROCEDURES Refer to the vehicle service information for specifications and special procedures. Then measure and adjust the camshaft timing in the engine provided by your instructor.

_____ 1. Write up a repair order.

_____ 2. Make sure you follow all procedures in the vehicle service information.

_____ 3. Inspect the timing belt or chain for signs of wear. **Note:** Belts will fray, crack, or lose teeth. (Wear on one side of a belt indicates a poorly aligned belt.) Chains will show signs of damage or wear at the guides or on the links.

_____ 4. Put a wrench on the crankshaft bolt and turn the crankshaft either clockwise or counter clockwise until the camshaft just begins to move.

_____ 5. Mark a timing sprocket tooth on the crankshaft and an index mark on the front of the engine.

_____ 6. Turn the crankshaft in the opposite direction until the camshaft just begins to move again. Mark the timing sprocket tooth location on the front of the engine. Using a protractor or a degree wheel, determine the difference in degrees between the marks. **Note:** The difference between the marks should be less than 5°. On engines with distributors, you can remove the distributor cap and watch distributor rotor movement while measuring the angle of play at the crankshaft using the timing mark scale. This can be done without disassembly. An engine will run with a difference of 10° to 15°, but this indicates that the chain or belt has stretched. You can also measure the chain deflection halfway between sprockets and compare to specifications. Timing belts may have provisions for adjustment.

(continued)

_____ 7. Adjust or replace the chain or belt according to the manufacturer's recommendations. If a chain is replaced, you should also replace all related sprockets. Make sure that all required index marks are properly aligned and that any tensioners (if used) are serviceable and properly adjusted. Many timing belts have a mileage replacement requirement.

_____ 8. Rotate the engine several times by hand in the normal rotation direction and verify that all index marks will line up again.

Performance ✓ Checklist

Name _____ Date _____ Class _____

PERFORMANCE STANDARDS:
Level 4–Performs skill without supervision and adapts to problem situations.
Level 3–Performs skill satisfactorily without assistance or supervision.
Level 2–Performs skill satisfactorily, but requires assistance/supervision.
Level 1–Performs parts of skill satisfactorily, but requires considerable assistance/supervision.

Attempt (circle one): **1 2 3 4**

Comments:

_____

_____

_____

_____

PERFORMANCE LEVEL ACHIEVED: _____

_____ 1. Safety rules and practices were followed at all times regarding this job.
_____ 2. Tools and equipment were used properly and stored upon completion of this job.
_____ 3. This completed job met the standards set and was done within the allotted time.
_____ 4. No injury or damage to property occurred during this job.
_____ 5. Upon completion of this job, the work area was cleaned correctly.

Instructor's Signature _____ Date _____

INSTALLING ENGINE COVERS

NATEF Standard(s) for Engine Repair:
A13 Install engine covers using gaskets, seals, and sealers as required.

Safety First
- ☐ Wear safety glasses at all times.
- ☐ Follow all safety rules when using common hand tools.

Tools & Equipment:
- Vehicle service information
- Sealer
- Engine oil
- Common hand tools
- Seal-driving tool
- Torque wrench
- Engine covers
- Soft-faced hammer

PROCEDURES Refer to the vehicle service information for specifications and special procedures. Then install engine covers on the engine of the vehicle provided by your instructor.

_____ 1. Write up a repair order.

_____ 2. Make sure you follow all procedures in the vehicle service information.

_____ 3. Refer to the vehicle service information for gasket installation.

_____ 4. Select the correct seal for the covers. Engine covers use either a conventional gasket or a "form-in-place" silicone sealer. Do not use an excessive amount of sealer. Some engines use a common seal at the bottom of the timing cover and the front of the oil pan. The timing cover may have a separate seal that fits around the end of the crankshaft. This seal has a metal ring that fits into a recess in the timing cover.

_____ 5. Use a seal-driving tool and a soft-faced hammer to install a new seal.

_____ 6. Apply a coat of engine oil to the seal lip. This will make it easier to install the crankshaft pulley without damaging the seal. This will provide lubricant between the seal and the crankshaft pulley.

_____ 7. Torque all cover mounting bolts to the specified values.

_____ 8. Install the crankshaft pulley and engine accessories and torque to specifications.

_____ 9. Install and tighten the oil drain plug. If the engine oil dipstick tube was removed during disassembly, it should be reinstalled now. Some dipstick tubes are pressed into place; others use an O-ring and a mounting bracket. Some engines have a separate lower tube that fits into the crankcase. When used, the lower tube must be installed first.

(continued)

Performance ✓ Checklist

Name _____ Date _____ Class _____

PERFORMANCE STANDARDS:
Level 4–Performs skill without supervision and adapts to problem situations.
Level 3–Performs skill satisfactorily without assistance or supervision.
Level 2–Performs skill satisfactorily, but requires assistance/supervision.
Level 1–Performs parts of skill satisfactorily, but requires considerable assistance/supervision.

Attempt (circle one): **1 2 3 4**

Comments:

_____

_____

_____

_____

PERFORMANCE LEVEL ACHIEVED: _____

_____ 1. Safety rules and practices were followed at all times regarding this job.

_____ 2. Tools and equipment were used properly and stored upon completion of this job.

_____ 3. This completed job met the standards set and was done within the allotted time.

_____ 4. No injury or damage to property occurred during this job.

_____ 5. Upon completion of this job, the work area was cleaned correctly.

Instructor's Signature _____ Date _____

INSTALLING AN ENGINE IN A FRONT-WHEEL DRIVE VEHICLE

NATEF Standard(s) for Engine Repair:

A12 Remove and reinstall engine in a late model front-wheel drive or rear-wheel drive vehicle (OBDII or newer); reconnect all attaching components and restore the vehicle to running condition.

Safety First

☐ Wear safety glasses at all times.

☐ Follow all safety rules when using common hand and air tools.

☐ Always connect a vehicle's exhaust to a vent hose before you run an engine in a closed shop. Unvented exhaust fumes in a closed shop can cause death.

☐ Keep jewelry, loose clothing, and hair away from moving parts while an engine is running.

☐ Be careful of hot engine parts.

☐ Use an assistant and lifting equipment when installing an engine.

☐ When testing the engine, set the parking brake and place automatic transmission in PARK or manual transmission in NEUTRAL.

Tools & Equipment:

- Lift equipment
- Vehicle service information
- Common hand tools
- Air tools
- Necessary related replacement parts and fluids

PROCEDURES Refer to the vehicle service information for specifications and special procedures. Then use the markings and tags you made during engine disassembly to install the engine provided by your instructor.

_____ 1. Write up a repair order.

_____ 2. Make sure you follow all procedures in the vehicle service information.

_____ 3. Clean and paint the engine compartment and allow enough drying time.

_____ 4. Spin-test the engine on the stand.

_____ 5. Reinstall the engine according to the manufacturer's recommendations for the make and model. Be sure to use an assistant. *Note:* The differences between front- and rear-wheel drive installations include that the transmissions on front-wheel drive vehicles are sometimes installed with the engine, the engine compartments will be more crowded, and the engines are usually mounted sideways to the vehicle.

_____ 6. Use the tags and marks from removal to make sure the linkages, hoses, belts, wires, and fasteners are installed correctly.

_____ 7. Prime the engine and make any adjustments that the manufacturer recommends.

(continued)

Engine Repair

_____ 8. Start the engine and make any necessary adjustments. **CAUTION:** Practice safety at all times.

_____ 9. Observe any previous scribe marks around hinges and replace the hood if it had to be removed.

Performance ✓ Checklist

Name _____ Date _____ Class _____

PERFORMANCE STANDARDS:
Level 4–Performs skill without supervision and adapts to problem situations.
Level 3–Performs skill satisfactorily without assistance or supervision.
Level 2–Performs skill satisfactorily, but requires assistance/supervision.
Level 1–Performs parts of skill satisfactorily, but requires considerable assistance/supervision.

Attempt (circle one): **1 2 3 4**

Comments:

_____

_____

_____

_____

PERFORMANCE LEVEL ACHIEVED: _____

_____ 1. Safety rules and practices were followed at all times regarding this job.
_____ 2. Tools and equipment were used properly and stored upon completion of this job.
_____ 3. This completed job met the standards set and was done within the allotted time.
_____ 4. No injury or damage to property occurred during this job.
_____ 5. Upon completion of this job, the work area was cleaned correctly.

Instructor's Signature _____ Date_____

INSTALLING AN ENGINE IN A REAR-WHEEL DRIVE VEHICLE

NATEF Standard(s) for Engine Repair:

A12 Remove and reinstall engine in a late model front-wheel drive or rear-wheel drive vehicle (OBDII or newer); reconnect all attaching components and restore the vehicle to running condition.

Safety First

☐ Wear safety glasses at all times.

☐ Follow all safety rules when using common hand and air tools.

☐ Always connect a vehicle's exhaust to a vent hose before you run an engine in a closed shop. Unvented exhaust fumes in a closed shop can cause death.

☐ Keep jewelry, loose clothing, and hair away from moving parts while an engine is running.

☐ Be careful of hot engine parts.

☐ Use an assistant and lifting equipment when installing an engine.

☐ When testing the engine, set the parking brake and place automatic transmission in PARK or manual transmission in NEUTRAL.

Tools & Equipment:

- Lift equipment
- Common hand tools
- Air tools
- Vehicle service information
- Necessary related replacement parts and tools

PROCEDURES Refer to the vehicle service information for specifications and special procedures. Then use the markings and tags you made during engine disassembly to install the engine provided by your instructor.

_____ 1. Write up a repair order.

_____ 2. Make sure you follow all procedures in the vehicle service information.

_____ 3. Clean and paint the engine compartment and allow enough drying time.

_____ 4. Spin-test the engine on the stand.

_____ 5. Reinstall the engine according to the manufacturer's recommendations for the make and model. Be sure to use an assistant.

_____ 6. Use the tags and marks from removal to make sure the linkages, hoses, belts, wires, and fasteners are installed correctly.

_____ 7. Prime the engine and make any adjustments that the manufacturer recommends.

_____ 8. Start the engine and make any necessary adjustments.

_____ 9. Observe any previous scribe marks around hinges and replace the hood if it had to be removed.

(continued)

Engine Repair

Performance ✓ Checklist

Name _____ Date _____ Class _____

PERFORMANCE STANDARDS:
Level 4–Performs skill without supervision and adapts to problem situations.
Level 3–Performs skill satisfactorily without assistance or supervision.
Level 2–Performs skill satisfactorily, but requires assistance/supervision.
Level 1–Performs parts of skill satisfactorily, but requires considerable assistance/supervision.

Attempt (circle one): **1 2 3 4**

Comments:

_____

_____

_____

_____

PERFORMANCE LEVEL ACHIEVED: _____

_____ **1.** Safety rules and practices were followed at all times regarding this job.
_____ **2.** Tools and equipment were used properly and stored upon completion of this job.
_____ **3.** This completed job met the standards set and was done within the allotted time.
_____ **4.** No injury or damage to property occurred during this job.
_____ **5.** Upon completion of this job, the work area was cleaned correctly.

Instructor's Signature _____ Date_____

Name _____ Date _____ Class _____

DIAGNOSING ENGINE COOLING AND HEATER SYSTEM HOSES AND BELTS

NATEF Standard(s) for Heating & Air Conditioning:

C3 Inspect engine cooling and heater system hoses and belts; perform necessary action.

DIRECTIONS: Fill in the blanks below by identifying (1) Safety First Practices that must be followed, (2) Tools and Equipment required, (3) three possible causes of engine cooling and heater system hose and belt problems, and (4) corrective actions that should be taken.

Safety First

- _____
- _____
- _____
- _____

Tools and Equipment Required:

- _____   • _____
- _____   • _____
- _____   • _____

QUICK ✔ Diagnostic for Cause(s)

List at least three common causes of engine cooling and heater system hose and belt problems:

1. _____
2. _____
3. _____
4. _____

QUICK ✔ Diagnostic for Corrective Action(s)

List possible corrective action(s):

1. _____
2. _____
3. _____
4. _____

Heating & Air Conditioning

DIAGNOSING HEATER CONTROL VALVES

NATEF Standard(s) for Heating & Air Conditioning:
C9 Inspect and test heater control valve(s); perform necessary action.

DIRECTIONS: Fill in the blanks below by identifying (1) Safety First Practices that must be followed, (2) Tools and Equipment required, (3) three possible causes of heater control valve problems, and (4) corrective actions that should be taken.

Safety First

- _____
- _____
- _____
- _____

Tools and Equipment Required:

- _____   • _____
- _____   • _____
- _____   • _____

QUICK ✔ Diagnostic for Cause(s)

List at least three common causes of heater control valve problems:

1. _____
2. _____
3. _____
4. _____
5. _____

QUICK ✔ Diagnostic for Corrective Action(s)

List possible corrective action(s):

1. _____
2. _____
3. _____
4. _____
5. _____

DIAGNOSING FANS

NATEF Standard(s) for Heating & Air Conditioning:
C7 Inspect and test cooling fan, fan clutch, fan shroud, and air dams; perform necessary action.

DIRECTIONS: Fill in the blanks below by identifying (1) Safety First Practices that must be followed, (2) Tools and Equipment required, (3) three possible causes of fan problems, and (4) corrective actions that should be taken.

Safety First
- _____
- _____
- _____
- _____

Tools and Equipment Required:
- _____   - _____
- _____   - _____
- _____   - _____

QUICK ✔ Diagnostic for Cause(s)

List at least three common causes of fan problems:

1. _____
2. _____
3. _____
4. _____
5. _____

QUICK ✔ Diagnostic for Corrective Action(s)

List possible corrective action(s):

1. _____
2. _____
3. _____
4. _____
5. _____

Heating & Air Conditioning

DIAGNOSING ELECTRICAL FAN CONTROL SYSTEM AND CIRCUITS

NATEF Standard(s) for Heating & Air Conditioning:
C8 Inspect and test electric cooling for control system and circuits; perform necessary action.

DIRECTIONS: Fill in the blanks below by identifying (1) Safety First Practices that must be followed, (2) Tools and Equipment required, (3) three possible causes of electrical fan control system and circuit problems, and (4) corrective actions that should be taken.

Safety First
- _____
- _____
- _____
- _____

Tools and Equipment Required:
- _____   • _____
- _____   • _____
- _____   • _____

QUICK ✔ Diagnostic for Cause(s)

List at least three common causes of electrical fan control system and circuit problems:
1. _____
2. _____
3. _____
4. _____
5. _____

QUICK ✔ Diagnostic for Corrective Action(s)

List possible corrective action(s):
1. _____
2. _____
3. _____
4. _____
5. _____

Name _____ Date _____ Class _____

DIAGNOSING A/C UNIT'S DISCHARGED OIL

NATEF Standard(s) for Heating & Air Conditioning:
A9 Inspect the condition of discharged oil; determine necessary action.

DIRECTIONS: Fill in the blanks below by identifying (1) Safety First Practices that must be followed, (2) Tools and Equipment required, (3) three possible causes of discharged oil problems, and (4) corrective actions that should be taken.

Safety First

- _____
- _____
- _____
- _____

Tools and Equipment Required:

- _____ • _____
- _____ • _____
- _____ • _____

Heating & Air Conditioning

QUICK ✔ Diagnostic for Cause(s)

List at least three common causes of discharged oil problems:

1. _____
2. _____
3. _____
4. _____
5. _____

QUICK ✔ Diagnostic for Corrective Action(s)

List possible corrective action(s):

1. _____
2. _____
3. _____
4. _____
5. _____

DIAGNOSING A/C SYSTEM CONDITIONS

NATEF Standard(s) for Heating & Air Conditioning:
B1-1 Diagnose A/C system conditions that cause the protection devices (pressure, thermal, and PCM) to interrupt system operation; determine necessary action.

DIRECTIONS: Fill in the blanks below by identifying (1) Safety First Practices that must be followed, (2) Tools and Equipment required, (3) three possible causes of A/C system problems, and (4) corrective actions that should be taken.

Safety First
- _____
- _____
- _____
- _____

Tools and Equipment Required:
- _____   - _____
- _____   - _____
- _____   - _____

QUICK ✔ Diagnostic for Cause(s)

List at least three common causes of A/C system problems:
1. _____
2. _____
3. _____
4. _____
5. _____

QUICK ✔ Diagnostic for Corrective Action(s)

List possible corrective action(s):
1. _____
2. _____
3. _____
4. _____
5. _____

DIAGNOSING A/C-HEATER CONTROL PANEL

NATEF Standard(s) for Heating & Air Conditioning:
D5 Inspect and test A/C-heater control panel assembly; determine necessary action.

DIRECTIONS: Fill in the blanks below by identifying (1) Safety First Practices that must be followed, (2) Tools and Equipment required, (3) three possible causes of A/C-heater control panel assembly problems, and (4) corrective actions that should be taken.

Safety First

- _____
- _____
- _____
- _____

Tools and Equipment Required:

- _____   - _____
- _____   - _____
- _____   - _____

QUICK ✔ Diagnostic for Cause(s)

List at least three common causes of A/C-heater control panel assembly problems:

1. _____
2. _____
3. _____
4. _____
5. _____

QUICK ✔ Diagnostic for Corrective Action(s)

List possible corrective action(s):

1. _____
2. _____
3. _____
4. _____
5. _____

Heating & Air Conditioning

DIAGNOSING A/C-HEATER CONTROL CABLES AND LINKAGES

NATEF Standard(s) for Heating & Air Conditioning:
D6 Inspect and test A/C-heater control cables, motors, and linkages; perform necessary action.

DIRECTIONS: Fill in the blanks below by identifying (1) Safety First Practices that must be followed, (2) Tools and Equipment required, (3) three possible causes of A/C-heater control cables and linkages problems, and (4) corrective actions that should be taken.

Safety First

- _____
- _____
- _____
- _____

Tools and Equipment Required:

- _____ • _____
- _____ • _____
- _____

QUICK ✔ Diagnostic for Cause(s)

List at least three common causes of A/C-heater control cables and linkages problems:

1. _____
2. _____
3. _____
4. _____
5. _____

QUICK ✔ Diagnostic for Corrective Action(s)

List possible corrective action(s):

1. _____
2. _____
3. _____
4. _____
5. _____

Name _____   Date _____   Class _____

DIAGNOSING A/C-HEATER DUCTS, HOSES, AND DOORS

NATEF Standard(s) for Heating & Air Conditioning:
D7 Inspect and test A/C-heater ducts, doors, hoses, cabin filters, and outlets; perform necessary action.

DIRECTIONS: Fill in the blanks below by identifying (1) Safety First Practices that must be followed, (2) Tools and Equipment required, (3) three possible causes of A/C-heater duct, hose, and door problems, and (4) corrective actions that should be taken.

Safety First
- _____
- _____
- _____
- _____

Tools and Equipment Required:
- _____   • _____
- _____   • _____
- _____

QUICK ✔ Diagnostic for Cause(s)

List at least three common causes of A/C-heater duct, hose, and door problems:

1. _____
2. _____
3. _____
4. _____
5. _____

QUICK ✔ Diagnostic for Corrective Action(s)

List possible corrective action(s):

1. _____
2. _____
3. _____
4. _____
5. _____

Heating & Air Conditioning

DIAGNOSING A/C COMPRESSOR LOAD CUT-OFF

NATEF Standard(s) for Heating & Air Conditioning:
D3 Test and diagnose A/C compressor clutch control systems; determine necessary action.

DIRECTIONS: Fill in the blanks below by identifying (1) Safety First Practices that must be followed, (2) Tools and Equipment required, (3) three possible causes of A/C compressor load cut-off systems problems, and (4) corrective actions that should be taken.

Safety First

- _____
- _____
- _____
- _____

Tools and Equipment Required:

- _____   - _____
- _____   - _____
- _____

QUICK ✔ Diagnostic for Cause(s)

List at least three common causes of A/C compressor load cut-off systems problems:

1. _____
2. _____
3. _____
4. _____
5. _____

QUICK ✔ Diagnostic for Corrective Action(s)

List possible corrective action(s):

1. _____
2. _____
3. _____
4. _____
5. _____

Name _____ Date _____ Class _____

DIAGNOSING OPERATING NOISES IN A/C SYSTEM

NATEF Standard(s) for Heating & Air Conditioning:
A6 Diagnose abnormal operating noises in the A/C system; determine necessary action.

DIRECTIONS: Fill in the blanks below by identifying (1) Safety First Practices that must be followed, (2) Tools and Equipment required, (3) three possible causes of noise problems in the A/C system, and (4) corrective actions that should be taken.

Safety First
- _____
- _____
- _____
- _____

Tools and Equipment Required:
- _____   • _____
- _____   • _____
- _____

QUICK ✔ Diagnostic for Cause(s)

List at least three common causes of noise problems in the A/C system:

1. _____
2. _____
3. _____
4. _____
5. _____

QUICK ✔ Diagnostic for Corrective Action(s)

List possible corrective action(s):

1. _____
2. _____
3. _____
4. _____
5. _____

Heating & Air Conditioning

DIAGNOSING EVAPORATOR HOUSING WATER DRAIN

NATEF Standard(s) for Heating & Air Conditioning:

B2-6 Inspect evaporator housing water drain; perform necessary action.

DIRECTIONS: Fill in the blanks below by identifying (1) Safety First Practices that must be followed, (2) Tools and Equipment required, (3) three possible causes of evaporator housing water drain problems, and (4) corrective actions that should be taken.

Safety First

- _____
- _____
- _____
- _____

Tools and Equipment Required:

- _____   - _____
- _____   - _____
- _____

QUICK ✔ Diagnostic for Cause(s)

List at least three common causes of evaporator housing water drain problems:

1. _____
2. _____
3. _____
4. _____

QUICK ✔ Diagnostic for Corrective Action(s)

List possible corrective action(s):

1. _____
2. _____
3. _____
4. _____

DIAGNOSING AUTOMATIC A/C AND HEATING

NATEF Standard(s) for Heating & Air Conditioning:
D8 Check operation of automatic and semi-automatic heating, ventilation, and air-conditioning (HVAC) control systems; determine necessary action.

DIRECTIONS: Fill in the blanks below by identifying (1) Safety First Practices that must be followed, (2) Tools and Equipment required, (3) three possible causes of automatic HVAC control system problems, and (4) corrective actions that should be taken.

Safety First

- _____
- _____
- _____
- _____

Tools and Equipment Required:

- _____   • _____
- _____   • _____
- _____

QUICK ✔ Diagnostic for Cause(s)

List at least three common causes of automatic HVAC control system problems:

1. _____
2. _____
3. _____
4. _____
5. _____

QUICK ✔ Diagnostic for Corrective Action(s)

List possible corrective action(s):

1. _____
2. _____
3. _____
4. _____
5. _____

Heating & Air Conditioning

DIAGNOSING TEMPERATURE CONTROL PROBLEMS IN HEATER/VENTILATION SYSTEM

NATEF Standard(s) for Heating & Air Conditioning:
C1 Diagnose temperature control problems in the heater/ventilation system; determine necessary action.

DIRECTIONS: Fill in the blanks below by identifying (1) Safety First Practices that must be followed, (2) Tools and Equipment required, (3) three possible causes of temperature control problems in the heater/ventilation system, and (4) corrective actions that should be taken.

Safety First
- _____
- _____
- _____
- _____

Tools and Equipment Required:
- _____   • _____
- _____   • _____
- _____   • _____

QUICK ✔ Diagnostic for Cause(s)

List at least three common causes of temperature control problems in the heater/ventilation system:
1. _____
2. _____
3. _____
4. _____

QUICK ✔ Diagnostic for Corrective Action(s)

List possible corrective action(s):
1. _____
2. _____
3. _____
4. _____

COMPLETING A VEHICLE REPAIR ORDER FOR A HEATING AND AIR CONDITIONING CONCERN

NATEF Standard(s) for Heating & Air Conditioning:

A1 Complete work order to include customer information, vehicle identifying information, customer concern, related service history, cause, and correction.

SAMPLE VEHICLE REPAIR ORDER　　　　Vehicle Repair Order # _____

Date ____/____/____

Customer Name & Phone #: _____　　Vehicle Make/Type: _____　　VIN: _____　　Mileage: _____

Service History: _____

Customer Concern: _____

Cause of Concern: _____

Suggested Repairs/Maintenance: _____

Services Performed: _____

	Parts			Labor	Time In:
Item	Description	Price		Diagnosis Time:	Time Complete:
1				Repair Time:	Total Hours:
2					
3					
4					
5					
6					

I hereby authorize the above repair work to be done using the necessary material, and hereby grant you and/or your employees permission to operate the vehicle herein described on streets, highways, or elsewhere for the purpose of testing and/or inspection. An express mechanic's lien is hereby acknowledged on above vehicle to secure the amount of repairs thereof.

X _____

Heating
& Air Conditioning

PROCEDURES Refer to the vehicle service information for specifications and special procedures. Then prepare a vehicle repair order for the vehicle provided by your instructor.

_____ 1. **Write legibly.** Others will be reading what you have written.

_____ 2. **Make sure all information is accurate.** Inaccurate information will slow the repair process.

_____ 3. **Complete every part of the Vehicle Repair Order.** Every part must be completed.

_____ 4. **Number the Vehicle Repair Order.** This will help others track the repair.

_____ 5. **Date the Vehicle Repair Order.** This will help document the service history.

_____ 6. **Enter the Customer Name and Phone Number.** Make sure you have spelled the Customer Name correctly. Double-check the Phone Number.

_____ 7. **Enter the Vehicle Make/Type.** This information is essential.

(continued)

_____ 8. **Enter the VIN (vehicle identification number).** This is a string of coded data that is unique to the vehicle. The location of the VIN depends on the manufacturer. It is usually found on the dashboard next to the windshield on the driver's side. The VIN is a rich source of information. It is needed to properly use a scan tool to read diagnostic trouble codes. Double-check the VIN to ensure accuracy.

_____ 9. **Enter the Mileage of the vehicle.** This information is part of the service history.

_____ 10. **Complete the Service History.** The service history is a history of all the service operations performed on a vehicle. The service history alerts the technician to previous problems with the vehicle. In the case of recurring problems, it helps the technician identify solutions that were ineffective.
   - A detailed service history is usually kept by the service facility where the vehicle is regularly serviced.
   - Information on service performed on the vehicle at other service centers is not available unless the customer makes it available. For this reason, ask the customer about service performed outside of the present service center.

_____ 11. **Identify the Customer Concern.** This should be a reasonably detailed and accurate description of the problem that the customer is having with the vehicle. The customer is usually the best source of information regarding the problem. This information can be used to perform the initial diagnosis. The customer has a passenger car. He says, "My car's heater doesn't produce enough heat, especially on cold days. Also, the engine is running rough and is difficult to start." Enter his concern on the Customer Concern line.

_____ 12. **Identify the Cause of Concern.** This will identify the problem. In this case, there may be several possible causes. Enter the possible causes on the Cause of Concern line.

_____ 13. Ask the customer to read the text at the bottom of the Labor box. By signing on the line at the bottom of this box, the customer authorizes repair work on the vehicle according to the terms specified.

_____ 14. **Identify Suggested Repairs/Maintenance.** This will identify what needs to be done to correct the problem.

_____ 15. **Identify Services Performed.** This will identify the specific maintenance and repair procedures that were performed to correct the problem.

_____ 16. **Provide Parts information.** This includes a numbered list of items used to complete the repair. It includes a description of each part with the price of the part.

_____ 17. **Provide Labor information.** The Diagnosis Time and the Repair Time are totaled to give the Total Hours.

Performance ✓ Checklist

Name _____ Date _____ Class _____

PERFORMANCE STANDARDS:
Level 4–Performs skill without supervision and adapts to problem situations.
Level 3–Performs skill satisfactorily without assistance or supervision.
Level 2–Performs skill satisfactorily, but requires assistance/supervision.
Level 1–Performs parts of skill satisfactorily, but requires considerable assistance/supervision.

Attempt (circle one): **1 2 3 4**

Comments:

_____

_____

_____

_____

PERFORMANCE LEVEL ACHIEVED: _____

_____ **1.** Safety rules and practices were followed at all times regarding this job.

_____ **2.** Tools and equipment were used properly and stored upon completion of this job.

_____ **3.** This completed job met the standards set and was done within the allotted time.

_____ **4.** No injury or damage to property occurred during this job.

_____ **5.** Upon completion of this job, the work area was cleaned correctly.

Instructor's Signature _____ Date_____

DIAGNOSING A/C COMPRESSOR CLUTCH CONTROL SYSTEMS

NATEF Standard(s) for Heating & Air Conditioning:
D3 Test and diagnose A/C compressor clutch control systems; determine necessary action.

Safety First
- ☐ Wear safety glasses at all times.
- ☐ Follow all safety rules when using common hand tools.
- ☐ Use exhaust vent if running engine in closed shop. Unvented exhaust fumes can cause death.
- ☐ Stay clear of cooling fan and hot surfaces.
- ☐ Keep jewelry, loose clothing, and hair away from moving parts while an engine is running.
- ☐ Make sure automatic transmission is in PARK or manual transmission is in NEUTRAL and set the parking brake.
- ☐ Follow all safety rules when using a lift or jack and jack stands.

Tools & Equipment:
- Vehicle service information
- Lift or jack and jack stands
- Common hand tools
- Scan tool
- DVOM

PROCEDURES Refer to the vehicle service information for specifications and special procedures. Then test and diagnose the air conditioner compressor clutch control system in the vehicle provided by your instructor.

_____ 1. Write up a repair order.

_____ 2. Make sure you follow all procedures in the vehicle service information.

_____ 3. If the clutch is slipping, check voltage to the clutch with a DVOM. Also use the meter to test clutch current draw and clutch resistance. Compare the readings against factory specifications.

_____ 4. Check the clamping diode. Disconnect the diode from the circuit. Use an ohmmeter to check the diode resistance. The diode should have little resistance to flow in one direction, but infinite resistance in the other direction.

_____ 5. If the clutch electrical circuit does not supply voltage to the clutch, electrically check the clutch switches with a DVOM. Many air conditioning systems have pressure and throttle position switches. These are known as control switches because they control the clutch.

_____ 6. If the clutch operates when directly energized with voltage but fails to work during system operation, check these electrical controls. Refer to the manufacturer's information for a wiring diagram that shows the switches in the circuit.

_____ 7. Some air conditioning control systems can be diagnosed with a scan tool. Use a scan tool on these systems to determine the status of all control switches and sensors.

(continued)

Performance ✓ Checklist

Name _____ Date _____ Class _____

PERFORMANCE STANDARDS:
Level 4–Performs skill without supervision and adapts to problem situations.
Level 3–Performs skill satisfactorily without assistance or supervision.
Level 2–Performs skill satisfactorily, but requires assistance/supervision.
Level 1–Performs parts of skill satisfactorily, but requires considerable assistance/supervision.

Attempt (circle one): **1 2 3 4**

Comments:

_____

_____

_____

_____

PERFORMANCE LEVEL ACHIEVED: _____

_____ **1.** Safety rules and practices were followed at all times regarding this job.

_____ **2.** Tools and equipment were used properly and stored upon completion of this job.

_____ **3.** This completed job met the standards set and was done within the allotted time.

_____ **4.** No injury or damage to property occurred during this job.

_____ **5.** Upon completion of this job, the work area was cleaned correctly.

Instructor's Signature _____ Date_____

TESTING, FLUSHING & HANDLING
ENGINE COOLANT

NATEF Standard(s) for Heating & Air Conditioning:

C5 Determine coolant condition and coolant type for vehicle application; drain and recover coolant.

C6 Flush system; refill system with recommended coolant; bleed system.

Safety First

- ☐ Wear safety glasses at all times.
- ☐ Follow all safety rules when using common hand tools.
- ☐ Always connect a vehicle's exhaust to a vent hose before you run an engine in a closed shop. Unvented exhaust fumes in a closed shop can cause death.
- ☐ Check Material Safety Data Sheets for chemical safety information.
- ☐ Disconnect electric fans from the power source before working in the vicinity of the fan blades.
- ☐ Keep jewelry, loose clothing, and hair away from moving parts while an engine is running.
- ☐ Be careful of hot engine parts.
- ☐ Set the parking brake and place automatic transmission in PARK or manual transmission in NEUTRAL.

Tools & Equipment:

- Coolant hydrometer
- Common hand tools
- Coolant refractometer
- Vehicle service information
- Coolant drain pan

PROCEDURES Refer to the vehicle service information for specifications and special procedures. Then test, drain, and replace the coolant in the engine provided by your instructor.

_____ 1. Write up a repair order.

_____ 2. Make sure you follow all procedures in the vehicle service information.

_____ 3. On a cool engine, remove the radiator cap. **CAUTION:** Remove the radiator cap slowly and then run your finger around the inside of the neck to check for rust, oil, and scale. If you find engine oil, it indicates an engine leak. Transmission oil in the coolant indicates an oil cooler leak. Light brown coolant is rusty and must be replaced.

_____ 4. Run the engine until it reaches normal operating temperature and then turn it off.

(continued)

_____ 5. Determine the coolant's age by checking maintenance records and check the protective qualities of the coolant by using the hydrometer or refractometer according to the manufacturer's instructions. *Note:* If the coolant fails any tests, you will have to replace it. See your instructor for special instructions to flush and drain the coolant system.

_____ 6. Replace coolant by first draining the old coolant into the coolant drain pan. *Note:* Be sure to dispose of coolant according to EPA regulations.

_____ 7. Look up the manufacturer's recommended coolant type and volume for the vehicle.

   Source: _____ *Page:*_____

_____ 8. If the system does not need to be flushed, close the engine drains, refill the system with the appropriate antifreeze, and add enough water to bring the level to "full." This should result in about a 50/50 solution. *Note:* Some technicians prefer to mix a 50/50 solution before putting it in the vehicle.

_____ 9. Fill the recovery reservoir to the hot or maximum level, leave the radiator cap off, and turn on the passenger compartment heater.

_____ 10. Run the engine until it reaches operating temperature.

_____ 11. Some manufacturers require bleeding of air from the highest point in the cooling system. Have your instructor supervise this operation. Follow the manufacturer's instructions. **CAUTION:** The coolant will be very hot. Do not allow it to touch your skin.

_____ 12. Stop the engine, add more coolant if necessary, and install the radiator cap.

_____ 13. Run the engine until it reaches operating temperature and turn it off.

_____ 14. Check for leaks. *Note:* Check especially around the thermostat housing, all the hoses, and the drains.

_____ 15. Repair any leaks.

Performance ✓ Checklist

Name _____   Date _____   Class _____

PERFORMANCE STANDARDS:
Level 4–Performs skill without supervision and adapts to problem situations.
Level 3–Performs skill satisfactorily without assistance or supervision.
Level 2–Performs skill satisfactorily, but requires assistance/supervision.
Level 1–Performs parts of skill satisfactorily, but requires considerable assistance/supervision.

Attempt (circle one): **1 2 3 4**

Comments:

_____

_____

_____

_____

PERFORMANCE LEVEL ACHIEVED: _____

_____ 1. Safety rules and practices were followed at all times regarding this job.
_____ 2. Tools and equipment were used properly and stored upon completion of this job.
_____ 3. This completed job met the standards set and was done within the allotted time.
_____ 4. No injury or damage to property occurred during this job.
_____ 5. Upon completion of this job, the work area was cleaned correctly.

Instructor's Signature _____   Date_____

DIAGNOSING TEMPERATURE CONTROL PROBLEMS

NATEF Standard(s) for Heating & Air Conditioning:

C1 Diagnose temperature control problems in the heater/ventilation system; determine necessary action.

Safety First

- ☐ Wear safety glasses at all times.
- ☐ Follow all safety rules when using common hand tools.
- ☐ Use exhaust vent if running engine in closed shop. Unvented exhaust fumes can cause death.
- ☐ Do not open a radiator cap while the engine is hot.
- ☐ Watch out for hot or moving parts.
- ☐ Make sure automatic transmission is in PARK or manual transmission is in NEUTRAL and set the parking brake.

Tools & Equipment:

- Vehicle service information
- Hand-held vacuum pump with gauge
- Common hand tools
- DVOM
- Container to heat water
- Thermometer
- Hot plate or stove

PROCEDURES Refer to the vehicle service information for specifications and special procedures. Then diagnose the temperature control problems in the vehicle provided by your instructor.

_____ 1. Write up a repair order.

_____ 2. Make sure you follow all procedures in the vehicle service information.

_____ 3. Check for inoperative outside air vent door. Adjust or repair outside air vent door.

_____ 4. Check for inoperative blend door. Adjust or repair blend door.

_____ 5. Check for inoperative heater control valve. Adjust or replace heater control valve.

_____ 6. Check for low coolant. Repair leaks and add coolant. Bleed air from the system as necessary.

_____ 7. If the thermostat is defective, test it. To do this:
- Suspend the thermostat in a pot of water. Be sure the thermostat is closed.
- Put a thermometer in the water.
- Use a hot plate or stove burner to heat the test water.
- Watch the thermometer. Note the temperature at which the thermostat begins to open as the water heats.

(continued)

Heating & Air Conditioning

- Note the temperature that causes a wide-open thermostat. That temperature should be equal to the thermostat rating.

  _____

- Measure the opening distance and compare it with specifications.

  _____

Note that some thermostats have accessory valves. These may allow coolant circulation through a bypass. They may also bleed air from the system. The small bypass valve may have a different opening temperature rating than the main radiator flow valve. Check the service information for specifics.

_____ 8. If a thermostat fails to open at the proper temperature or does not open far enough, replace it. Also, replace the gasket or O-ring.

_____ 9. It is good to test a thermostat to prove it is defective. However, thermostats are not expensive and the labor rate is low. Most technicians will simply replace a thermostat if there is any doubt about its operation.

Performance ✓ Checklist

Name _____ Date _____ Class _____

PERFORMANCE STANDARDS:
Level 4–Performs skill without supervision and adapts to problem situations.
Level 3–Performs skill satisfactorily without assistance or supervision.
Level 2–Performs skill satisfactorily, but requires assistance/supervision.
Level 1–Performs parts of skill satisfactorily, but requires considerable assistance/supervision.

Attempt (circle one): **1 2 3 4**

Comments:

_____

_____

_____

_____

PERFORMANCE LEVEL ACHIEVED: _____

_____ 1. Safety rules and practices were followed at all times regarding this job.
_____ 2. Tools and equipment were used properly and stored upon completion of this job.
_____ 3. This completed job met the standards set and was done within the allotted time.
_____ 4. No injury or damage to property occurred during this job.
_____ 5. Upon completion of this job, the work area was cleaned correctly.

Instructor's Signature _____ Date _____

PERFORMING SYSTEM PRESSURE TESTS

NATEF Standard(s) for Heating & Air Conditioning:

C2 Perform cooling system pressure tests; (check coolant condition, inspect and test radiator, pressure cap, coolant recovery tank, and hoses); determine necessary action.

Safety First

- ☐ Wear safety glasses at all times.
- ☐ Follow all safety rules when using common hand tools.
- ☐ Always connect a vehicle's exhaust to a vent hose before you run an engine in a closed shop. Unvented exhaust fumes in a closed shop can cause death.
- ☐ Always follow all safety rules for removing radiator caps.
- ☐ Keep jewelry, loose clothing, and hair away from moving parts while an engine is running.
- ☐ Be careful of hot engine parts.
- ☐ Set the parking brake and place automatic transmission in PARK or manual transmission in NEUTRAL.

Tools & Equipment:
- Cooling system analyzer
- Common hand tools
- Coolant thermometer
- Vehicle service information
- Voltmeter

PROCEDURES Refer to the vehicle service information for specifications and special procedures. Then test the system pressures in the engine provided by your instructor.

_____ 1. Write up a repair order.

_____ 2. Make sure you follow all procedures in the vehicle service information.

_____ 3. Look up the recommended system pressure and temperature specifications in the vehicle service information.

Source: _____ Page:_____

_____ 4. Test the radiator cap with the cooling system pressure tester by following the directions with the kit. **Note:** Refer to the kit's directions to perform all the pressure tests.

_____ 5. Pump the tester slowly while noting when the pressure reading on the gauge stops. **Note:** This is where the relief valve on the cap opens.

_____ 6. Release the tester and repeat Steps #4 and #5. If the radiator cap's pressure reading is below the minimum specified—or 2 to 3 psi above the maximum—replace the cap.

_____ 7. Test the engine's cooling system with the cooling system tester kit by attaching the tester to the radiator filler neck and pumping the tester until the pressure on the gauge reaches the pressure specified by the manufacturer's recommendations.

(*continued*)

_____ 8. Keep pressure on the system for at least two minutes while watching the gauge for any pressure drop. ***Note:*** If the pressure holds steady, go to Step #10.

_____ 9. If the pressure drops, inspect the system for external leaks. Repair any leaks you find.

_____ 10. Check for milky looking oil on the end of the oil dipstick. This is a sign of an internal coolant leak. ***Note:*** If coolant is in the oil, the cylinder head, gasket, or block is damaged. See your instructor if you find contaminated oil.

_____ 11. Test the internal combustion components by first removing the radiator cap and allowing the engine to run until it reaches normal operating temperature. Then turn off the engine. **CAUTION:** Be careful to follow all safety rules when removing a radiator cap from a warm engine. Remove it slowly. Never attempt to remove a radiator cap from a hot engine.

_____ 12. Attach the pressure tester to the radiator according to the manufacturer's recommended procedure. If there is a leak, isolate the source with the power balance test. **CAUTION:** Do not exceed the cap pressure rating.

_____ 13. Check the coolant temperature by first allowing the engine to cool and then checking and correcting the coolant level if necessary.

_____ 14. Run the engine until the thermostat opens and check the coolant temperature by placing the coolant thermometer in the neck of the radiator. **CAUTION:** If the temperature is above or below the recommended temperature, the thermostat may be faulty.

_____ 15. With the engine off, conduct an electrolysis test by grounding the positive probe of the voltmeter to the radiator and inserting the negative probe into the coolant.

_____ 16. If the reading is over 0.5 volts, flush the system and replace the coolant.

Performance ✓ Checklist

Name _____ Date _____ Class _____

PERFORMANCE STANDARDS:
Level 4–Performs skill without supervision and adapts to problem situations.
Level 3–Performs skill satisfactorily without assistance or supervision.
Level 2–Performs skill satisfactorily, but requires assistance/supervision.
Level 1–Performs parts of skill satisfactorily, but requires considerable assistance/supervision.

Attempt (circle one): **1 2 3 4**

Comments:

_____

_____

_____

_____

PERFORMANCE LEVEL ACHIEVED: _____

_____ 1. Safety rules and practices were followed at all times regarding this job.

_____ 2. Tools and equipment were used properly and stored upon completion of this job.

_____ 3. This completed job met the standards set and was done within the allotted time.

_____ 4. No injury or damage to property occurred during this job.

_____ 5. Upon completion of this job, the work area was cleaned correctly.

Instructor's Signature _____ Date_____

INSPECTING COOLING AND HEATING SYSTEM HOSES AND BELTS

NATEF Standard(s) for Heating & Air Conditioning:

C3 Inspect engine cooling and heater system hoses and belts; perform necessary action.

Safety First

- ☐ Wear safety glasses at all times.
- ☐ Follow all safety rules when using common hand tools.
- ☐ Use exhaust vent if running engine in closed shop. Unvented exhaust fumes can cause death.
- ☐ Make sure that the engine is cold. Do not open radiator cap on a hot engine.
- ☐ Follow all safety rules when working with automatic belt tensions.
- ☐ Make sure automatic transmission is in PARK or manual transmission is in NEUTRAL and set the parking brake.
- ☐ Keep jewelry, loose clothing, and hair away from moving parts while an engine is running.
- ☐ Stay clear of cooling fan and hot surfaces.

Tools & Equipment:

- Vehicle service information
- Belt tension gauge
- Hose-clamp pliers
- Common hand tools
- Coolant drain pan
- Antifreeze hydrometer
- Pry bars

PROCEDURES Refer to the vehicle service information for specifications and special procedures. Then inspect the belts and hoses of the cooling and heating system in the vehicle provided by your instructor.

Inspecting Belts

_____ 1. Write up a repair order.

_____ 2. Make sure you follow all procedures in the vehicle service information.

_____ 3. Examine the drive belt(s) closely.

_____ 4. Check belt tension with a belt tension gauge. A high-pitched squeal is a sign of a loose and slipping belt.

_____ 5. To check the V-belt, examine the inner friction surface for cracks, grease, glazing, scars, tears, splits, and ply separation.

_____ 6. Examine how deeply into the pulley the belt runs. A V-belt that rides too low bottoms in the pulley valley and loses traction. If the V-belt has a companion belt running in tandem, replace both belts at the same time.

(continued)

Heating & Air Conditioning

_____ 7. Do not turn a flat serpentine belt more than 90° to examine the belt traction surface. Excess rotation can damage internal belt cords. Most flat serpentine belts have an automatic belt tensioner. Back off the tensioner strength if you need viewing access to see the belt traction surface. Inspect the serpentine belt carefully.

_____ 8. Check the serpentine belt for excessive wear and cracking, splits, glazing, chunking, and broken cords. Replace the belt if any of these conditions are found. Serpentine belts often have numerous tiny cracks in their surface. This is considered normal and is not cause for replacement.

_____ 9. After installing a new belt, use an appropriate tension gauge and adjust the belt tension. Make sure the belt seats properly. Ensure that it mates properly with the pulleys.

_____ 10. Ribbed serpentine belts run with the belt ridges mating with grooves on the pulley. Check the pulley position so the belt sits true and straight on the pulley grooves.

_____ 11. If the pulleys are not properly aligned, shim the component mountings to align the pulleys with the belt. New V-belts will seat and loosen after the engine is first run. Therefore, after a new V-belt is installed, tension should be checked after the system operates for several minutes.

Inspecting Hoses

_____ 1. Look for hose bulging caused by pressure and heat.

_____ 2. Check each hose for hardening.

_____ 3. Examine each for interference.

_____ 4. Check for abrasion.

_____ 5. Squeeze the hoses, feeling for softness that indicates weakness.

_____ 6. Inspect system tubing. In many cooling systems, tubes are used rather than hoses. Tubes often have sealing O-rings or gaskets at the ends. If the initial inspection or vehicle mileage indicates that seals or hoses should be serviced, replace the hoses.

_____ 7. To replace a hose, do not attempt to pull it from the connecting fitting. Instead, push on the hose while twisting. If the hose seal does not break loose, it may be necessary to cut the hose from its connecting fitting. If so, use a sharp knife blade to carefully cut through the hose connecting area at a 45° angle around the joint.

_____ 8. A lower radiator hose often contains an inner coil spring. This spring prevents hose collapse from the suction of the water pump. Check the condition of the support spring. If the spring is deteriorated, replace the hose. Make sure the spring does not extend into the water pump. If the spring protrudes into the pump, it can cause interference.

Installing a New Hose

_____ 1. Make sure the hose fits the application. Compare its shape and length to the old hose.

_____ 2. Slide new clamps loosely onto the hose.

_____ 3. Slide the hose ends in place. Most receiving nipples bulge the hose slightly. Place the clamps slightly beyond this bulge.

_____ 4. Tighten the clamps snugly. When tightening hose clamps, leave about 1/8"–1/4" of the hose extending beyond the clamp. Tighten clamps securely, but do not overtighten. A clamp must be snug, but not so tight that it cuts into the hose.

Performance ✓ Checklist

Name _____ Date _____ Class _____

PERFORMANCE STANDARDS:
Level 4–Performs skill without supervision and adapts to problem situations.
Level 3–Performs skill satisfactorily without assistance or supervision.
Level 2–Performs skill satisfactorily, but requires assistance/supervision.
Level 1–Performs parts of skill satisfactorily, but requires considerable assistance/supervision.

Attempt (circle one): **1 2 3 4**

Comments:

_____

_____

_____

_____

PERFORMANCE LEVEL ACHIEVED: _____

_____ **1.** Safety rules and practices were followed at all times regarding this job.

_____ **2.** Tools and equipment were used properly and stored upon completion of this job.

_____ **3.** This completed job met the standards set and was done within the allotted time.

_____ **4.** No injury or damage to property occurred during this job.

_____ **5.** Upon completion of this job, the work area was cleaned correctly.

Instructor's Signature _____ Date _____

Heating
& Air Conditioning

INSPECTING & REPLACING THERMOSTATS

NATEF Standard(s) for Heating & Air Conditioning:
C4 Inspect, test, and replace thermostat and gasket.

Safety First

☐ Wear safety glasses at all times.
☐ Follow all safety rules when using common hand tools.
☐ Always connect a vehicle's exhaust to a vent hose before you run an engine in a closed shop. Unvented exhaust fumes in a closed shop can cause death.
☐ Check Material Safety Data Sheets for chemical safety information.
☐ Disconnect electric fans from the power source before working in the vicinity of the fan blades.
☐ Keep jewelry, loose clothing, and hair away from moving parts while an engine is running.
☐ Be careful of hot engine parts.
☐ Set the parking brake and place automatic transmission in PARK or manual transmission in NEUTRAL.

Tools & Equipment:
- Syringe
- Cooling pot
- Sealant or gasket
- Vehicle service information
- Hot plate
- Wire
- Torque wrench
- Thermometer
- Putty knife
- Common hand tools

PROCEDURES Refer to the vehicle service information for specifications and special procedures. Then test and replace the thermostat in the engine provided by your instructor.

_____ 1. Write up a repair order.

_____ 2. Make sure you follow all procedures in the vehicle service information.

_____ 3. On a cool engine, remove the radiator cap. **CAUTION:** Remove the radiator cap slowly.

_____ 4. Using the syringe, decrease the coolant level to beneath the thermostat housing (or partially drain the radiator). *Note:* Save the coolant if it is clean. If it needs to be discarded, be sure to dispose of it according to EPA regulations.

_____ 5. Remove the thermostat housing or cap. *Note:* You may need to tap the housing free with a rubber hammer.

_____ 6. Remove and rinse the thermostat.

_____ 7. Inspect the thermostat for damage such as cracks, breaks, corrosion, and damaged seals.

_____ 8. Replace any defective or damaged thermostats.

(continued)

_____ 9. Look up the manufacturer's recommended thermostat temperature rating for the vehicle.

Source: _____ *Page:* _____

_____ 10. Replace the thermostat if its temperature rating (stamped on the engine side of the thermostat) does not match the manufacturer's recommended rating.

_____ 11. Refer to your instructor for performance testing an undamaged thermostat with water, the hot plate, and the thermometer.

_____ 12. Look up the manufacturer's recommended thermostat selection.

Source: _____ *Page:* _____

_____ 13. Install the correct thermostat and then reinstall the thermostat housing with a new gasket. *Note:* A high-temperature silicon sealant can also be used. See your instructor for the procedure.

_____ 14. Torque the bolts to specification, reconnect the radiator hose, and replace the coolant that you removed.

_____ 15. Start the engine and run it until it reaches operating temperature and the thermostat opens.

_____ 16. Check for leaks and turn off the engine.

_____ 17. Repair any leaks.

Performance ✓ Checklist

Name _____ Date _____ Class _____

PERFORMANCE STANDARDS:
Level 4–Performs skill without supervision and adapts to problem situations.
Level 3–Performs skill satisfactorily without assistance or supervision.
Level 2–Performs skill satisfactorily, but requires assistance/supervision.
Level 1–Performs parts of skill satisfactorily, but requires considerable assistance/supervision.

Attempt (circle one): **1 2 3 4**

Comments:

_____

_____

_____

PERFORMANCE LEVEL ACHIEVED: _____

_____ 1. Safety rules and practices were followed at all times regarding this job.

_____ 2. Tools and equipment were used properly and stored upon completion of this job.

_____ 3. This completed job met the standards set and was done within the allotted time.

_____ 4. No injury or damage to property occurred during this job.

_____ 5. Upon completion of this job, the work area was cleaned correctly.

Instructor's Signature _____ Date_____

DIAGNOSING HEATER CONTROL VALVES

NATEF Standard(s) for Heating & Air Conditioning:

C9 Inspect and test heater control valve(s); perform necessary action.

Safety First

- ☐ Wear safety glasses at all times.
- ☐ Follow all safety rules when using common hand tools.
- ☐ Use exhaust vent if running engine in closed shop. Unvented exhaust fumes can cause death.
- ☐ Watch out for hot or moving parts.
- ☐ Make sure engine is cool and pressure is relieved from cooling system before removing heater control valve. Do not open the radiator cap on a hot engine.
- ☐ Make sure automatic transmission is in PARK or manual transmission is in NEUTRAL and set the parking brake.

Tools & Equipment:

- Vehicle service information
- Radiator pressure tester
- Hose-clamp pliers
- Common hand tools
- Hand-held vacuum pump with a gauge
- DVOM
- Coolant drain pan

PROCEDURES Refer to the vehicle service information for specifications and special procedures. Then diagnose the heater control valves in the vehicle provided by your instructor.

_____ 1. Write up a repair order.

_____ 2. Make sure you follow all procedures in the vehicle service information.

_____ 3. Check for control valve leaking coolant. Use pressure tester if necessary. Replace valve.

_____ 4. Check for damaged or inoperative control valve. Replace valve.

_____ 5. Check for improper vacuum signal to control valve. Repair vacuum circuit.

_____ 6. Check for control cable that is misadjusted. Adjust control cable.

(continued)

Heating & Air Conditioning

Performance ✓ Checklist

Name _____ Date _____ Class _____

PERFORMANCE STANDARDS:
Level 4–Performs skill without supervision and adapts to problem situations.
Level 3–Performs skill satisfactorily without assistance or supervision.
Level 2–Performs skill satisfactorily, but requires assistance/supervision.
Level 1–Performs parts of skill satisfactorily, but requires considerable assistance/supervision.

Attempt (circle one): **1 2 3 4**

Comments:

_____

_____

_____

_____

PERFORMANCE LEVEL ACHIEVED: _____

_____ **1.** Safety rules and practices were followed at all times regarding this job.

_____ **2.** Tools and equipment were used properly and stored upon completion of this job.

_____ **3.** This completed job met the standards set and was done within the allotted time.

_____ **4.** No injury or damage to property occurred during this job.

_____ **5.** Upon completion of this job, the work area was cleaned correctly.

Instructor's Signature _____ Date_____

DIAGNOSING FAN PROBLEMS

NATEF Standard(s) for Heating & Air Conditioning:

C7 Inspect and test cooling fan, fan clutch, fan shroud, and air dams; perform necessary action.

Safety First

- ☐ Wear safety glasses at all times.
- ☐ Follow all safety rules when using common hand tools.
- ☐ Use exhaust vent if running engine in closed shop. Unvented exhaust fumes can cause death.
- ☐ Watch out for hot or moving parts.
- ☐ Make sure automatic transmission is in PARK or manual transmission is in NEUTRAL and set the parking brake.

Tools & Equipment:

- Vehicle service information
- Common hand tools

PROCEDURES Refer to the vehicle service information for specifications and special procedures. Then diagnose the fan problems in the vehicle provided by your instructor.

_____ 1. Write up a repair order.

_____ 2. Make sure you follow all procedures in the vehicle service information.

_____ 3. Check for loose fan belt. Tighten or replace fan belt.

_____ 4. Check for seized fan clutch. Replace fan clutch.

_____ 5. Check for freewheeling fan clutch. Replace fan clutch.

_____ 6. Check fan blade for damage or stress cracks. Replace defective fan blade.

_____ 7. Check fan shroud for damage and proper mounting. Replace damaged fan shroud.

_____ 8. Check condition of air dam and proper mounting. Replace damaged air dam.

(continued)

Heating & Air Conditioning

Performance ✓ Checklist

Name _____ Date _____ Class _____

PERFORMANCE STANDARDS:
Level 4–Performs skill without supervision and adapts to problem situations.
Level 3–Performs skill satisfactorily without assistance or supervision.
Level 2–Performs skill satisfactorily, but requires assistance/supervision.
Level 1–Performs parts of skill satisfactorily, but requires considerable assistance/supervision.

Attempt (circle one): **1 2 3 4**

Comments:

_____

_____

_____

_____

PERFORMANCE LEVEL ACHIEVED: _____

_____ 1. Safety rules and practices were followed at all times regarding this job.

_____ 2. Tools and equipment were used properly and stored upon completion of this job.

_____ 3. This completed job met the standards set and was done within the allotted time.

_____ 4. No injury or damage to property occurred during this job.

_____ 5. Upon completion of this job, the work area was cleaned correctly.

Instructor's Signature _____ Date_____

DIAGNOSING COOLING FAN CONTROL SYSTEMS

NATEF Standard(s) for Heating & Air Conditioning:

C8 Inspect and test electric cooling fan control system and circuits; determine necessary action.

Safety First

☐ Wear safety glasses at all times.

☐ Follow all safety rules when using common hand tools.

☐ Use exhaust vent if running engine in closed shop. Unvented exhaust fumes can cause death.

☐ Watch out for hot or moving parts.

☐ Electrical cooling fans may start with engine off. Use caution around fans.

☐ Make sure automatic transmission is in PARK or manual transmission is in NEUTRAL and set the parking brake.

Tools & Equipment:

- Vehicle service information
- Thermometer
- Jumper wire
- Common hand tools
- DVOM
- 12-volt test light
- Scan tool

PROCEDURES Refer to the vehicle service information for specifications and special procedures. Then diagnose the cooling fan control systems in the vehicle provided by your instructor.

_____ 1. Write up a repair order.

_____ 2. Make sure you follow all procedures in the vehicle service information.

_____ 3. Before checking the fan or the fan circuit, be sure the problem is not simply a replaceable fuse or a fusible link. If the electric fan does not run, check for continuity through the fuse using a 12-volt test light or multimeter.

_____ 4. If the coolant fan system can be tested with a scan tool, connect a scan tool and check for codes. Some systems allow the fan to be operated by the scan tool to test the operation of the fan motor.

_____ 5. If the engine is hot and the coolant fan is off, turn on the air conditioner. If the coolant fan now operates, the problem is likely the coolant temperature sensor or switch. Check out the sensor or switch and wiring to the relay. A jumper wire can be used if the circuit is fully understood. Improper use of a jumper wire can lead to problems. You may be unknowingly bypassing other components in the circuit.

_____ 6. Most testing can be done at the relay. A voltmeter can verify battery voltage available to the relay. This terminal jumped to the fan motor terminal can verify wiring continuity and motor operation.

(continued)

_____ 7 The trigger circuit to the relay can be tested with a voltmeter. This may be a ground trigger from the PCM. The PCM may require inputs from several other sensors as well as the engine coolant temperature sensor before triggering the relay.

_____ 8. The relay can be properly tested with the use of a voltmeter and ohmmeter.

_____ 9. Once the necessary parts are installed or wiring repaired, the current draw of the fan motor should be checked. A defective fan motor can draw excessive amperage and damage a relay or switch. Many times the failed part is replaced and the fan works, but the root cause of the problem has not been corrected and the problem occurs again later.

_____ 10. Many technicians replace the relay when a failed fan motor is replaced because they're afraid that the relay contacts are burned and may cause a failure of the system later.

Performance ✓ Checklist

Name _____ Date _____ Class _____

PERFORMANCE STANDARDS:
Level 4–Performs skill without supervision and adapts to problem situations.
Level 3–Performs skill satisfactorily without assistance or supervision.
Level 2–Performs skill satisfactorily, but requires assistance/supervision.
Level 1–Performs parts of skill satisfactorily, but requires considerable assistance/supervision.

Attempt (circle one): **1 2 3 4**

Comments:

_____

_____

_____

_____

PERFORMANCE LEVEL ACHIEVED: _____

_____ 1. Safety rules and practices were followed at all times regarding this job.
_____ 2. Tools and equipment were used properly and stored upon completion of this job.
_____ 3. This completed job met the standards set and was done within the allotted time.
_____ 4. No injury or damage to property occurred during this job.
_____ 5. Upon completion of this job, the work area was cleaned correctly.

Instructor's Signature _____ Date _____

PERFORMING A SYSTEM TEST WITH A MANIFOLD GAUGE

NATEF Standard(s) for Heating & Air Conditioning:

A2 Identify and interpret heating and air conditioning concern; determine necessary action.

A5 Performance test A/C system; diagnose A/C system malfunctions using principles of refrigeration.

A6 Diagnose abnormal operating noises in the A/C system; determine necessary action.

A7 Identify refrigerant type; select and connect proper gauge set; record pressure readings.

Safety First

- ☐ Wear safety glasses at all times.
- ☐ Follow all safety rules when using common hand tools.
- ☐ Always connect a vehicle's exhaust to a vent hose before you run an engine in a closed shop. Unvented exhaust fumes in a closed shop can cause death.
- ☐ Keep jewelry, loose clothing, and hair away from moving parts while an engine is running.
- ☐ Be careful of hot engine parts.

Tools & Equipment:

- Proper refrigerant storage container
- Manifold gauge set
- Common hand tools
- Thermometer
- Vehicle service information

PROCEDURES Refer to the vehicle service information for specifications and special procedures. Then perform the system test on the A/C system provided by your instructor.

_____ 1. Write up a repair order.

_____ 2. Make sure you follow all procedures in the vehicle service information.

_____ 3. Perform a visual check and place a thermometer in the appropriate A/C dash outlet.

_____ 4. Be sure the engine is not running.

_____ 5. Locate the low- and high-side service valves. The type of service valves and placards in the engine compartment should help you identify the type of refrigerant.

_____ 6. Slowly remove the protective caps from the low-side and high-side service valves.

_____ 7. Close both hand valves on the manifold gauge.

_____ 8. Connect the center hose to the proper storage container.

_____ 9. Properly connect the high-side gauge to the high-side service valve using the high-side hose and open hose valve, if appropriate.

(continued)

Heating & Air Conditioning

_____ **10.** Connect the low-side gauge to the low-side service valve using the low-side hose and open hose valve, if appropriate.

_____ **11.** Start engine, turn air conditioner to maximum cold, and turn blower speed to high.

_____ **12.** Follow performance test instructions. Open the windows if instructed, close the car doors, and wait for 5 minutes. *Note:* By allowing the air conditioner to run for 5 minutes, the system becomes stable.

_____ **13.** Note the high and low pressure and compare them to normal pressure for the vehicle you are testing. *Note:* Normal pressure can be found in the vehicle service information. Differences can be used for diagnostic purposes. Ambient temperature is important when comparing pressure. Also compare the thermometer reading at the dash outlet.

_____

_____ **14.** If the test indicates low refrigerant, the system must be fully charged and re-tested. Compare test results to troubleshooting data to determine needed repairs, if necessary.

_____ **15.** If the system is low on refrigerant, conduct a refrigerant leak test to determine the cause.

_____ **16.** Make sure the manifold valves are closed, close hose valves if appropriate, and remove the hoses from the service valves.

_____ **17.** Replace the caps on the service valves.

Performance ✓ Checklist

Name _____ Date _____ Class _____

PERFORMANCE STANDARDS:
Level 4–Performs skill without supervision and adapts to problem situations.
Level 3–Performs skill satisfactorily without assistance or supervision.
Level 2–Performs skill satisfactorily, but requires assistance/supervision.
Level 1–Performs parts of skill satisfactorily, but requires considerable assistance/supervision.

Attempt (circle one): **1 2 3 4**

Comments:

_____

_____

_____

PERFORMANCE LEVEL ACHIEVED: _____

_____ **1.** Safety rules and practices were followed at all times regarding this job.
_____ **2.** Tools and equipment were used properly and stored upon completion of this job.
_____ **3.** This completed job met the standards set and was done within the allotted time.
_____ **4.** No injury or damage to property occurred during this job.
_____ **5.** Upon completion of this job, the work area was cleaned correctly.

Instructor's Signature _____ Date_____

INSPECTING DISCHARGED OIL

NATEF Standard(s) for Heating & Air Conditioning:

A9 Inspect the condition of discharged oil; determine necessary action.

Safety First
- ☐ Wear safety glasses at all times.
- ☐ Follow all safety rules when using common hand tools.
- ☐ Follow all safety rules when working with refrigerant. Be careful to avoid open flame as refrigerant is flammable and dangerous if breathed once ignited.

Tools & Equipment:
- Vehicle service information
- A/C recovery machine
- Common hand tools
- A/C gauge set

PROCEDURES Refer to the vehicle service information for specifications and special procedures. Then inspect the discharged oil from the vehicle provided by your instructor.

_____ 1. Write up a repair order.

_____ 2. Make sure you follow all procedures in the vehicle service information.

_____ 3. Determine what type of refrigerant is used in your vehicle.

_____ 4. Using the appropriate recovery machine, recover any remaining refrigerant from the vehicle's A/C system.

_____ 5. Following procedure, remove the collected oil from the recovery process.

_____ 6. Examine the oil for signs of contamination due to moisture, corrosion, desiccant, or internal parts failure.

_____ 7. If any contaminants are found, further diagnosis must be made to determine if the system needs parts replaced or just a flush to remove the contamination.

(continued)

Heating & Air Conditioning

Performance ✓ Checklist

Name _____ Date _____ Class _____

PERFORMANCE STANDARDS:
Level 4–Performs skill without supervision and adapts to problem situations.
Level 3–Performs skill satisfactorily without assistance or supervision.
Level 2–Performs skill satisfactorily, but requires assistance/supervision.
Level 1–Performs parts of skill satisfactorily, but requires considerable assistance/supervision.

Attempt (circle one): **1 2 3 4**

Comments:

_____
_____
_____
_____

PERFORMANCE LEVEL ACHIEVED: _____

_____ **1.** Safety rules and practices were followed at all times regarding this job.
_____ **2.** Tools and equipment were used properly and stored upon completion of this job.
_____ **3.** This completed job met the standards set and was done within the allotted time.
_____ **4.** No injury or damage to property occurred during this job.
_____ **5.** Upon completion of this job, the work area was cleaned correctly.

Instructor's Signature _____ Date_____

INSPECTING & REPLACING THE A/C CONDENSER

NATEF Standard(s) for Heating & Air Conditioning:

B2-3 Inspect A/C condenser for airflow restrictions; perform necessary action.

B2-8 Remove, inspect, and reinstall condenser; determine required oil quantity.

Safety First

☐ Wear safety glasses at all times.

☐ Follow all safety rules when using common hand tools.

☐ When replacing the condenser, be careful not to cut yourself on its fins.

☐ Always connect a vehicle's exhaust to a vent hose before you run an engine in a closed shop. Unvented exhaust fumes in a closed shop can cause death.

☐ Keep jewelry, loose clothing, and hair away from moving parts while an engine is running.

☐ Be careful of hot engine parts.

Tools & Equipment:

- Refrigerant oil and measuring device
- Leak detector
- New condenser
- Recovery and charge equipment
- Common hand tools
- Thermometer
- Necessary O-rings, gaskets, or seals
- Vehicle service information

PROCEDURES Refer to the vehicle service information for specifications and special procedures. Then inspect and replace the A/C condenser in the vehicle provided by your instructor.

_____ 1. Write up a repair order.

_____ 2. Make sure you follow all procedures in the vehicle service information.

_____ 3. Inspect the condenser for damaged or worn parts, damaged or worn bolts, or bent fins.

_____ 4. Remove leaves, bugs, or other debris from the condenser coils.

_____ 5. If anything is damaged, replace the condenser. *Note:* The refrigerant must be recovered before this procedure, and the system must be evacuated and charged after it. Before charging, add oil as needed and replace the receiver/drier if required.

_____ 6. Disconnect the inlet hose with the proper tools.

_____ 7. Use the proper tools to disconnect the outlet hose.

_____ 8. Remove the bolts holding the condenser to the vehicle.

_____ 9. Remove the condenser from the vehicle.

(continued)

Heating & Air Conditioning

_____ **10.** Place the new condenser into the vehicle. **CAUTION:** Be careful not to damage the condenser's fins.

_____ **11.** Insert bolts into the condenser and mount.

_____ **12.** Tighten the bolts with the proper tools. **CAUTION:** Be careful not to over-tighten the bolts.

_____ **13.** Add the proper amount of refrigerant oil to the condenser as required.

_____ **14.** Reconnect the outlet hose.

_____ **15.** Reconnect the inlet hose.

_____ **16.** Properly evacuate and charge the system with the correct type and amount of refrigerant.

_____ **17.** Leak test and performance test the system for proper operation.

Performance ✓ Checklist

Name _____ Date _____ Class _____

PERFORMANCE STANDARDS:
Level 4–Performs skill without supervision and adapts to problem situations.
Level 3–Performs skill satisfactorily without assistance or supervision.
Level 2–Performs skill satisfactorily, but requires assistance/supervision.
Level 1–Performs parts of skill satisfactorily, but requires considerable assistance/supervision.

Attempt (circle one): **1 2 3 4**

Comments:

_____

_____

_____

_____

PERFORMANCE LEVEL ACHIEVED: _____

_____ **1.** Safety rules and practices were followed at all times regarding this job.

_____ **2.** Tools and equipment were used properly and stored upon completion of this job.

_____ **3.** This completed job met the standards set and was done within the allotted time.

_____ **4.** No injury or damage to property occurred during this job.

_____ **5.** Upon completion of this job, the work area was cleaned correctly.

Instructor's Signature _____ Date _____

INSPECTING & REPLACING THE RECEIVER/DRIER

NATEF Standard(s) for Heating & Air Conditioning:

B2-3 Inspect A/C condenser for airflow restrictions; perform necessary action.

Safety First

- ☐ Wear safety glasses at all times.
- ☐ Follow all safety rules when using common hand tools.
- ☐ If plugs are not installed on the new receiver/drier, do not use it.
- ☐ Always connect a vehicle's exhaust to a vent hose before you run an engine in a closed shop. Unvented exhaust fumes in a closed shop can cause death.
- ☐ Keep jewelry, loose clothing, and hair away from moving parts while an engine is running.
- ☐ Be careful of hot engine parts.

Tools & Equipment:

- New receiver/drier
- Recovery and charge equipment
- Common hand tools
- Leak detector
- Necessary O-rings, gaskets, or seals
- Vehicle service information
- Thermometer
- Refrigerant oil and measuring device

PROCEDURES Refer to the vehicle service information for specifications and special procedures. Then inspect and replace the receiver/drier in the vehicle provided by your instructor.

_____ 1. Write up a repair order.

_____ 2. Make sure you follow all procedures in the vehicle service information.

_____ 3. Inspect the receiver/drier for damaged or worn parts or bolts.

_____ 4. If anything is damaged, replace the receiver/drier. **CAUTION:** The refrigerant must be recovered before this procedure, and the system must be evacuated and charged after it. Before charging, add oil as needed.

_____ 5. Disconnect the inlet hose with the proper tools.

_____ 6. Use the proper tools to disconnect the outlet hose.

_____ 7. Remove the bolts holding the receiver/drier to the vehicle.

_____ 8. Remove the receiver/drier from the vehicle.

_____ 9. Place the new receiver/drier into the vehicle. _Note:_ Try to install it at the same angle as before. Add the appropriate amount of proper refrigerant oil to the receiver/drier.

(continued)

_____ **10.** Insert bolts into the receiver/drier and mount.

_____ **11.** Tighten the bolts with the proper wrench. **CAUTION:** Be careful not to over-tighten the bolts.

_____ **12.** Reconnect the outlet hose.

_____ **13.** Reconnect the inlet hose.

_____ **14.** Properly evacuate and charge the system with the correct type and amount of refrigerant.

_____ **15.** Leak test and performance test the system for proper operation.

Performance ✓ Checklist

Name _____ Date _____ Class _____

PERFORMANCE STANDARDS:
Level 4–Performs skill without supervision and adapts to problem situations.
Level 3–Performs skill satisfactorily without assistance or supervision.
Level 2–Performs skill satisfactorily, but requires assistance/supervision.
Level 1–Performs parts of skill satisfactorily, but requires considerable assistance/supervision.

Attempt (circle one): **1 2 3 4**

Comments:

_____

_____

_____

_____

PERFORMANCE LEVEL ACHIEVED: _____

_____ **1.** Safety rules and practices were followed at all times regarding this job.

_____ **2.** Tools and equipment were used properly and stored upon completion of this job.

_____ **3.** This completed job met the standards set and was done within the allotted time.

_____ **4.** No injury or damage to property occurred during this job.

_____ **5.** Upon completion of this job, the work area was cleaned correctly.

Instructor's Signature _____ Date_____

DETERMINING THE SOURCE OF LOST VACUUM IN VACUUM LINES

NATEF Standard(s) for Heating & Air Conditioning:

D4 Diagnose malfunctions in the vacuum, mechanical, and electrical components and controls of the heating, ventilation, and A/C (HVAC) system; determine necessary action.

Safety First

- ☐ Wear safety glasses at all times.
- ☐ Follow all safety rules when using common hand tools.
- ☐ Always connect a vehicle's exhaust to a vent hose before you run an engine in a closed shop. Unvented exhaust fumes in a closed shop can cause death.
- ☐ Keep jewelry, loose clothing, and hair away from moving parts while an engine is running.
- ☐ Be careful of hot engine parts.

Tools & Equipment:

- Vacuum tee and gauge
- Common hand tools
- Vehicle service information

PROCEDURES Refer to the vehicle service information for specifications and special procedures. Then determine the source of lost vacuum in the HVAC system provided by your instructor.

_____ 1. Write up a repair order.

_____ 2. Make sure you follow all procedures in the vehicle service information.

_____ 3. Insert the vacuum tee in the larger vacuum supply line in front of the vacuum reservoir and connect the gauge.

_____ 4. Start the vehicle.

_____ 5. Have another person move the function control switch to all possible positions ("A/C Max," "A/C Norm," "Vent," etc.).

_____ 6. Read the gauge at all positions of the function control switch.

_____ 7. Vacuum is being lost in the lines if the gauge reads "Insufficient Vacuum" at all positions of the function control switch.

_____ 8. If the gauge reads "Insufficient Vacuum" at only some positions of the function control switch, vacuum is being lost in lines active at that position of the function control switch.

_____ 9. Move the gauge past the vacuum reservoir if the gauge reads "Below Normal" at all positions of the function control switch and connect the gauge directly to the air conditioning system side of the reservoir.

(continued)

_____ **10.** If the gauge reads "Insufficient Vacuum" at the vacuum reservoir, vacuum is being lost in the supply line. Vacuum is being lost in the lines after the vacuum reservoir if the gauge now reads "Engine Manifold Vacuum."

_____ **11.** Trace out vacuum circuits and locate the source of the vacuum leak. Record the source of this vacuum leak.

_____

_____ **12.** Repair the vacuum hose or replace the leaking component and verify proper system operation.

Performance ✓ Checklist

Name _____ Date _____ Class _____

PERFORMANCE STANDARDS:
Level 4–Performs skill without supervision and adapts to problem situations.
Level 3–Performs skill satisfactorily without assistance or supervision.
Level 2–Performs skill satisfactorily, but requires assistance/supervision.
Level 1–Performs parts of skill satisfactorily, but requires considerable assistance/supervision.

Attempt (circle one): **1 2 3 4**

Comments:

_____

_____

_____

_____

PERFORMANCE LEVEL ACHIEVED: _____

_____ **1.** Safety rules and practices were followed at all times regarding this job.
_____ **2.** Tools and equipment were used properly and stored upon completion of this job.
_____ **3.** This completed job met the standards set and was done within the allotted time.
_____ **4.** No injury or damage to property occurred during this job.
_____ **5.** Upon completion of this job, the work area was cleaned correctly.

Instructor's Signature _____ Date_____

REPLACING VACUUM LINES

NATEF Standard(s) for Heating & Air Conditioning:
D4 Diagnose malfunctions in the vacuum, mechanical, and electrical components and controls of the heating, ventilation, and A/C (HVAC) system; determine necessary action.

Safety First

☐ Wear safety glasses at all times.
☐ Follow all safety rules when using common hand tools.
☐ Always connect a vehicle's exhaust to a vent hose before you run an engine in a closed shop. Unvented exhaust fumes in a closed shop can cause death.
☐ Be careful not to cut yourself while using the knife.

Tools & Equipment:
- Common hand tools
- Knife
- Neoprene tubing (same diameter as vacuum lines)
- Vehicle service information

PROCEDURES Refer to the vehicle service information for specifications and special procedures. Then replace the vacuum lines in the HVAC system provided by your instructor.

_____ 1. Write up a repair order.

_____ 2. Make sure you follow all procedures in the vehicle service information.

_____ 3. Locate the vacuum line that is the source of lost vacuum. (See Job Sheet HA-15.)

_____ 4. Check the entire line for cracks or bad connections that may be causing the system to lose vacuum.

_____ 5. Remove the damaged vacuum line.

_____ 6. Cut a new vacuum line about ½ inch longer than the section removed.

_____ 7. Replace the bad section of vacuum line from fitting to fitting.

_____ 8. Verify proper operation of the system.

(continued)

Performance ✓ Checklist

Name _____ Date _____ Class _____

PERFORMANCE STANDARDS:
Level 4–Performs skill without supervision and adapts to problem situations.
Level 3–Performs skill satisfactorily without assistance or supervision.
Level 2–Performs skill satisfactorily, but requires assistance/supervision.
Level 1–Performs parts of skill satisfactorily, but requires considerable assistance/supervision.

Attempt (circle one): **1 2 3 4**

Comments:

_____
_____
_____
_____

PERFORMANCE LEVEL ACHIEVED: _____

_____ **1.** Safety rules and practices were followed at all times regarding this job.

_____ **2.** Tools and equipment were used properly and stored upon completion of this job.

_____ **3.** This completed job met the standards set and was done within the allotted time.

_____ **4.** No injury or damage to property occurred during this job.

_____ **5.** Upon completion of this job, the work area was cleaned correctly.

Instructor's Signature _____ Date _____

INSPECTING, REPLACING & ADJUSTING A/C COMPRESSOR DRIVE BELTS

NATEF Standard(s) for Heating & Air Conditioning:
B1-2 Inspect A/C compressor drive belts; determine necessary action.

Safety First

- ☐ Wear safety glasses at all times.
- ☐ Follow all safety rules when using common hand tools.
- ☐ Always connect a vehicle's exhaust to a vent hose before you run an engine in a closed shop. Unvented exhaust fumes in a closed shop can cause death.
- ☐ Keep jewelry, loose clothing, and hair away from moving parts while an engine is running.
- ☐ Be careful of hot engine parts.

Tools & Equipment:

- Belt tension gauge
- Common hand tools
- Vehicle service information

PROCEDURES Refer to the vehicle service information for specifications and special procedures. Then inspect, replace, and adjust the compressor drive belts on the A/C system provided by your instructor.

_____ 1. Write up a repair order.

_____ 2. Make sure you follow all procedures in the vehicle service information.

_____ 3. Visually inspect the belts for cracks, glazing, or the presence of oil. If any of these are found, replace the belt.

_____ 4. Check belt tension using a belt tension gauge.

_____ 5. If a belt requires adjustment, determine the method of adjustment. It may be an adjustable idler or a spring-loaded tensioner.

_____ 6. Adjust the belt until the proper tension is shown on the gauge.

_____ 7. If a new belt was installed, run the engine until it reaches normal operating temperature, and then re-tighten the belt. **CAUTION:** Overtightening the belt can cause compressor and crankshaft damage.

(continued)

Performance ✓ Checklist

Name _____ Date _____ Class _____

PERFORMANCE STANDARDS:
Level 4–Performs skill without supervision and adapts to problem situations.
Level 3–Performs skill satisfactorily without assistance or supervision.
Level 2–Performs skill satisfactorily, but requires assistance/supervision.
Level 1–Performs parts of skill satisfactorily, but requires considerable assistance/supervision.

Attempt (circle one): **1 2 3 4**

Comments:

_____

_____

_____

_____

PERFORMANCE LEVEL ACHIEVED: _____

_____ 1. Safety rules and practices were followed at all times regarding this job.

_____ 2. Tools and equipment were used properly and stored upon completion of this job.

_____ 3. This completed job met the standards set and was done within the allotted time.

_____ 4. No injury or damage to property occurred during this job.

_____ 5. Upon completion of this job, the work area was cleaned correctly.

Instructor's Signature _____ Date_____

DIAGNOSING AIR CONDITIONING PROTECTION DEVICES

NATEF Standard(s) for Heating & Air Conditioning:

B1-1 Diagnose A/C system conditions that cause the protection devices (pressure, thermal, and PCM) to interrupt system operation; determine necessary action.

Safety First

☐ Wear safety glasses at all times.

☐ Follow all safety rules when using common hand tools.

☐ Use exhaust vent if running engine in closed shop. Unvented exhaust fumes can cause death.

☐ Stay clear of cooling fan and hot surfaces.

☐ Keep jewelry, loose clothing, and hair away from moving parts while an engine is running.

☐ Make sure automatic transmission is in PARK or manual transmission is in NEUTRAL and set the parking brake.

Tools & Equipment:

- Vehicle service information
- Refrigerant recovery, recycle, and charge equipment with manifold gauge set

- Common hand tools
- Scan tool

- DVOM
- Refrigerant leak detector

PROCEDURES Refer to the vehicle service information for specifications and special procedures. Then diagnose the air conditioning protection devices on the vehicle provided by your instructor.

_____ 1. Write up a repair order.

_____ 2. Make sure you follow all procedures in the vehicle service information.

_____ 3. If the A/C system operation is interrupted, check out the condition that would activate a particular protection device.

_____ 4. If the high-pressure switch is open, install a manifold gauge set and verify a high-pressure condition.

    a. Check for an overheated condenser. Is the engine cooling fan operating? Is the engine overheating?

    b. Is the system overcharged? Check for a proper refrigerant charge.

_____ 5. If the low-pressure or clutch-cycling pressure switch is open, install a manifold gauge set and verify a low-pressure condition.

    a. Is the system empty or does it have a low charge of refrigerant?

(continued)

Heating & Air Conditioning

_____ 6. If the ambient temperature switch is open, is the outside temperature below the open value of the switch?

_____ 7. If the power steering pressure switch is open, is the power steering pump pressure excessive?

_____ 8. Use a scan tool to see if there are any codes set that would cause the PCM to deny A/C operation. Any of the above conditions may cause the PCM to deny A/C operation. A defective MAP or TPS sensor, indicating information that the engine is operating under a heavy load, could also possibly cause the PCM to deny A/C operation.

Performance ✓ Checklist

Name _____ Date _____ Class _____

PERFORMANCE STANDARDS:
Level 4–Performs skill without supervision and adapts to problem situations.
Level 3–Performs skill satisfactorily without assistance or supervision.
Level 2–Performs skill satisfactorily, but requires assistance/supervision.
Level 1–Performs parts of skill satisfactorily, but requires considerable assistance/supervision.

Attempt (circle one): **1 2 3 4**

Comments:

_____

_____

_____

_____

PERFORMANCE LEVEL ACHIEVED: _____

_____ 1. Safety rules and practices were followed at all times regarding this job.

_____ 2. Tools and equipment were used properly and stored upon completion of this job.

_____ 3. This completed job met the standards set and was done within the allotted time.

_____ 4. No injury or damage to property occurred during this job.

_____ 5. Upon completion of this job, the work area was cleaned correctly.

Instructor's Signature _____ Date_____

INSPECTING, TESTING & REPLACING
AN A/C COMPRESSOR CLUTCH

NATEF Standard(s) for Heating & Air Conditioning:
B1-3 Inspect, test, and replace A/C compressor clutch components and/or assembly.

Safety First

☐ Wear safety glasses at all times.
☐ Follow all safety rules when using common hand tools.
☐ Follow all safety rules when checking electrical connections.
☐ Always connect a vehicle's exhaust to a vent hose before you run an engine in a closed shop. Unvented exhaust fumes in a closed shop can cause death.
☐ Keep jewelry, loose clothing, and hair away from moving parts while an engine is running.
☐ Be careful of hot engine parts.

Tools & Equipment:
- Digital multimeter
- Feeler gauge
- Snap-ring pliers
- Common hand tools
- Clutch puller
- Vehicle service information

PROCEDURES Refer to the vehicle service information for specifications and special procedures. Then inspect, test, and replace the compressor clutch on the A/C system provided by your instructor.

_____ 1. Write up a repair order.

_____ 2. Make sure you follow all procedures in the vehicle service information.

_____ 3. Check clutch operation by turning the A/C on and observing the clutch.

_____ 4. If the clutch does not operate, check the electrical connections and refrigerant level.

_____ 5. Using a digital multimeter, check the clutch coil for opens or shorts or excessive current draw.

_____ 6. Remove the belt.

_____ 7. Check the condition of the bearing by rotating the pulley. It should spin freely.

_____ 8. Remove the compressor clutch using a proper puller, if required. *Note:* This can be done without recovering the system refrigerant if the clutch is accessible without removing the compressor.

_____ 9. Replace the clutch assembly.

_____ 10. Using a feeler gauge, check the clutch clearance.

(continued)

Heating & Air Conditioning

_____ **11.** Adjust clearance to specifications using shims or whatever procedure is required by the manufacturer.

Source: _____ *Page:* _____

_____ **12.** Reinstall and adjust the belt.

_____ **13.** Check for proper operation.

Performance ✓ Checklist

Name _____ Date _____ Class _____

PERFORMANCE STANDARDS:
Level 4–Performs skill without supervision and adapts to problem situations.
Level 3–Performs skill satisfactorily without assistance or supervision.
Level 2–Performs skill satisfactorily, but requires assistance/supervision.
Level 1–Performs parts of skill satisfactorily, but requires considerable assistance/supervision.

Attempt (circle one): **1 2 3 4**

Comments:

_____

_____

_____

_____

PERFORMANCE LEVEL ACHIEVED: _____

_____ **1.** Safety rules and practices were followed at all times regarding this job.

_____ **2.** Tools and equipment were used properly and stored upon completion of this job.

_____ **3.** This completed job met the standards set and was done within the allotted time.

_____ **4.** No injury or damage to property occurred during this job.

_____ **5.** Upon completion of this job, the work area was cleaned correctly.

Instructor's Signature _____ Date_____

INSPECTING & REPLACING A/C SYSTEM HOSES AND/OR GASKETS AND O-RINGS

NATEF Standard(s) for Heating & Air Conditioning:

B2-2 Remove and inspect A/C system mufflers, hoses, lines, fittings, O-rings, seals, and service valves; perform necessary action.

Safety First

- ☐ Wear safety glasses at all times.
- ☐ Follow all safety rules when using common hand tools.
- ☐ Always connect a vehicle's exhaust to a vent hose before you run an engine in a closed shop. Unvented exhaust fumes in a closed shop can cause death.
- ☐ Keep jewelry, loose clothing, and hair away from moving parts while an engine is running.
- ☐ Be careful of hot engine parts.

Tools & Equipment:

- Refrigeration oil and measuring device
- Leak detector
- Common hand tools
- Recovery and charge equipment
- Thermometer
- Vehicle service information
- Replacement refrigerant hoses
- Necessary O-rings, gaskets, or seals

PROCEDURES Refer to the vehicle service information for specifications and special procedures. Then inspect and replace the hoses and/or gaskets and O-rings in the A/C system provided by your instructor.

_____ 1. Write up a repair order.

_____ 2. Make sure you follow all procedures in the vehicle service information.

_____ 3. Look carefully for cracks in each hose in the refrigeration system. *Note:* A leak detector can be used to find a leaking hose. Also, a cracked hose will usually have oil leaking out of it.

_____ 4. If any cracks are found, replace the system hose and/or gaskets and O-rings. *Note:* The refrigerant must be recovered before this procedure, and the system must be evacuated and charged after it. Before charging, add oil as needed.

_____ 5. Use an appropriate tool to loosen or remove hose connections.

_____ 6. Compare the replacement hose to the original hose. If there is any difference, see your instructor.

_____ 7. Make sure that the proper gaskets and/or O-rings are attached. **CAUTION:** Replacement O-rings and gaskets must be exactly the same quality and size as the old ones.

(continued)

_____ 8. Inspect the new hose for any restrictions or foreign material inside.

_____ 9. Apply refrigerant oil on the ends of the insert fittings and/or gaskets and O-rings. Add refrigerant oil as necessary to the system.

_____ 10. Hand start the hose connections to be sure they are properly fitted.

_____ 11. Tighten all connections to the proper torque. **CAUTION:** Overtightening may damage hose fittings.

_____ 12. Tighten the clamps, if equipped.

_____ 13. Evacuate the system.

_____ 14. Charge the system with the proper amount of refrigerant.

_____ 15. Leak test and performance test the system for proper operation.

Performance ✓ Checklist

Name _____ Date _____ Class _____

PERFORMANCE STANDARDS:
Level 4–Performs skill without supervision and adapts to problem situations.
Level 3–Performs skill satisfactorily without assistance or supervision.
Level 2–Performs skill satisfactorily, but requires assistance/supervision.
Level 1–Performs parts of skill satisfactorily, but requires considerable assistance/supervision.

Attempt (circle one): **1 2 3 4**

Comments:

_____

_____

_____

_____

PERFORMANCE LEVEL ACHIEVED: _____

_____ 1. Safety rules and practices were followed at all times regarding this job.

_____ 2. Tools and equipment were used properly and stored upon completion of this job.

_____ 3. This completed job met the standards set and was done within the allotted time.

_____ 4. No injury or damage to property occurred during this job.

_____ 5. Upon completion of this job, the work area was cleaned correctly.

Instructor's Signature _____ Date _____

DIAGNOSING ABNORMAL OPERATING NOISES IN THE AIR CONDITIONING SYSTEM

NATEF Standard(s) for Heating & Air Conditioning:
A6 Diagnose abnormal operating noises in the A/C system; determine necessary action.

Safety First
- ☐ Wear safety glasses at all times.
- ☐ Follow all safety rules when using common hand tools.
- ☐ Use exhaust vent if running engine in closed shop. Unvented exhaust fumes can cause death.
- ☐ Watch out for hot or moving parts.
- ☐ Make sure automatic transmission is in PARK or manual transmission is in NEUTRAL and set the parking brake.

Tools & Equipment:
- Vehicle service information
- Stethoscope
- Common hand tools
- Fender cover

PROCEDURES Refer to the vehicle service information for specifications and special procedures. Then diagnose the abnormal operating noise in the heating and air conditioning system of the vehicle provided by your instructor.

_____ 1. Write up a repair order.

_____ 2. Make sure you follow all procedures in the vehicle service information.

_____ 3. Check for loose belts. Tighten belts.

_____ 4. Check for slipping compressor clutch. Replace clutch.

_____ 5. Check for defective compressor clutch bearing. Replace bearing.

_____ 6. Check for defective compressor. Replace compressor.

_____ 7. Check for low refrigerant charge, causing the clutch to cycle too often. Correct reason for refrigerant loss and recharge to proper level.

_____ 8. Check for foreign material in blower squirrel cage. Remove foreign materials from blower squirrel cage.

_____ 9. Check for defective blower motor bearing. Replace blower motor.

(continued)

Heating & Air Conditioning

Performance ✓ Checklist

Name _____   Date _____   Class _____

PERFORMANCE STANDARDS:
Level 4–Performs skill without supervision and adapts to problem situations.
Level 3–Performs skill satisfactorily without assistance or supervision.
Level 2–Performs skill satisfactorily, but requires assistance/supervision.
Level 1–Performs parts of skill satisfactorily, but requires considerable assistance/supervision.

Attempt (circle one): **1 2 3 4**

Comments:

_____

_____

_____

_____

PERFORMANCE LEVEL ACHIEVED: _____

_____ **1.** Safety rules and practices were followed at all times regarding this job.

_____ **2.** Tools and equipment were used properly and stored upon completion of this job.

_____ **3.** This completed job met the standards set and was done within the allotted time.

_____ **4.** No injury or damage to property occurred during this job.

_____ **5.** Upon completion of this job, the work area was cleaned correctly.

Instructor's Signature _____   Date_____

PERFORMING A LEAK TEST WITH AN ELECTRONIC LEAK DETECTOR

NATEF Standard(s) for Heating & Air Conditioning:
A8 Leak test A/C system; determine necessary action.

Safety First
☐ Wear safety glasses at all times.
☐ Follow all safety rules when working with hoses and connections.

Tools & Equipment:
- Manual for leak detector
- Vehicle service information
- Electronic (halogen) leak detector
- Charging equipment and refrigerant (if necessary)

PROCEDURES Refer to the vehicle service information for specifications and special procedures. Then perform the leak test on the A/C system provided by your instructor.

_____ 1. Write up a repair order.

_____ 2. Make sure you follow all procedures in the vehicle service information. This test should be performed with the engine off.

_____ 3. Visually check the hoses and connections. If the system is totally empty, refrigerant will have to be added before leak testing. There must be at least 50 psi of pressure in the system.

_____ 4. Turn on the leak detector and let it stabilize for about 30 seconds.

_____ 5. Make adjustments. *Note:* Refer to the electronic leak detector manual for the exact time and operation procedure.

   Source: _____ *Page:* _____

_____ 6. Turn the sensitivity switch to SEARCH.

_____ 7. Move the search probe UNDER all connections and hoses. *Note:* Refrigerant is heavier than air, so the probe must be kept lower than the connections and hoses.

_____ 8. The leak detector will tick faster when a leak is detected.

_____ 9. Note the location of any leaks.

_____ 10. Turn off the leak detector and report any leaks to your instructor.

(continued)

Heating & Air Conditioning

Performance ✓ Checklist

Name _____ Date _____ Class _____

PERFORMANCE STANDARDS:
Level 4–Performs skill without supervision and adapts to problem situations.
Level 3–Performs skill satisfactorily without assistance or supervision.
Level 2–Performs skill satisfactorily, but requires assistance/supervision.
Level 1–Performs parts of skill satisfactorily, but requires considerable assistance/supervision.

Attempt (circle one): **1 2 3 4**

Comments:

_____

_____

_____

_____

PERFORMANCE LEVEL ACHIEVED: _____

_____ **1.** Safety rules and practices were followed at all times regarding this job.
_____ **2.** Tools and equipment were used properly and stored upon completion of this job.
_____ **3.** This completed job met the standards set and was done within the allotted time.
_____ **4.** No injury or damage to property occurred during this job.
_____ **5.** Upon completion of this job, the work area was cleaned correctly.

Instructor's Signature _____ Date_____

IDENTIFYING & RECOVERING A/C SYSTEM REFRIGERANT

NATEF Standard(s) for Heating & Air Conditioning:
E2 Identify (by label application or use of a refrigerant identifier) and recover
A/C system refrigerant.

☐ Wear safety glasses at all times.

Tools & Equipment:
• Recovery station   • Refrigerant identifier   • Vehicle service information

PROCEDURES Refer to the vehicle service information for specifications and special procedures. Then identify and recover the refrigerant in the A/C system provided by your instructor.

_____ 1. Write up a repair order.

_____ 2. Make sure you follow all procedures in the vehicle service information.

_____ 3. Identify the refrigerant type by system service port design or placard.

_____ 4. A refrigerant sample can be tested for purity using a refrigerant identifier. *Note:* Some recovery stations come equipped with a built-in refrigerant identifier.

_____ 5. Attach hoses to the system.

_____ 6. Program the recovery unit and press the start button.

_____ 7. When recovery is complete, a vacuum will show on the gauges. If a positive pressure is indicated within 5 minutes, the recovery process should be continued.

(continued)

Heating & Air Conditioning

Performance ✓ Checklist

Name _____ Date _____ Class _____

PERFORMANCE STANDARDS:
Level 4–Performs skill without supervision and adapts to problem situations.
Level 3–Performs skill satisfactorily without assistance or supervision.
Level 2–Performs skill satisfactorily, but requires assistance/supervision.
Level 1–Performs parts of skill satisfactorily, but requires considerable assistance/supervision.

Attempt (circle one): **1 2 3 4**

Comments:

_____

_____

_____

_____

PERFORMANCE LEVEL ACHIEVED: _____

_____ **1.** Safety rules and practices were followed at all times regarding this job.

_____ **2.** Tools and equipment were used properly and stored upon completion of this job.

_____ **3.** This completed job met the standards set and was done within the allotted time.

_____ **4.** No injury or damage to property occurred during this job.

_____ **5.** Upon completion of this job, the work area was cleaned correctly.

Instructor's Signature _____ Date _____

USING & MAINTAINING REFRIGERANT HANDLING EQUIPMENT

NATEF Standard(s) for Heating & Air Conditioning:
E1 Perform correct use and maintenance of refrigerant handling equipment.

Safety First

- ☐ Wear safety glasses at all times.
- ☐ Follow all safety rules when using common hand tools.
- ☐ Use exhaust vent if running engine in closed shop. Unvented exhaust fumes can cause death.
- ☐ Keep jewelry, loose clothing, and hair away from moving parts while an engine is running.
- ☐ Stay clear of cooling fan and hot surfaces.
- ☐ Make sure automatic transmission is in PARK or manual transmission is in NEUTRAL and set the parking brake.

Tools & Equipment:
- Vehicle service information
- Appropriate refrigerant as necessary
- Common hand tools
- Manufacturer's instructions for refrigerant recovery, recycle, and charging equipment
- Refrigerant recovery, recycle, and charging equipment
- Appropriate refrigerant oil as necessary

PROCEDURES Refer to the vehicle service information for specifications and special procedures. Then correctly use and maintain the refrigerant handling equipment provided by your instructor.

_____ 1. Make sure that all recovery and recycling equipment is UL-certified. It must carry a label or tag indicating that the equipment is in compliance with government regulations. Typical wording for such a label is "Design certified by Underwriters Laboratories for compliance with SAEJ-1991." SAEJ-1991 is the document that provides the design standards adopted by the EPA for air conditioning service equipment.

_____ 2. Make sure that all service hoses have shutoff valves located within 12" [30 cm] of the hose ends. This prevents the unnecessary release of refrigerant when the hoses are disconnected. The hoses have unique service fittings.

_____ 3. Check that hoses are also color-coded with stripes of different colors for R12 and R134a.

_____ 4. Properly attach refrigerant recovery, recycle, and charging equipment to a vehicle provided by your instructor. Follow all vehicle manufacturer and equipment instructions.

_____ 5. Properly recover refrigerant from the vehicle A/C system.

(continued)

Heating & Air Conditioning

_____ 6. Properly recycle the refrigerant as necessary on this equipment.

_____ 7. Program the equipment as necessary for proper evacuation and charge of the vehicle system.

_____ 8. Properly add refrigerant oil to the vehicle A/C as necessary.

_____ 9. Following equipment instructions, properly charge and test the vehicle A/C system.

Performance ✓ Checklist

Name _____ Date _____ Class _____

PERFORMANCE STANDARDS:
Level 4–Performs skill without supervision and adapts to problem situations.
Level 3–Performs skill satisfactorily without assistance or supervision.
Level 2–Performs skill satisfactorily, but requires assistance/supervision.
Level 1–Performs parts of skill satisfactorily, but requires considerable assistance/supervision.

Attempt (circle one): **1 2 3 4**

Comments:

_____

_____

_____

_____

PERFORMANCE LEVEL ACHIEVED: _____

_____ 1. Safety rules and practices were followed at all times regarding this job.

_____ 2. Tools and equipment were used properly and stored upon completion of this job.

_____ 3. This completed job met the standards set and was done within the allotted time.

_____ 4. No injury or damage to property occurred during this job.

_____ 5. Upon completion of this job, the work area was cleaned correctly.

Instructor's Signature _____ Date_____

TESTING RECYCLED REFRIGERANT FOR NONCONDENSABLE GASES

NATEF Standard(s) for Heating & Air Conditioning:
E5 Test recycled refrigerant for noncondensable gases.

Safety First

☐ Wear safety glasses at all times.

Tools & Equipment:
- Pressure gauge
- Vehicle service information
- Recovery/recycle machine
- Precise thermometer
- Refrigerant in storage container

PROCEDURES Refer to the vehicle service information for specifications and special procedures. Then test the recycled refrigerant provided by your instructor for noncondensable gases.

Note: Many recovery/recycle machines will automatically remove noncondensable gases. Others will alert the operator and allow manual venting of these gases. Follow the equipment manufacturer's instructions. The following test can also identify noncondensable gases.

_____ 1. Write up a repair order.

_____ 2. Make sure you follow all procedures in the vehicle service information.

_____ 3. Store the refrigerant container at 65°F (18.3°C) or above for at least 12 hours, out of direct sunlight.

_____ 4. Connect an appropriate pressure gauge to the container and read the pressure.

_____ 5. Measure the air temperature within 4 inches of the container with a precise thermometer.

_____ 6. Compare the temperature and pressure to the appropriate standard temperature/pressure chart for the refrigerant in question.

_____ 7. If the pressure is at or below the pressure indicated on the chart for that temperature, the refrigerant does not contain excessive noncondensable gases.

_____ 8. If the pressure is higher than the pressure indicated on the chart, slowly vent the vapor from the top of the container into a recovery/recycle unit until the pressure falls below the appropriate value.

(continued)

Heating & Air Conditioning

Performance ✓ Checklist

Name _____ Date _____ Class _____

PERFORMANCE STANDARDS:
Level 4–Performs skill without supervision and adapts to problem situations.
Level 3–Performs skill satisfactorily without assistance or supervision.
Level 2–Performs skill satisfactorily, but requires assistance/supervision.
Level 1–Performs parts of skill satisfactorily, but requires considerable assistance/supervision.

Attempt (circle one): **1 2 3 4**

Comments:

_____

_____

_____

_____

PERFORMANCE LEVEL ACHIEVED: _____

_____ 1. Safety rules and practices were followed at all times regarding this job.
_____ 2. Tools and equipment were used properly and stored upon completion of this job.
_____ 3. This completed job met the standards set and was done within the allotted time.
_____ 4. No injury or damage to property occurred during this job.
_____ 5. Upon completion of this job, the work area was cleaned correctly.

Instructor's Signature _____ Date_____

USING A REFRIGERANT RECYCLING CENTER

NATEF Standard(s) for Heating & Air Conditioning:
E3 Recycle refrigerant.
E6 Evacuate and charge A/C system.

Safety First
- ☐ Wear safety glasses at all times.
- ☐ Follow all safety rules when working with refrigerant.

Tools & Equipment:
- Refrigerant recycling center
- Refrigerant recycling center operation manual
- Vehicle service information

PROCEDURES Refer to the vehicle service information for specifications and special procedures. Then recycle the refrigerant in the A/C system provided by your instructor.

Note: There are many different types of recycling equipment. Some machines recover refrigerant from numerous vehicles until a certain volume is reached. Then the machine is programmed to recycle the bulk refrigerant. This refrigerant is then transferred to a charge station for future use. Other machines recover refrigerant and recycle it while the machine is in the evacuation mode. This recycled refrigerant is then used to charge the vehicle being serviced. Appropriate instructions must be followed.

_____ 1. Write up a repair order.

_____ 2. Make sure you follow all procedures in the vehicle service information.

_____ 3. Connect the recycling center to the air conditioning system as outlined in the vehicle operation manual.

   Source: _____ *Page:* _____

_____ 4. Following the directions in the recycling center operating manual, recover, evacuate, recycle, and/or charge the system.

   Source: _____ *Page:* _____

_____ 5. Disconnect the recycling center from the air conditioning system as outlined in the operation manual.

   Source: _____ *Page:* _____

Heating & Air Conditioning

(continued)

Performance ✓ Checklist

Name _____ Date _____ Class _____

PERFORMANCE STANDARDS:
Level 4–Performs skill without supervision and adapts to problem situations.
Level 3–Performs skill satisfactorily without assistance or supervision.
Level 2–Performs skill satisfactorily, but requires assistance/supervision.
Level 1–Performs parts of skill satisfactorily, but requires considerable assistance/supervision.

Attempt (circle one): **1 2 3 4**

Comments:

_____

_____

_____

_____

PERFORMANCE LEVEL ACHIEVED: _____

_____ 1. Safety rules and practices were followed at all times regarding this job.
_____ 2. Tools and equipment were used properly and stored upon completion of this job.
_____ 3. This completed job met the standards set and was done within the allotted time.
_____ 4. No injury or damage to property occurred during this job.
_____ 5. Upon completion of this job, the work area was cleaned correctly.

Instructor's Signature _____ Date_____

LABELING & STORING REFRIGERANT

NATEF Standard(s) for Heating & Air Conditioning:
E4 Label and store refrigerant.

☐ Wear safety glasses at all times.

Tools & Equipment:
- Container labels
- Vehicle service information
- Refrigerant storage containers

PROCEDURES Refer to the vehicle service information for specifications and special procedures. Then label and store the refrigerant provided by your instructor.

_____ 1. Write up a repair order.

_____ 2. Make sure you follow all procedures in the vehicle service information.

_____ 3. Add refrigerant only to appropriate DOT-approved storage containers. **CAUTION:** Never add more than the allowed percentage of capacity. Also, never add refrigerant to nonrefillable containers.

_____ 4. Make sure that the storage containers are properly labeled.

_____ 5. Store the containers in an area away from heat or direct sunlight.

Heating & Air Conditioning

(continued)

Performance ✓ Checklist

Name _____ Date _____ Class _____

PERFORMANCE STANDARDS:
Level 4–Performs skill without supervision and adapts to problem situations.
Level 3–Performs skill satisfactorily without assistance or supervision.
Level 2–Performs skill satisfactorily, but requires assistance/supervision.
Level 1–Performs parts of skill satisfactorily, but requires considerable assistance/supervision.

Attempt (circle one): **1 2 3 4**

Comments:

_____

_____

_____

_____

PERFORMANCE LEVEL ACHIEVED: _____

_____ **1.** Safety rules and practices were followed at all times regarding this job.
_____ **2.** Tools and equipment were used properly and stored upon completion of this job.
_____ **3.** This completed job met the standards set and was done within the allotted time.
_____ **4.** No injury or damage to property occurred during this job.
_____ **5.** Upon completion of this job, the work area was cleaned correctly.

Instructor's Signature _____ Date_____

CHARGING THE A/C SYSTEM

NATEF Standard(s) for Heating & Air Conditioning:

E6 Evacuate and charge A/C system.

Safety First

- ☐ Wear safety glasses at all times.
- ☐ Follow all safety rules when using common hand tools.
- ☐ Always connect a vehicle's exhaust to a vent hose before you run an engine in a closed shop. Unvented exhaust fumes in a closed shop can cause death.
- ☐ Keep jewelry, loose clothing, and hair away from moving parts while an engine is running.
- ☐ Be careful of hot engine parts.

Tools & Equipment:

- Can tap and individual refrigerant cans, bulk refrigerant and scales, or charging cylinder
- Manifold gauge set
- Vehicle service information
- Vacuum pump
- Common hand tools

PROCEDURES Refer to the vehicle service information for specifications and special procedures. Then charge the A/C system provided by your instructor.

Note: Although most repair facilities use an electronic charging station, the following procedure is still valid when only minimum equipment is available.

_____ 1. Write up a repair order.

_____ 2. Make sure you follow all procedures in the vehicle service information.

_____ 3. Visually inspect the air conditioning system.

_____ 4. Connect the manifold gauge to the system.

_____ 5. Close both hand valves on the manifold gauge.

_____ 6. Attach the vacuum pump and evacuate the system (see Job Sheet HA-29).

_____ 7. Close both hand valves on the manifold gauge and disconnect the vacuum pump.

_____ 8. Note the weight of the bulk container. Place the bulk container on scales and calculate its weight when charging is complete. You could also fill a charging cylinder with enough refrigerant to charge the system or calculate the number of individual 12-oz. cans needed to charge the system.

_____ 9. Connect the refrigerant supply to the center hose on the manifold gauge. **CAUTION:** When connecting or disconnecting cans, check with your instructor to make sure that you are doing it correctly.

(continued)

Heating & Air Conditioning

_____ **10.** Open the low-side hand valve on the manifold gauge. **CAUTION:** Keep the high-side hand valve closed.

_____ **11.** Hold the refrigerant can upright to allow refrigerant vapor to flow into the system. *Note:* Place the can in lukewarm tap water to assist in vaporization of the refrigerant.

_____ **12.** Start the engine and turn on the A/C. The low-pressure switch may have to be jumped to engage the compressor clutch.

_____ **13.** Watch the pressure on the manifold gauge. Fill the A/C system with refrigerant until the determined amount has been dispersed. *Note:* See the vehicle service information for specifications.

Source: _____ Page:_____

_____ **14.** Close the low-side hand valve on the manifold gauge when the system is charged with the appropriate amount of refrigerant. Verify appropriate operating pressures.

_____ **15.** Remove the refrigerant can or shut off the valve on the bulk container. **CAUTION:** When removing the can, check with your instructor to make sure that you are doing it correctly.

_____ **16.** Properly disconnect the manifold gauge set from the vehicle.

_____ **17.** Reconnect the protective caps on the service valves.

_____ **18.** Leak test and performance test the system to verify proper operation.

Performance ✓ Checklist

Name _____ Date _____ Class _____

PERFORMANCE STANDARDS:
Level 4–Performs skill without supervision and adapts to problem situations.
Level 3–Performs skill satisfactorily without assistance or supervision.
Level 2–Performs skill satisfactorily, but requires assistance/supervision.
Level 1–Performs parts of skill satisfactorily, but requires considerable assistance/supervision.

Attempt (circle one): **1 2 3 4**

Comments:

_____

_____

_____

PERFORMANCE LEVEL ACHIEVED: _____

_____ **1.** Safety rules and practices were followed at all times regarding this job.

_____ **2.** Tools and equipment were used properly and stored upon completion of this job.

_____ **3.** This completed job met the standards set and was done within the allotted time.

_____ **4.** No injury or damage to property occurred during this job.

_____ **5.** Upon completion of this job, the work area was cleaned correctly.

Instructor's Signature _____ Date_____

EVACUATING THE A/C SYSTEM

NATEF Standard(s) for Heating & Air Conditioning:
E6 Evacuate and charge A/C system.

Safety First
☐ Wear safety glasses at all times.
☐ Follow all safety rules when using common hand tools.

Tools & Equipment:
- Vacuum pump with manual
- Manifold gauge set
- Vehicle service information
- Common hand tools

PROCEDURES Refer to the vehicle service information for specifications and special procedures. Then evacuate the A/C system provided by your instructor.

Note: To evacuate the system, it must be empty of all refrigerant. If the system contains refrigerant, a recovery machine must be used before evacuating the system (see Job Sheet HA-23).

_____ 1. Write up a repair order.

_____ 2. Make sure you follow all procedures in the vehicle service information.

_____ 3. Visually inspect the air conditioning system.

_____ 4. Connect the manifold gauge to the system.

_____ 5. If required, check the vacuum pump manual for proper servicing procedures.

   Source: _____ *Page:* _____

_____ 6. Close all manifold hand valves.

_____ 7. Use the wrench (if needed) to take off the caps to the inlet and exhaust ports on the vacuum pump.

_____ 8. Connect the center hose of the manifold gauge to the vacuum pump inlet port and turn on the vacuum pump.

_____ 9. Open the high- and low-side hand valves on the manifold gauge.

_____ 10. Watch the pressure gauges on the manifold gauge. They should begin to drop.

_____ 11. Pump until the low-side pressure gauge reads 26 to 30 inches.

_____ 12. Close the high- and low-side hand valves on the manifold gauge and turn off the vacuum pump.

_____ 13. Wait 2-5 minutes to see if the system will hold a vacuum. If not, the system has a leak and needs additional service. If there is no leakage, turn on the vacuum pump and re-open the manifold hand valves.

(continued)

Heating & Air Conditioning

_____ **14.** Continue pumping for a minimum of 30 minutes.

_____ **15.** Close the hand valves and turn off the vacuum pump.

_____ **16.** Disconnect the vacuum pump from the manifold gauge.

_____ **17.** Reconnect the caps to the inlet and exhaust ports on the vacuum pump.

_____ **18.** The system is now ready to be charged (see Job Sheet HA-28).

Performance ✓ Checklist

Name _____ Date _____ Class _____

PERFORMANCE STANDARDS:
Level 4–Performs skill without supervision and adapts to problem situations.
Level 3–Performs skill satisfactorily without assistance or supervision.
Level 2–Performs skill satisfactorily, but requires assistance/supervision.
Level 1–Performs parts of skill satisfactorily, but requires considerable assistance/supervision.

Attempt (circle one): **1 2 3 4**

Comments:

_____

_____

_____

_____

PERFORMANCE LEVEL ACHIEVED: _____

_____ **1.** Safety rules and practices were followed at all times regarding this job.

_____ **2.** Tools and equipment were used properly and stored upon completion of this job.

_____ **3.** This completed job met the standards set and was done within the allotted time.

_____ **4.** No injury or damage to property occurred during this job.

_____ **5.** Upon completion of this job, the work area was cleaned correctly.

Instructor's Signature _____ Date_____

SELECTING, MEASURING & ADDING OIL TO THE A/C SYSTEM

NATEF Standard(s) for Heating & Air Conditioning:
A10 Determine recommended oil for system application.

Safety First
- ☐ Wear safety glasses at all times.
- ☐ Follow all safety rules when using common hand tools.
- ☐ Follow all safety rules when recovering, evacuating, and recharging an A/C system.

Tools & Equipment:
- Common hand tools
- A/C recovery/ charging station
- Appropriate refrigerant oil
- Clean graduated measuring container
- Vehicle service information

PROCEDURES Refer to the vehicle service information for specifications and special procedures. Then select, measure, and add oil to the A/C system provided by your instructor.

_____ 1. Write up a repair order.

_____ 2. Make sure you follow all procedures in the vehicle service information.

_____ 3. If the refrigerant is being recovered, use an appropriate recovery station. Measure and record the amount of oil removed.

_____

_____ 4. Select the correct oil. R12 systems use mineral oil, and R134a or retrofitted systems use either PAG or ester oil. **CAUTION:** These oils are not interchangeable.

_____ 5. If a component is being replaced, add the amount of oil specified in the vehicle service information plus any oil removed during the recovery process. **CAUTION:** Most systems hold 6 to 9 ounces of oil. An overcharge or undercharge of oil will adversely affect system operation.

Source: _____ _Page:_ _____

_____ 6. Reassemble the system as required. If the system was not disassembled, most charging stations have a provision to inject needed oil between the evacuation and charge mode.

_____ 7 Evacuate and recharge the system.

(continued)

Performance ✓ Checklist

Name _____ Date _____ Class _____

PERFORMANCE STANDARDS:
Level 4–Performs skill without supervision and adapts to problem situations.
Level 3–Performs skill satisfactorily without assistance or supervision.
Level 2–Performs skill satisfactorily, but requires assistance/supervision.
Level 1–Performs parts of skill satisfactorily, but requires considerable assistance/supervision.

Attempt (circle one): **1 2 3 4**

Comments:

_____

_____

_____

_____

PERFORMANCE LEVEL ACHIEVED: _____

_____ 1. Safety rules and practices were followed at all times regarding this job.

_____ 2. Tools and equipment were used properly and stored upon completion of this job.

_____ 3. This completed job met the standards set and was done within the allotted time.

_____ 4. No injury or damage to property occurred during this job.

_____ 5. Upon completion of this job, the work area was cleaned correctly.

Instructor's Signature _____ Date _____

INSPECTING & REPLACING THE ORIFICE TUBE OR EXPANSION VALVE

NATEF Standard(s) for Heating & Air Conditioning:

B2-5 Remove and install expansion valve or orifice (expansion) tube.

Safety First

- ☐ Wear safety glasses at all times.
- ☐ Follow all safety rules when using common hand tools.
- ☐ Always connect a vehicle's exhaust to a vent hose before you run an engine in a closed shop. Unvented exhaust fumes in a closed shop can cause death.
- ☐ Keep jewelry, loose clothing, and hair away from moving parts while an engine is running.
- ☐ Be careful of hot engine parts.

Tools & Equipment:

- Common hand tools
- Recovery and charge equipment
- Thermometer
- Vehicle service information
- Necessary O-rings, gaskets, or seals
- Refrigerant and measuring device
- Leak detector
- New orifice tube or expansion valve

PROCEDURES Refer to the vehicle service information for specifications and special procedures. Then inspect and replace the orifice tube or expansion valve in the vehicle provided by your instructor.

_____ 1. Write up a repair order.

_____ 2. Make sure you follow all procedures in the vehicle service information.

_____ 3. Locate the orifice tube or expansion valve and inspect it for damaged connections or parts.

_____ 4. If anything is damaged, replace the orifice tube or expansion valve.

_____ 5. If you need to replace the orifice tube or expansion valve, recover the refrigerant first.

_____ 6. Disconnect hoses from the orifice tube or expansion valve.

_____ 7. Remove the orifice tube or expansion valve.

_____ 8. Apply refrigerant oil on the new orifice tube or expansion valve connections.

_____ 9. Put the new orifice tube or expansion valve in place. If necessary, add refrigerant oil to the system.

_____ 10. Reconnect hoses to the orifice tube or expansion valve using the proper O-rings, gaskets, or seals.

(*continued*)

Heating & Air Conditioning

_____ **11.** Evacuate the system.

_____ **12.** Charge the system.

_____ **13.** Leak test and performance test the system for proper operation.

Performance ✓ Checklist

Name _____ Date _____ Class _____

PERFORMANCE STANDARDS:
Level 4–Performs skill without supervision and adapts to problem situations.
Level 3–Performs skill satisfactorily without assistance or supervision.
Level 2–Performs skill satisfactorily, but requires assistance/supervision.
Level 1–Performs parts of skill satisfactorily, but requires considerable assistance/supervision.

Attempt (circle one): **1 2 3 4**

Comments:

_____

_____

_____

_____

PERFORMANCE LEVEL ACHIEVED: _____

_____ **1.** Safety rules and practices were followed at all times regarding this job.

_____ **2.** Tools and equipment were used properly and stored upon completion of this job.

_____ **3.** This completed job met the standards set and was done within the allotted time.

_____ **4.** No injury or damage to property occurred during this job.

_____ **5.** Upon completion of this job, the work area was cleaned correctly.

Instructor's Signature _____ Date_____

INSPECTING & REPLACING
THE ACCUMULATOR

NATEF Standard(s) for Heating & Air Conditioning:

B2-4 Remove, inspect, and install receiver/drier or accumulator/drier; determine required oil quantity.

Safety First

- ☐ Wear safety glasses at all times.
- ☐ Follow all safety rules when using common hand tools.
- ☐ Follow all safety rules when disconnecting and reconnecting electrical connections.
- ☐ Always connect a vehicle's exhaust to a vent hose before you run an engine in a closed shop. Unvented exhaust fumes in a closed shop can cause death.
- ☐ Keep jewelry, loose clothing, and hair away from moving parts while an engine is running.
- ☐ Be careful of hot engine parts.

Tools & Equipment:

- New accumulator
- Recovery and charge equipment
- Common hand tools
- Thermometer
- Necessary O-rings, gaskets, or seals
- Vehicle service information
- Leak detector
- Refrigerant oil and measuring device

PROCEDURES Refer to the vehicle service information for specifications and special procedures. Then inspect and replace the accumulator in the vehicle provided by your instructor.

_____ 1. Write up a repair order.

_____ 2. Make sure you follow all procedures in the vehicle service information.

_____ 3. Locate the accumulator and inspect it for damaged connections or parts.

_____ 4. If anything is damaged, replace the accumulator.

_____ 5. To do this, recover the refrigerant in the system.

_____ 6. Disconnect the power source.

_____ 7. Disconnect hoses from the accumulator.

_____ 8. Disconnect electrical connections to the accumulator.

_____ 9. Remove the bolts connecting accumulator to accumulator mounting hardware, if equipped.

_____ 10. Remove accumulator from accumulator mounting hardware.

(continued)

_____ 11. Record the amount of refrigerant oil in the old accumulator so that you'll know how much oil to add to the new accumulator.

_____

_____ 12. Add the proper amount of oil to the new accumulator.

_____ 13. Place the new accumulator in the accumulator mounting hardware.

_____ 14. Reconnect bolts connecting the accumulator to the accumulator mounting hardware.

_____ 15. Reconnect electrical connections.

_____ 16. Reconnect hoses to the accumulator.

_____ 17. Reconnect the power source.

_____ 18. Properly evacuate and charge the system with the correct type and amount of refrigerant.

_____ 19. Leak test and performance test the system for proper operation.

Performance ✓ Checklist

Name _____ Date _____ Class _____

PERFORMANCE STANDARDS:
Level 4–Performs skill without supervision and adapts to problem situations.
Level 3–Performs skill satisfactorily without assistance or supervision.
Level 2–Performs skill satisfactorily, but requires assistance/supervision.
Level 1–Performs parts of skill satisfactorily, but requires considerable assistance/supervision.

Attempt (circle one): **1 2 3 4**

Comments:

_____

_____

_____

_____

PERFORMANCE LEVEL ACHIEVED: _____

_____ 1. Safety rules and practices were followed at all times regarding this job.

_____ 2. Tools and equipment were used properly and stored upon completion of this job.

_____ 3. This completed job met the standards set and was done within the allotted time.

_____ 4. No injury or damage to property occurred during this job.

_____ 5. Upon completion of this job, the work area was cleaned correctly.

Instructor's Signature _____ Date_____

INSPECTING & REPLACING THE A/C COMPRESSOR AND MOUNTINGS

NATEF Standard(s) for Heating & Air Conditioning:

B1-4 Remove, inspect, and reinstall A/C compressor and mountings; determine required oil quantity; determine necessary action.

Safety First

- ☐ Wear safety glasses at all times.
- ☐ Follow all safety rules when working with hoses and connections.
- ☐ Follow all safety rules when disconnecting and reconnecting electrical wires.
- ☐ Always connect a vehicle's exhaust to a vent hose before you run an engine in a closed shop. Unvented exhaust fumes in a closed shop can cause death.
- ☐ Keep jewelry, loose clothing, and hair away from moving parts while an engine is running.
- ☐ Be careful of hot engine parts.

Tools & Equipment:

- Assorted line plugs
- New A/C recovery and charge equipment
- Appropriate refrigerant oil and graduated measuring device
- New compressor
- New receiver/drier or accumulator (if required)
- Common hand tools
- Thermometer
- New compressor mounting hardware
- Vehicle service information

Heating & Air Conditioning

PROCEDURES Refer to the vehicle service information for specifications and special procedures. Then inspect and replace (if necessary) the A/C compressor and mountings provided by your instructor.

_____ 1. Write up a repair order.

_____ 2. Make sure you follow all procedures in the vehicle service information.

_____ 3. Look for damaged or worn parts, bolts, or wires on the compressor and mounting hardware.

_____ 4. If anything is damaged, replace the compressor and mountings. *Note:* The refrigerant must be recovered before this procedure, and the system must be evacuated and charged after it. Before charging, add oil as needed and change the receiver/drier if required.

_____ 5. Disconnect the power source.

_____ 6. Using the proper wrench, disconnect the two refrigerant lines connected to the compressor. *Note:* On some vehicles, it may be easier to wait until Step #11 to disconnect the lines.

_____ 7. Press a line plug into the end of each refrigerant line. *Note:* This will keep anything from getting into the hose.

(continued)

_____ 8. Loosen the drive belt pulley and tension adjustors.

_____ 9. Loosen the bolts on the mounting hardware of the compressor.

_____ 10. Remove the drive belt from the compressor and disconnect any electrical lines to it.

_____ 11. Using the proper wrenches, remove all remaining bolts that hold the compressor to the compressor mount.

_____ 12. Lift the compressor from the engine. Drain the refrigerant oil from the old compressor and measure it to help determine how much new oil to add to the system.

_____ 13. Disconnect the mounting hardware and connect the new mounting hardware.

_____ 14. Place the new compressor into the compressor's mounting hardware and replace the compressor mounting hardware bolts. *Note:* Tighten the bolts finger-tight.

_____ 15. Add refrigerant oil as necessary and attach suction and discharge hoses to the compressor using new O-rings or gaskets.

_____ 16. Slip the drive belt back onto the compressor drive pulley, and then tighten all the bolts between the compressor and compressor mounting hardware.

_____ 17. Adjust the belt pulley and tension adjustors to the manufacturer's specifications.

_____ 18. Check the alignment of the compressor pulley with other pulleys.

_____ 19. Replace the receiver/drier or accumulator if required.

_____ 20. Reconnect all the electrical wires to the compressor.

_____ 21. Properly evacuate the system, charge it, and then check it for leaks.

_____ 22. Performance test the system to verify that it is operating properly.

Performance ✓ Checklist

Name _____ Date _____ Class _____

PERFORMANCE STANDARDS:
Level 4–Performs skill without supervision and adapts to problem situations.
Level 3–Performs skill satisfactorily without assistance or supervision.
Level 2–Performs skill satisfactorily, but requires assistance/supervision.
Level 1–Performs parts of skill satisfactorily, but requires considerable assistance/supervision.

Attempt (circle one): **1 2 3 4**

Comments:

_____

_____

_____

_____

PERFORMANCE LEVEL ACHIEVED: _____

_____ 1. Safety rules and practices were followed at all times regarding this job.

_____ 2. Tools and equipment were used properly and stored upon completion of this job.

_____ 3. This completed job met the standards set and was done within the allotted time.

_____ 4. No injury or damage to property occurred during this job.

_____ 5. Upon completion of this job, the work area was cleaned correctly.

Instructor's Signature _____ Date_____

INSPECTING & REPLACING
ELECTRICAL CONTROLS

NATEF Standard(s) for Heating & Air Conditioning:

D1 Diagnose malfunctions in the electrical controls of heating, ventilation, and A/C (HVAC) systems; determine necessary action.

Safety First

- ☐ Wear safety glasses at all times.
- ☐ Follow all safety rules when using common hand tools.
- ☐ Follow all safety rules when working with wiring and components.
- ☐ Always connect a vehicle's exhaust to a vent hose before you run an engine in a closed shop. Unvented exhaust fumes in a closed shop can cause death.
- ☐ Keep jewelry, loose clothing, and hair away from moving parts while an engine is running.
- ☐ Be careful of hot engine parts.

Tools & Equipment:

- Common hand tools
- Necessary replacement electrical controls
- Vehicle service information
- Schematic of air conditioning electrical system
- Volt ohmmeter (VOM)

PROCEDURES Refer to the vehicle service information for specifications and special procedures. Then inspect and replace the electrical controls in the HVAC system provided by your instructor.

_____ 1. Write up a repair order.

_____ 2. Make sure you follow all procedures in the vehicle service information.

_____ 3. Locate electrical control to be inspected.

_____ 4. Check the wires for damaged insulation.

_____ 5. Check the wire attachments to the electrical control for solid connections.

_____ 6. Attach the VOM to the electrical control and compare the reading on the VOM to service specifications.

   VOM Reading: _____   _Service Specification:_____

_____ 7. If any wires or connections are bad, replace the electrical controls or repair the wiring or terminals.

_____ 8. To do this, disconnect the power source first.

(continued)

_____   **9.** Record the wiring connector position.

_____

_____ **10.** Disconnect the wiring connector from the electrical control.

_____ **11.** Remove the electrical control from its mounting hardware.

_____ **12.** Place the new electrical control in the mounting hardware and reconnect the electrical plug.
CAUTION: Be sure to connect the wires to the correct terminal on the electrical controls.

_____ **13.** Reconnect the power source.

_____ **14.** Test the system for proper operation.

Performance ✓ Checklist

Name _____ Date _____ Class _____

PERFORMANCE STANDARDS:
Level 4–Performs skill without supervision and adapts to problem situations.
Level 3–Performs skill satisfactorily without assistance or supervision.
Level 2–Performs skill satisfactorily, but requires assistance/supervision.
Level 1–Performs parts of skill satisfactorily, but requires considerable assistance/supervision.

Attempt (circle one): **1 2 3 4**

Comments:

_____

_____

_____

_____

PERFORMANCE LEVEL ACHIEVED: _____

_____ **1.** Safety rules and practices were followed at all times regarding this job.

_____ **2.** Tools and equipment were used properly and stored upon completion of this job.

_____ **3.** This completed job met the standards set and was done within the allotted time.

_____ **4.** No injury or damage to property occurred during this job.

_____ **5.** Upon completion of this job, the work area was cleaned correctly.

Instructor's Signature _____ Date_____

OBSERVING & RECORDING RELATED HVAC TROUBLE CODES

NATEF Standard(s) for Heating & Air Conditioning:
A11 Using scan tool, observe and record related HVAC data and trouble codes.

Safety First
- ☐ Wear safety glasses at all times.
- ☐ Follow all safety rules when using common hand tools.

Tools & Equipment:
- Vehicle service information
- Common hand tools
- Scan tool

PROCEDURES Refer to the vehicle service information for specifications and special procedures. Then use a scan tool to observe and record related HVAC data and trouble codes on the vehicle provided by your instructor.

_____ 1. Write up a repair order.

_____ 2. Make sure you follow all procedures in the vehicle service information.

_____ 3. Identify the data link connector for the A/C system on your vehicle.

_____ 4. Connect the scan tool and establish communication with the appropriate module responsible for HVAC system operation.

_____ 5. While the vehicle is running and the A/C system is in operation, observe the sensor display.

_____ 6. How many sensors does your display list for the air conditioning system?

_____

_____ 7. Locate and retrieve any stored trouble codes.

_____ 8. List the codes displayed on the scan tool for your vehicle.

_____

Heating & Air Conditioning

(continued)

Performance ✓ Checklist

Name _____ Date _____ Class _____

PERFORMANCE STANDARDS:
Level 4–Performs skill without supervision and adapts to problem situations.
Level 3–Performs skill satisfactorily without assistance or supervision.
Level 2–Performs skill satisfactorily, but requires assistance/supervision.
Level 1–Performs parts of skill satisfactorily, but requires considerable assistance/supervision.

Attempt (circle one): **1 2 3 4**

Comments:

_____

_____

_____

_____

PERFORMANCE LEVEL ACHIEVED: _____

_____ 1. Safety rules and practices were followed at all times regarding this job.

_____ 2. Tools and equipment were used properly and stored upon completion of this job.

_____ 3. This completed job met the standards set and was done within the allotted time.

_____ 4. No injury or damage to property occurred during this job.

_____ 5. Upon completion of this job, the work area was cleaned correctly.

Instructor's Signature _____ Date _____

INSPECTING & REPLACING A SWITCH
OR ELECTRICAL COMPONENT

NATEF Standard(s) for Heating & Air Conditioning:
D2 Inspect and test A/C-heater blower, motors, resistors, switches, relays, wiring, and protection devices; perform necessary action.

Safety First

- ☐ Wear safety glasses at all times.
- ☐ Follow all safety rules when using common hand tools.
- ☐ Follow all safety rules when working with wiring and components.
- ☐ Always connect a vehicle's exhaust to a vent hose before you run an engine in a closed shop. Unvented exhaust fumes in a closed shop can cause death.
- ☐ Keep jewelry, loose clothing, and hair away from moving parts while an engine is running.
- ☐ Be careful of hot engine parts.

Tools & Equipment:
- New switch or electrical component
- Vehicle service information
- Common hand tools
- Volt ohmmeter (VOM)

PROCEDURES Refer to the vehicle service information for specifications and special procedures. Then inspect and replace the switch or electrical component in the A/C-heater system provided by your instructor.

_____ 1. Write up a repair order.

_____ 2. Make sure you follow all procedures in the vehicle service information.

_____ 3. Locate the switch or component to be inspected. *Note:* If the switch or component is mounted in the dash, parts of the dash may have to be removed.

_____ 4. Check the wires for damaged insulation.

_____ 5. Check the wire attachments to the switch or component for solid connections.

_____ 6. Attach the VOM to the switch or component and compare the reading on the VOM to service specification.

   VOM Reading: _____   *Service Specification:*_____

_____ 7. If any wires or connections are bad, replace the switch or component or repair the wiring and terminals.

_____ 8. To do this, disconnect the power source.

(continued)

Heating & Air Conditioning

_____ 9. Record where the wiring harness is attached to the switch or component.

_____

_____ 10. Disconnect the wiring harness from the switch or component.

_____ 11. Remove the switch or component from its mounting hardware.

_____ 12. Place the switch or component in the mounting hardware and reconnect the wiring harness.
CAUTION: Be sure to connect the wires to the correct terminals on the switch.

_____ 13. Reconnect the power source.

_____ 14. Test the system for proper operation.

Performance ✓ Checklist

Name _____   Date _____   Class _____

PERFORMANCE STANDARDS:
Level 4–Performs skill without supervision and adapts to problem situations.
Level 3–Performs skill satisfactorily without assistance or supervision.
Level 2–Performs skill satisfactorily, but requires assistance/supervision.
Level 1–Performs parts of skill satisfactorily, but requires considerable assistance/supervision.

Attempt (circle one): **1 2 3 4**

Comments:

_____

_____

_____

_____

PERFORMANCE LEVEL ACHIEVED: _____

_____ 1. Safety rules and practices were followed at all times regarding this job.
_____ 2. Tools and equipment were used properly and stored upon completion of this job.
_____ 3. This completed job met the standards set and was done within the allotted time.
_____ 4. No injury or damage to property occurred during this job.
_____ 5. Upon completion of this job, the work area was cleaned correctly.

Instructor's Signature _____   Date_____

DIAGNOSING FLUID USAGE, LEVEL, AND CONDITION

NATEF Standard(s) for Automatic Transmission & Transaxle:
A5 Diagnose fluid loss and condition concerns; check fluid level on transmissions with and without dipstick; determine necessary action.

DIRECTIONS: Fill in the blanks below by identifying (1) Safety First Practices that must be followed, (2) Tools and Equipment required, (3) three possible causes of fluid usage, level, and condition problems, and (4) corrective actions that should be taken.

Safety First

- _____
- _____
- _____
- _____

Tools and Equipment Required:

- _____   - _____
- _____   - _____
- _____   - _____

QUICK ✔ Diagnostic for Cause(s)

List at least three common causes of fluid usage, level, and condition problems:

1. _____
2. _____
3. _____
4. _____
5. _____

QUICK ✔ Diagnostic for Corrective Action(s)

List possible corrective action(s):

1. _____
2. _____
3. _____
4. _____
5. _____

Automatic Transmission & Transaxle

DIAGNOSING THE TORQUE CONVERTER AND TRANSMISSION/TRANSAXLE COOLING SYSTEM

DIRECTIONS: Fill in the blanks below by identifying (1) Safety First Practices that must be followed, (2) Tools and Equipment required, (3) three possible causes of torque converter and transmission/transaxle cooling system problems, and (4) corrective actions that should be taken.

Safety First

- _____
- _____
- _____
- _____

Tools and Equipment Required:

- _____   • _____
- _____   • _____
- _____   • _____

QUICK ✔ Diagnostic for Cause(s)

List at least three common causes of torque converter and transmission/transaxle cooling system problems:

1. _____
2. _____
3. _____
4. _____
5. _____

QUICK ✔ Diagnostic for Corrective Action(s)

List possible corrective action(s):

1. _____
2. _____
3. _____
4. _____
5. _____

DIAGNOSING NOISES AND VIBRATIONS

NATEF Standard(s) for Automatic Transmission & Transaxle:
A10 Diagnose noise and vibration concerns; determine necessary action.

DIRECTIONS: Fill in the blanks below by identifying (1) Safety First Practices that must be followed, (2) Tools and Equipment required, (3) three possible causes of noise and vibration problems, and (4) corrective actions that should be taken.

Safety First

- _____
- _____
- _____
- _____

Tools and Equipment Required:

- _____   - _____
- _____   - _____
- _____

QUICK ✔ Diagnostic for Cause(s)

List at least three common causes of noise and vibration problems:

1. _____
2. _____
3. _____
4. _____
5. _____

QUICK ✔ Diagnostic for Corrective Action(s)

List possible corrective action(s):

1. _____
2. _____
3. _____
4. _____
5. _____

Automatic Transmission & Transaxle

DIAGNOSING GEAR TRAIN, SHAFTS, BUSHINGS, AND CASE

NATEF Standard(s) for Automatic Transmission & Transaxle:
D3-6 Inspect case bores, passages, bushings, vents, and mating surfaces; determine necessary action.

DIRECTIONS: Fill in the blanks below by identifying (1) Safety First Practices that must be followed, (2) Tools and Equipment required, (3) three possible causes of gear train, shafts, bushings, and case problems, and (4) corrective actions that should be taken.

Safety First

- _____
- _____
- _____
- _____

Tools and Equipment Required:

- _____ · _____
- _____ · _____
- _____

QUICK ✔ Diagnostic for Cause(s)

List at least three common causes of gear train, shafts, bushings, and case problems:

1. _____
2. _____
3. _____
4. _____

QUICK ✔ Diagnostic for Corrective Action(s)

List possible corrective action(s):

1. _____
2. _____
3. _____
4. _____

Name _____ Date _____ Class _____

DIAGNOSING FRICTION AND REACTION UNITS

NATEF Standard(s) for Automatic Transmission & Transaxle:
D4-1 Inspect clutch drum, piston, check-balls, springs, retainers, seals, and friction and pressure plates; determine necessary action.

DIRECTIONS: Fill in the blanks below by identifying (1) Safety First Practices that must be followed, (2) Tools and Equipment required, (3) three possible causes of friction and reaction units problems, and (4) corrective actions that should be taken.

Safety First
- _____
- _____
- _____
- _____

Tools and Equipment Required:

- _____ • _____
- _____ • _____
- _____ • _____

QUICK ✔ Diagnostic for Cause(s)

List at least three common causes of friction and reaction units problems:

1. _____
2. _____
3. _____
4. _____
5. _____

QUICK ✔ Diagnostic for Corrective Action(s)

List possible corrective action(s):

1. _____
2. _____
3. _____
4. _____
5. _____

Automatic Transmission & Transaxle

Name _____ Date _____ Class _____

DIAGNOSING ELECTRONIC, MECHANICAL, AND HYDRAULIC VACUUM CONTROL SYSTEM

NATEF Standard(s) for Automatic Transmission & Transaxle:
A9 Diagnose electronic, mechanical, hydraulic, vacuum control system concerns; determine necessary action.

DIRECTIONS: Fill in the blanks below by identifying (1) Safety First Practices that must be followed, (2) Tools and Equipment required, (3) three possible causes of electronic, mechanical, and hydraulic vacuum control system problems, and (4) corrective actions that should be taken.

Safety First
- _____
- _____
- _____
- _____

Tools and Equipment Required:
- _____   • _____
- _____   • _____
- _____   • _____

QUICK ✔ Diagnostic for Cause(s)

List at least three common causes of electronic, mechanical, and hydraulic vacuum control system problems:
1. _____
2. _____
3. _____
4. _____

QUICK ✔ Diagnostic for Corrective Action(s)

List possible corrective action(s):
1. _____
2. _____
3. _____
4. _____

COMPLETING A VEHICLE REPAIR ORDER FOR AN AUTOMATIC TRANSMISSION AND TRANSAXLE CONCERN

NATEF Standard(s) for Automatic Transmission & Transaxle:
A1 Complete work order to include customer information, vehicle identifying information, customer concern, related service history, cause, and correction.

SAMPLE VEHICLE REPAIR ORDER   Vehicle Repair Order # _____

Date ____/____/____

Customer Name & Phone #:   Vehicle Make/Type: VIN: Mileage:

Service History: _____

Customer Concern: _____

Cause of Concern: _____

Suggested Repairs/Maintenance: _____

Services Performed: _____

Parts				Labor	Time In:
Item	Description	Price		Diagnosis Time:	Time Complete:
1				Repair Time:	Total Hours:
2					
3					
4					
5					
6					

I hereby authorize the above repair work to be done using the necessary material, and hereby grant you and/or your employees permission to operate the vehicle herein described on streets, highways, or elsewhere for the purpose of testing and/or inspection. An express mechanic's lien is hereby acknowledged on above vehicle to secure the amount of repairs thereof.

X _____

PROCEDURES Refer to the vehicle service information for specifications and special procedures. Then prepare a vehicle repair order for the vehicle provided by your instructor.

_____ 1. **Write legibly.** Others will be reading what you have written.

_____ 2. **Make sure all information is accurate.** Inaccurate information will slow the repair process.

_____ 3. **Complete every part of the Vehicle Repair Order.** Every part must be completed.

_____ 4. **Number the Vehicle Repair Order.** This will help others track the repair.

_____ 5. **Date the Vehicle Repair Order.** This will help document the service history.

_____ 6. **Enter the Customer Name and Phone Number.** Make sure you have spelled the Customer Name correctly. Double-check the Phone Number.

_____ 7. **Enter the Vehicle Make/Type.** This information is essential.

(*continued*)

Automatic Transmission & Transaxle

_____ 8. **Enter the VIN (vehicle identification number).** This is a string of coded data that is unique to the vehicle. The location of the VIN depends on the manufacturer. It is usually found on the dashboard next to the windshield on the driver's side. The VIN is a rich source of information. It is needed to properly use a scan tool to read diagnostic trouble codes. Double-check the VIN to ensure accuracy.

_____ 9. **Enter the Mileage of the vehicle.** This information is part of the service history.

_____ 10. **Complete the Service History.** The service history is a history of all the service operations performed on a vehicle. The service history alerts the technician to previous problems with the vehicle. In the case of recurring problems, it helps the technician identify solutions that were ineffective.

   • A detailed service history is usually kept by the service facility where the vehicle is regularly serviced.

   • Information on service performed on the vehicle at other service centers is not available unless the customer makes it available. For this reason, ask the customer about service performed outside of the present service center.

_____ 11. **Identify the Customer Concern.** This should be a reasonably detailed and accurate description of the problem that the customer is having with the vehicle. The customer is usually the best source of information regarding the problem. This information can be used to perform the initial diagnosis. The customer has a passenger car. He says, "My transmission is not shifting properly. I had the same problem recently, and an attendant at a local service station added a quart of transmission fluid. It worked fine for a while. Now the problem has started again." Enter his concern on the Customer Concern line.

_____ 12. **Identify the Cause of Concern.** This will identify the problem. In this case, there may be several possible causes. Enter the possible causes on the Cause of Concern line.

_____ 13. Ask the customer to read the text at the bottom of the Labor box. By signing on the line at the bottom of this box, the customer authorizes repair work on the vehicle according to the terms specified.

_____ 14. **Identify Suggested Repairs/Maintenance.** This will identify what needs to be done to correct the problem.

_____ 15. **Identify Services Performed.** This will identify the specific maintenance and repair procedures that were performed to correct the problem.

_____ 16. **Provide Parts information.** This includes a numbered list of items used to complete the repair. It includes a description of each part with the price of the part.

_____ 17. **Provide Labor information.** The Diagnosis Time and the Repair Time are totaled to give the Total Hours.

Performance ✓ Checklist

Name _____ Date _____ Class _____

PERFORMANCE STANDARDS:
Level 4–Performs skill without supervision and adapts to problem situations.
Level 3–Performs skill satisfactorily without assistance or supervision.
Level 2–Performs skill satisfactorily, but requires assistance/supervision.
Level 1–Performs parts of skill satisfactorily, but requires considerable assistance/supervision.

Attempt (circle one): **1 2 3 4**

Comments:

_____

_____

_____

_____

PERFORMANCE LEVEL ACHIEVED: _____

_____ 1. Safety rules and practices were followed at all times regarding this job.
_____ 2. Tools and equipment were used properly and stored upon completion of this job.
_____ 3. This completed job met the standards set and was done within the allotted time.
_____ 4. No injury or damage to property occurred during this job.
_____ 5. Upon completion of this job, the work area was cleaned correctly.

Instructor's Signature _____ Date_____

Automatic Transmission
& Transaxle

INSPECTING THE PARKING PAWL ASSEMBLY

NATEF Standard(s) for Automatic Transmission & Transaxle:
D3-9 Inspect and reinstall parking pawl, shaft, spring, and retainer; determine necessary action.

Safety First
- ☐ Wear safety glasses at all times.
- ☐ Follow all safety rules when using common hand tools.

Tools & Equipment:
- Common hand tools
- Vehicle service information

PROCEDURES Refer to the vehicle service information for specifications and special procedures. Then inspect the parking pawl assembly in the vehicle provided by your instructor.

_____ 1. Write up a repair order.

_____ 2. Make sure you follow all procedures in the vehicle service information.

_____ 3. Inspect the gear teeth, spring, and pawl for wear or damage.

_____ 4. If applicable, inspect the manual shaft for freedom of movement and other damage.

_____ 5. Check the shaft hole in the pawl for wear and concentricity (roundness).

_____ 6. Following the directions in the vehicle service information, replace any damaged parts.

Source: _____ Page: _____

(continued)

Performance ✓ Checklist

Name _____ Date _____ Class _____

PERFORMANCE STANDARDS:
Level 4–Performs skill without supervision and adapts to problem situations.
Level 3–Performs skill satisfactorily without assistance or supervision.
Level 2–Performs skill satisfactorily, but requires assistance/supervision.
Level 1–Performs parts of skill satisfactorily, but requires considerable assistance/supervision.

Attempt (circle one): **1 2 3 4**

Comments:

_____

_____

_____

_____

PERFORMANCE LEVEL ACHIEVED: _____

_____ 1. Safety rules and practices were followed at all times regarding this job.

_____ 2. Tools and equipment were used properly and stored upon completion of this job.

_____ 3. This completed job met the standards set and was done within the allotted time.

_____ 4. No injury or damage to property occurred during this job.

_____ 5. Upon completion of this job, the work area was cleaned correctly.

Instructor's Signature _____ Date _____

DIAGNOSING FLUID CONDITIONS

NATEF Standard(s) for Automatic Transmission & Transaxle:
A5 Diagnose fluid loss and condition concerns; check fluid level on transmissions with and without dipstick; determine necessary action.

Safety First

- ☐ Wear safety glasses at all times.
- ☐ Follow all safety rules when using common hand tools.
- ☐ Follow all safety rules when using an automotive lift.
- ☐ Keep jewelry, loose clothing, and hair away from moving parts when an engine is running.
- ☐ Stay clear of cooling fan and hot surfaces.
- ☐ Make sure automatic transmission or transaxle is in proper gear selection and the parking brake is set when checking fluid levels with a dipstick.

Tools & Equipment:
- Vehicle service information
- Common hand tools
- Automotive lift

PROCEDURES Refer to the vehicle service information for specifications and special procedures. Then diagnose the condition of the transmission/transaxle fluid in the vehicle provided by your instructor.

_____ 1. Write up a repair order.

_____ 2. Make sure you follow all procedures in the vehicle service information.

_____ 3. Check for external transmission fluid leakage around all gaskets and seals.

_____ 4. Check the fluid's small. Damaged fluid tends to smell burned (like burned popcorn). Burned fluid is usually dark and may contain material worn from clutches and bands. This condition indicates that the transmission may need an overhaul. Some fluids, such as Dexron III may smell burned even when they are not. If the fluid is not dark, there is a good chance it is not burned. If in doubt, smell some fresh Dexron fluid. Then compare the smell to that of the suspect fluid.

_____ 5. Check for dark fluid. ATF is usually dyed red. ATF may darken slightly with use. If the fluid appears very dark or brown, it is oxidized and contaminated. Oxidation causes varnish to form. The varnish may cause valves in the transmission to stick. This may result in poor shift quality, improper lubrication, and damage to the parts.

_____ 6. Check for milky brown fluid. If coolant (antifreeze) contaminates oxidized ATF, the fluid appears milky brown. Non-oxidized fluid will appear milky pink. This means there is probably a leak in the transmission fluid cooler, which is located in the radiator of the vehicle. If the ATF is contaminated by antifreeze, the transaxle or transmission must be overhauled. The lockup torque converter, if the unit has one, should also be replaced. If there is a leak in the transmission fluid cooler, the cooler must also be replaced or repaired. The cooler lines must then be properly flushed.

(continued)

Automatic Transmission & Transaxle

_____ **7.** Check for foamy fluid. Aeration makes the fluid appear foamy. Aeration can lead to transmission failure, varnish formation, or excessive venting of the transmission fluid from the unit. Aeration usually results from overfilling.

_____ **8.** To check the fluid level on automatic transmissions or transaxles using dipsticks, make sure that the vehicle has been driven and the fluid is at normal temperature. The vehicle must be level and the gear selector must be in the proper position. This may be PARK on some models and NEUTRAL on others. Be sure to properly apply the parking brake. With the engine running, dip the fluid level and make necessary adjustments by adding a little fluid at a time. Be careful not to overfill.

_____ **9.** To check the fluid level on automatic transmissions or transaxles that do not have dipsticks, follow the specific instructions of the manufacturer. These units are sealed and have special procedures for verifying proper fluid level when serviced. This may require the vehicle to be lifted on an automotive lift, a plug to be removed, and the fluid adjusted to a particular level. Some vehicles require the use of a scan tool to determine fluid temperature and a chart to determine the appropriate fluid level.

Performance ✓ Checklist

Name _____   Date _____   Class _____

PERFORMANCE STANDARDS:
Level 4–Performs skill without supervision and adapts to problem situations.
Level 3–Performs skill satisfactorily without assistance or supervision.
Level 2–Performs skill satisfactorily, but requires assistance/supervision.
Level 1–Performs parts of skill satisfactorily, but requires considerable assistance/supervision.

Attempt (circle one): **1 2 3 4**

Comments:

_____

_____

_____

_____

PERFORMANCE LEVEL ACHIEVED: _____

_____ **1.** Safety rules and practices were followed at all times regarding this job.
_____ **2.** Tools and equipment were used properly and stored upon completion of this job.
_____ **3.** This completed job met the standards set and was done within the allotted time.
_____ **4.** No injury or damage to property occurred during this job.
_____ **5.** Upon completion of this job, the work area was cleaned correctly.

Instructor's Signature _____   Date_____

PERFORMING VISUAL INSPECTION OF AUTOMATIC TRANSMISSION FLUID

NATEF Standard(s) for Automatic Transmission & Transaxle:
B2 Service transmission; perform visual inspection; replace fluids and filters.

Safety First

☐ Wear safety glasses at all times.

☐ Always connect a vehicle's exhaust to a vent hose before you run an engine in a closed shop. Unvented exhaust fumes in a closed shop can cause death.

☐ Keep jewelry, loose clothing, and hair away from moving parts while an engine is running.

☐ Be careful of hot engine parts.

☐ Make sure that the parking brake is firmly set before inspecting transmission fluids.

Tools & Equipment:
• Shop towels
• Vehicle service information
• Manufacturer-recommended transmission fluid

PROCEDURES Refer to the vehicle service information for specifications and special procedures. Then perform the visual inspection of the automatic transmission fluid in the engine provided by your instructor.

_____ 1. Write up a repair order.

_____ 2. Make sure you follow all procedures in the vehicle service information.

Fluid Level Inspection

_____ 1. Position the vehicle on a flat surface, engage the parking brake, start the engine, and let it warm to normal operating temperature.

_____ 2. With the engine at curb idle, move the transmission shift lever slowly through all gear ranges, ending in the gear (PARK or NEUTRAL) recommended by the vehicle manufacturer.

_____ 3. Open the hood, remove the transmission-fluid dipstick, and carefully touch the wet end to determine whether the fluid is cool, warm, or hot.

_____ 4. Wipe the dipstick clean with a shop towel and reinsert the dipstick into its tube until seated.

_____ 5. Remove the dipstick and read the fluid level. If the fluid felt cool or warm, the level should be between dimples above the FULL mark. If the fluid felt hot, the level should be at or in the area marked HOT.

_____ 6. If needed, add just enough of the manufacturer-recommended transmission fluid to raise the level from the ADD mark to the FULL mark. *Note:* Remember the marks on the dipstick indicate pints–not quarts as on the engine-oil dipstick.

(continued)

Automatic Transmission & Transaxle

_____ 7. **CAUTION:** Do not overfill the transmission. Refill capacities listed in the vehicle service information are approximate. Correct fluid level should be determined by checking the mark on the dipstick rather than by the amount added.

_____ 8. Reinsert and seat the dipstick.

Automatic Transmission Fluid Inspection

_____ 1. Remove the transmission-fluid dipstick.

_____ 2. Visually inspect the fluid for discoloration and sniff it for a burned smell.

_____ 3. If the fluid is discolored or has a burned smell, check with your instructor.

Performance ✓ Checklist

Name _____ Date _____ Class _____

PERFORMANCE STANDARDS:
Level 4–Performs skill without supervision and adapts to problem situations.
Level 3–Performs skill satisfactorily without assistance or supervision.
Level 2–Performs skill satisfactorily, but requires assistance/supervision.
Level 1–Performs parts of skill satisfactorily, but requires considerable assistance/supervision.

Attempt (circle one): 1 2 3 4

Comments:

_____

_____

_____

_____

PERFORMANCE LEVEL ACHIEVED: _____

_____ 1. Safety rules and practices were followed at all times regarding this job.

_____ 2. Tools and equipment were used properly and stored upon completion of this job.

_____ 3. This completed job met the standards set and was done within the allotted time.

_____ 4. No injury or damage to property occurred during this job.

_____ 5. Upon completion of this job, the work area was cleaned correctly.

Instructor's Signature _____ Date_____

REPLACING AUTOMATIC TRANSMISSION FLUIDS AND FILTERS

NATEF Standard(s) for Automatic Transmission & Transaxle:
B2 Service transmission; perform visual inspection; replace fluids and filters.

Safety First

- ☐ Wear safety glasses at all times.
- ☐ Follow all safety rules when using common hand tools.
- ☐ Check Material Safety Data Sheets for chemical safety information.
- ☐ Transmission fluid in a recently driven vehicle is hot. Remove bolts carefully to avoid splashing transmission fluid.
- ☐ Follow all safety rules when using a lift.
- ☐ Always connect a vehicle's exhaust to a vent hose before you run an engine in a closed shop. Unvented exhaust fumes in a closed shop can cause death.
- ☐ Be careful of hot exhaust parts.

Tools & Equipment:

- Lift
- Parts-cleaning solution
- Transmission fluid
- Replacement pan gasket or sealer
- Large, 5-quart drain pan
- Gasket scraper
- Common hand tools
- Replacement filter as required
- Stiff brush
- Inch-pound torque wrench
- Vehicle service information
- Replacement filter O-ring or seal

PROCEDURES Refer to the vehicle service information for specifications and special procedures. Then replace the fluids and filters in the automatic transmission provided by your instructor.

_____ 1. Write up a repair order.

_____ 2. Make sure you follow all procedures in the vehicle service information.

_____ 3. Safely raise the vehicle on a lift.

_____ 4. Place the drain pan under the transmission pan.

_____ 5. Remove all but two corner pan bolts.

_____ 6. Loosen the two remaining pan bolts.

_____ 7. Carefully pry the pan loose with a pry bar, allowing fluid to drain.

_____ 8. Remove the pan and filter.

_____ 9. Remove any dirt and gasket material from the pan and transmission case.

_____ 10. Replace or clean the filter according to the manufacturer's procedure.

(continued)

Automatic Transmission & Transaxle

_____ 11. Check pan flanges for distortion and straighten, if necessary.

_____ 12. Install a new pan gasket or sealer as required.

_____ 13. Torque pan bolts in a crisscross pattern to manufacturer's specifications.

_____ 14. Safely lower the vehicle to the ground.

_____ 15. Refill the transmission with transmission fluid.

_____ 16. Start the engine and move the shift lever through all gear positions.

_____ 17. Check fluid level and correct as necessary.

_____ 18. Check for leaks.

Performance ✓ Checklist

Name _____ Date _____ Class _____

PERFORMANCE STANDARDS:
Level 4–Performs skill without supervision and adapts to problem situations.
Level 3–Performs skill satisfactorily without assistance or supervision.
Level 2–Performs skill satisfactorily, but requires assistance/supervision.
Level 1–Performs parts of skill satisfactorily, but requires considerable assistance/supervision.

Attempt (circle one): **1 2 3 4**

Comments:

_____

_____

_____

_____

PERFORMANCE LEVEL ACHIEVED: _____

_____ 1. Safety rules and practices were followed at all times regarding this job.

_____ 2. Tools and equipment were used properly and stored upon completion of this job.

_____ 3. This completed job met the standards set and was done within the allotted time.

_____ 4. No injury or damage to property occurred during this job.

_____ 5. Upon completion of this job, the work area was cleaned correctly.

Instructor's Signature _____ Date_____

REMOVING & REPLACING EXTERNAL GASKETS AND SEALS (COVERS WITH FLAT GASKETS OR O-RINGS)

NATEF Standard(s) for Automatic Transmission & Transaxle:
C3 Inspect and replace external seals and gaskets.

Safety First
- ☐ Wear safety glasses at all times.
- ☐ Follow all safety rules when using common hand tools.
- ☐ Follow all safety rules when using a lift.
- ☐ Check Material Safety Data Sheets for chemical safety information.

Tools & Equipment:
- Transmission fluid
- Lift
- Common hand tools
- Cleaning solution
- Special seal removers and installers
- Drain pan
- Vehicle service information

PROCEDURES Refer to the vehicle service information for specifications and special procedures. Then remove and replace the gaskets and seals in the transmission provided by your instructor.

_____ 1. Write up a repair order.

_____ 2. Make sure you follow all procedures in the vehicle service information.

_____ 3. Safely raise the vehicle on a lift.

_____ 4. Remove all dirt from around the mating surfaces.

_____ 5. Place a drain pan under the cover that is to be removed.

_____ 6. Loosen the bolts or clips that hold the cover in place. **CAUTION:** Some accumulator and servo covers contain a strong spring. Be careful that it does not fly off and hit you.

_____ 7. Remove the cover, gasket, and/or O-ring.

_____ 8. Place the gaskets and O-rings with their proper components.

_____ 9. Scrape the old gasket material from both mating surfaces. **CAUTION:** Be careful not to scratch the surfaces.

_____ 10. Clean the cover with solvent and blow dry it.

_____ 11. Make sure the new seal is the same type, size, and thickness as the old one.

_____ 12. Check the new seal for fit.

(continued)

Automatic Transmission & Transaxle

_____ 13. Following the manufacturer's instructions, use an appropriate seal installer to install the new seal.

Source: _____ Page: _____

_____ 14. Install the cover. Make sure you torque bolts to the manufacturer's specifications.

_____ 15. Safely lower the vehicle to the ground.

_____ 16. Replace any lost fluid.

Performance ✓ Checklist

Name _____ Date _____ Class _____

PERFORMANCE STANDARDS:
Level 4–Performs skill without supervision and adapts to problem situations.
Level 3–Performs skill satisfactorily without assistance or supervision.
Level 2–Performs skill satisfactorily, but requires assistance/supervision.
Level 1–Performs parts of skill satisfactorily, but requires considerable assistance/supervision.

Attempt (circle one): **1 2 3 4**

Comments:

_____

_____

_____

_____

PERFORMANCE LEVEL ACHIEVED: _____

_____ 1. Safety rules and practices were followed at all times regarding this job.

_____ 2. Tools and equipment were used properly and stored upon completion of this job.

_____ 3. This completed job met the standards set and was done within the allotted time.

_____ 4. No injury or damage to property occurred during this job.

_____ 5. Upon completion of this job, the work area was cleaned correctly.

Instructor's Signature _____ Date _____

REMOVING & REPLACING EXTERNAL GASKETS AND SEALS (SHAFTS WITH METAL-CLAD SEALS)

NATEF Standard(s) for Automatic Transmission & Transaxle:
C3 Inspect and replace external seals and gaskets.

Safety First

☐ Wear safety glasses at all times.
☐ Follow all safety rules when using common hand tools.
☐ Follow all safety rules when working with compressed air.
☐ Follow all safety rules when using a lift.

Tools & Equipment:
- Lubricant
- Drain pan
- Compressed air
- Other tools and equipment as required
- Transaxle fluid
- Lift
- Vehicle service information
- Special seal removers and installers
- Seals
- Common hand tools

PROCEDURES Refer to the vehicle service information for specifications and special procedures. Then remove and replace the gaskets and seals in the transmission provided by your instructor.

_____ 1. Write up a repair order.

_____ 2. Make sure you follow all procedures in the vehicle service information.

_____ 3. Safely raise the vehicle on a lift.

_____ 4. Use compressed air to remove all dirt from around the seal.

_____ 5. Place a drain pan under the seal that is to be replaced.

_____ 6. Carefully pry the old seal out of the housing. **CAUTION:** Be careful not to nick the housing surface.

_____ 7. Look for damage on the sealing surface and in the housing.

_____ 8. Place the seal with its proper component.

_____ 9. Clean the area where the new seal will be installed.

_____ 10. Make sure the new seal is the same type, size, and thickness as the old one.

_____ 11. Check the new seal for fit.

(continued)

Automatic Transmission & Transaxle

_____ **12.** Following the manufacturer's instructions, use an apropriate seal installer to install the new seal. **CAUTION:** Make sure that you drive against the outer edge of the new seal only. Driving against the inner edge can damage the seal and cause continued leakage.

Source: _____ *Page:* _____

_____ **13.** Install the linkage.

_____ **14.** Safely lower the vehicle to the ground.

_____ **15.** Replace any lost fluid.

Performance ✓ Checklist

Name _____ Date _____ Class _____

PERFORMANCE STANDARDS:
Level 4–Performs skill without supervision and adapts to problem situations.
Level 3–Performs skill satisfactorily without assistance or supervision.
Level 2–Performs skill satisfactorily, but requires assistance/supervision.
Level 1–Performs parts of skill satisfactorily, but requires considerable assistance/supervision.

Attempt (circle one): **1 2 3 4**

Comments:

_____

_____

_____

_____

PERFORMANCE LEVEL ACHIEVED: _____

_____ **1.** Safety rules and practices were followed at all times regarding this job.

_____ **2.** Tools and equipment were used properly and stored upon completion of this job.

_____ **3.** This completed job met the standards set and was done within the allotted time.

_____ **4.** No injury or damage to property occurred during this job.

_____ **5.** Upon completion of this job, the work area was cleaned correctly.

Instructor's Signature _____ Date_____

INSPECTING THE TRANSMISSION HOUSING

NATEF Standard(s) for Automatic Transmission & Transaxle:
C4 Inspect extension housing, bushings, and seals; perform necessary action.

Safety First

- ☐ Wear safety glasses at all times.
- ☐ Follow all safety rules when using common hand tools.
- ☐ Follow all safety rules when using a lift.
- ☐ Check Material Safety Data Sheets for chemical safety information.

Tools & Equipment:
- Thread inserts
- Common hand tools
- Automatic transmission overhaul kit
- Seal installers and replacers
- Vehicle service information
- Other tools and equipment as required
- Carburetor cleaner
- Lift

PROCEDURES Refer to the vehicle service information for specifications and special procedures. Then inspect the transmission housing in the transmission provided by your instructor.

_____ 1. Write up a repair order.

_____ 2. Make sure you follow all procedures in the vehicle service information.

_____ 3. Safely raise the vehicle on a lift.

Transmission Case Inspection

_____ 1. Inspect the transmission case for porosity or other visible damage.

_____ 2. Check all threaded holes for damaged or stripped threads. Pay careful attention to the pump threads.

_____ 3. Using the appropriate vehicle service information as a guide, repair any stripped threads.

   Source: _____ *Page:*_____

_____ 4. Check all sleeves and bores for scratches or wear.

_____ 5. Check the cooler inlet and outlet fittings for damage. *Note:* If the fittings are the plastic quick-disconnect type, check for cracks or breakage.

_____ 6. After the case is cleaned, install a new manual lever shaft seal.

Automatic Transmission & Transaxle

(continued)

Extension Housing Inspection

_____ 1. Remove the rear seal.

_____ 2. Remove the bushing from the extension housing. Be careful not to damage the housing where the bushing rides. The bushing can be carefully cut from the housing with a bushing-cutter, or it can be collapsed with a chisel and hammer and removed.

_____ 3. Use carburetor cleaner to clean the housing. Then let it air dry.

_____ 4. Inspect the housing for porosity or any other visible damage.

_____ 5. Install a new extension housing bushing.

_____ 6. Install a new rear seal. Make sure you don't distort the seal during installation.

_____ 7. Safely lower the vehicle to the ground.

Performance ✓ Checklist

Name _____ Date _____ Class _____

PERFORMANCE STANDARDS:
Level 4–Performs skill without supervision and adapts to problem situations.
Level 3–Performs skill satisfactorily without assistance or supervision.
Level 2–Performs skill satisfactorily, but requires assistance/supervision.
Level 1–Performs parts of skill satisfactorily, but requires considerable assistance/supervision.

Attempt (circle one): **1 2 3 4**

Comments:

_____

_____

_____

_____

PERFORMANCE LEVEL ACHIEVED: _____

_____ 1. Safety rules and practices were followed at all times regarding this job.
_____ 2. Tools and equipment were used properly and stored upon completion of this job.
_____ 3. This completed job met the standards set and was done within the allotted time.
_____ 4. No injury or damage to property occurred during this job.
_____ 5. Upon completion of this job, the work area was cleaned correctly.

Instructor's Signature _____ Date_____

REMOVING & REPLACING THE EXTENSION HOUSING BUSHING AND SEAL

NATEF Standard(s) for Automatic Transmission & Transaxle:
C4 Inspect extension housing, bushings, and seals; perform necessary action.

Safety First

☐ Wear safety glasses at all times.
☐ Follow all safety rules when using common hand tools.
☐ Follow all safety rules when using a lift.

Tools & Equipment:
- Lift
- Other tools and equipment as required
- Common hand tools
- Appropriate seal and bushing removing and installing tools
- Vehicle service information
- Crocus cloth

PROCEDURES Refer to the vehicle service information for specifications and special procedures. Then remove and replace the extension housing bushing and seal in the transmission provided by your instructor.

_____ 1. Write up a repair order.

_____ 2. Make sure you follow all procedures in the vehicle service information.

_____ 3. Safely raise the vehicle on a lift.

_____ 4. In order to replace the extension housing seal, the drive shaft has to be removed. To aid in reassembly, always keep subassembly parts grouped together in a logical sequence and make marks on the rear universal joints and the drive shaft.

_____ 5. Using the correct tool, remove the bushing and install a new one in the correct position.

_____ 6. Before installing a new seal, inspect the sealing surface of the universal joint yoke for scoring. Also inspect the counterbore of the housing for burrs. Use a crocus cloth to remove any burrs.

_____ 7. Using a suitable seal driver, install the new seal in the proper direction.

_____ 8. After installation, check for leaks in the housing seal areas and make sure the transmission fluid is at the proper level.

_____ 9. Lube the seal lip with transmission fluid.

_____ 10. If the sealing surface of the component (yoke, torque converter input hub) is grooved or scarred, the component should be replaced because the seal will not seal around a damaged surface.

_____ 11. Reassemble the drive shaft using the marks you made during removal.

_____ 12. Safely lower the vehicle to the ground.

(continued)

Performance ✓ Checklist

Name _____ Date _____ Class _____

PERFORMANCE STANDARDS:
Level 4–Performs skill without supervision and adapts to problem situations.
Level 3–Performs skill satisfactorily without assistance or supervision.
Level 2–Performs skill satisfactorily, but requires assistance/supervision.
Level 1–Performs parts of skill satisfactorily, but requires considerable assistance/supervision.

Attempt (circle one): **1 2 3 4**

Comments:

_____

_____

_____

_____

PERFORMANCE LEVEL ACHIEVED: _____

_____ **1.** Safety rules and practices were followed at all times regarding this job.
_____ **2.** Tools and equipment were used properly and stored upon completion of this job.
_____ **3.** This completed job met the standards set and was done within the allotted time.
_____ **4.** No injury or damage to property occurred during this job.
_____ **5.** Upon completion of this job, the work area was cleaned correctly.

Instructor's Signature _____ Date_____

REMOVING & REPLACING OIL DELIVERY SEAL RINGS

NATEF Standard(s) for Automatic Transmission & Transaxle:
D3-3 Inspect oil delivery seal rings, ring grooves, and sealing surface areas.

Safety First
- ☐ Wear safety glasses at all times.
- ☐ Follow all safety rules when using common hand tools.

Tools & Equipment:
- Wax lubricant
- Common hand tools
- Automatic transmission overhaul kit
- Special removal and installation tools
- Feeler gauge
- Vehicle service information

PROCEDURES Refer to the vehicle service information for specifications and special procedures. Then remove and replace the oil delivery seal rings in the transmission provided by your instructor.

_____ 1. Write up a repair order.

_____ 2. Make sure you follow all procedures in the vehicle service information.

_____ 3. Locate and remove the seal rings. Make sure you use the special removal tool that is recommended in the vehicle service information.

   Source: _____ *Page:* _____

_____ 4. Carefully check the new seal ring against the removed seal ring to make sure that it is the same size (both in cross section and total diameter) and type.

_____ 5. Check each seal ring for fit prior to installation.

_____ 6. Install the new seal rings using the special installation tool that is recommended in the vehicle service information.

   Source: _____ *Page:* _____

_____ 7. If you are installing metal sealing rings, allow a slight gap between the ring ends to allow for expansion.

_____ 8. If you are installing a lipped seal ring, be careful not to catch the seal lip on the edge of the bore. This can cut the seal lip and cause a fluid leak.

_____ 9. To ease the installation of lipped seal rings, use a wax lubricant, the proper tool, a seal guide, and precompress the seal lip.

(continued)

Automatic Transmission & Transaxle

_____ 10. Always install lipped seal rings with the lips facing the direction recommended in the vehicle service information.

Source: _____ *Page:* _____

_____ 11. Be careful when installing the next unit on top of a scarf-cut Teflon® sealing ring. The ends tend to stick out and they can get caught and cut during installation.

_____ 12. Make sure you always lap the ends of scarf-cut Teflon® seal rings on the correct sides.

Performance ✓ Checklist

Name _____ Date _____ Class _____

PERFORMANCE STANDARDS:
Level 4–Performs skill without supervision and adapts to problem situations.
Level 3–Performs skill satisfactorily without assistance or supervision.
Level 2–Performs skill satisfactorily, but requires assistance/supervision.
Level 1–Performs parts of skill satisfactorily, but requires considerable assistance/supervision.

Attempt (circle one): **1 2 3 4**

Comments:

_____

_____

_____

_____

PERFORMANCE LEVEL ACHIEVED: _____

_____ 1. Safety rules and practices were followed at all times regarding this job.
_____ 2. Tools and equipment were used properly and stored upon completion of this job.
_____ 3. This completed job met the standards set and was done within the allotted time.
_____ 4. No injury or damage to property occurred during this job.
_____ 5. Upon completion of this job, the work area was cleaned correctly.

Instructor's Signature _____ Date_____

INSPECTING THE CONVERTER
FLEX PLATE ASSEMBLY

NATEF Standard(s) for Automatic Transmission & Transaxle:
D2-1 Inspect converter flex plate, attaching parts, pilot, pump drive, and seal areas.

Safety First

- ☐ Wear safety glasses at all times.
- ☐ Follow all safety rules when using common hand tools.
- ☐ Follow all safety rules when using a lift.

Tools & Equipment:
- Lift
- Common hand tools
- Vehicle service information

PROCEDURES Refer to the vehicle service information for specifications and special procedures. Then inspect the converter flex plate assembly in the vehicle provided by your instructor.

_____ 1. Write up a repair order.

_____ 2. Make sure you follow all procedures in the vehicle service information.

_____ 3. Safely raise the vehicle on a lift.

_____ 4. Remove the torque converter splash shield. *Note:* This may require removal of the engine mount or the starter motor, or both.

_____ 5. Inspect the flex plate for metal-fatigue cracks, especially in the areas around the crankshaft flange bolts.

_____ 6. Inspect the starter drive teeth for wear or broken teeth.

_____ 7. If any cracks or broken teeth are found, replace the flex plate.

_____ 8. Inspect the bellhousing for evidence of oil or ATF. If oil or ATF is found, transmission removal may be necessary.

_____ 9. Safely lower the vehicle to the ground.

Automatic Transmission & Transaxle

(continued)

Performance ✓ Checklist

Name _____ Date _____ Class _____

PERFORMANCE STANDARDS:
Level 4–Performs skill without supervision and adapts to problem situations.
Level 3–Performs skill satisfactorily without assistance or supervision.
Level 2–Performs skill satisfactorily, but requires assistance/supervision.
Level 1–Performs parts of skill satisfactorily, but requires considerable assistance/supervision.

Attempt (circle one): **1 2 3 4**

Comments:

_____

_____

_____

_____

PERFORMANCE LEVEL ACHIEVED: _____

_____ 1. Safety rules and practices were followed at all times regarding this job.

_____ 2. Tools and equipment were used properly and stored upon completion of this job.

_____ 3. This completed job met the standards set and was done within the allotted time.

_____ 4. No injury or damage to property occurred during this job.

_____ 5. Upon completion of this job, the work area was cleaned correctly.

Instructor's Signature _____ Date_____

MEASURING TORQUE CONVERTER ENDPLAY AND CHECKING FOR INTERFERENCE

NATEF Standard(s) for Automatic Transmission & Transaxle:
D2-2 Measure torque converter endplay and check for interference; check stator clutch.

Safety First
- ☐ Wear safety glasses at all times.
- ☐ Follow all safety rules when using common hand tools.

Tools & Equipment:
- Dial indicator
- Holding device specialty tool
- Common hand tools
- Vehicle service information

PROCEDURES Refer to the vehicle service information for specifications and special procedures. Then check the torque converter provided by your instructor.

_____ 1. Write up a repair order.

_____ 2. Make sure you follow all procedures in the vehicle service information.

Measure Torque Converter Endplay

_____ 1. Insert the holding tool into the converter hub until it bottoms.

_____ 2. Tighten the tool in place to lock it into the turbine spline.

_____ 3. Attach a dial indicator so that the indicator button is on the converter hub.

_____ 4. Zero the indicator reading, and then lift the holding tool upward as far as it will go.

_____ 5. Read the dial indicator. Replace the converter if the reading exceeds the endplay specified in the vehicle service information. *Note:* Lockup converters generally have a maximum internal endplay of 0.050″ or 1.27 mm.

Source: _____ Page: _____

Check for Stator/Turbine Interference

_____ 1. Place the converter downward on the workbench.

_____ 2. Using snap-ring pliers, engage the turbine hub.

_____ 3. While holding the converter, rotate the turbine shaft. Replace the converter if the turbine shaft does not turn freely in both directions.

(continued)

Automatic Transmission & Transaxle

Check Torque Converter Stator Clutch Operation

_____ **1.** Inspect the stator clutch splines for wear or damage.

_____ **2.** Using snap-ring pliers, manually rotate the stator. The splines should freewheel in one direction and lock in the other.

Performance ✓ Checklist

Name _____ Date _____ Class _____

PERFORMANCE STANDARDS:
Level 4–Performs skill without supervision and adapts to problem situations.
Level 3–Performs skill satisfactorily without assistance or supervision.
Level 2–Performs skill satisfactorily, but requires assistance/supervision.
Level 1–Performs parts of skill satisfactorily, but requires considerable assistance/supervision.

Attempt (circle one): **1 2 3 4**

Comments:

_____

_____

_____

_____

PERFORMANCE LEVEL ACHIEVED: _____

_____ **1.** Safety rules and practices were followed at all times regarding this job.

_____ **2.** Tools and equipment were used properly and stored upon completion of this job.

_____ **3.** This completed job met the standards set and was done within the allotted time.

_____ **4.** No injury or damage to property occurred during this job.

_____ **5.** Upon completion of this job, the work area was cleaned correctly.

Instructor's Signature _____ Date_____

PERFORMING A STALL TEST

NATEF Standard(s) for Automatic Transmission & Transaxle:
A7 Perform stall test; determine necessary action.

Safety First
- [ ] Wear safety glasses at all times.
- [ ] Follow all safety rules when using common hand tools.
- [ ] To avoid personal and material damage during a stall test, follow all safety precautions and operational procedures in vehicle service information.
- [ ] Do not allow anyone to stand near the vehicle during a stall test.
- [ ] Do not perform a stall test for longer than 5 seconds at a time. Extreme heat is generated.
- [ ] Following a stall test, place the transmission in NEUTRAL and run the engine at 2,000 rpm for about a minute to allow the fluid to cool.

Tools & Equipment:
- Vehicle service information
- Wheel chocks
- Common hand tools
- Tachometer

PROCEDURES Refer to the vehicle service information for specifications and special procedures. Then conduct a stall test on the vehicle provided by your instructor.

_____ 1. Write up a repair order.

_____ 2. Make sure you follow all procedures in the vehicle service information.

_____ 3. The stall test is not recommended by all manufacturers. Conduct a stall test only when a malfunctioning stator one-way clutch problem is suspected (stator clutch not holding).

_____ 4. Locate the stall testing specifications in the vehicle service information.

_____ 5. Check and inspect ATF level and condition.

_____ 6. Chock all wheels and set the parking brake.

_____ 7. Attach a tachometer to the vehicle.

_____ 8. Apply and hold the service brake by stepping on the brake pedal.

_____ 9. Start the engine and place the transmission in a forward range.

_____ 10. Increase engine speed to wide-open throttle and note the maximum indicated engine rpm.

_____ 11. Conduct the stall test in all forward and reverse ranges. The indicated engine rpm should be the same in all gear ranges.

(continued)

Automatic Transmission & Transaxle

_____ 12. If the engine rpm is lower than specified in the vehicle service information, the problem may be:
- The engine is not properly tuned.
- The torque converter stator one-way clutch is slipping.

_____ 13. If engine rpm is higher than specified in the vehicle service information, the problem may be:
- Low fluid level.
- A restricted fluid filter.
- A sticking pressure regulator valve.
- Slipping clutches, bands, shaft splines, or one-way clutch.

Performance ✓ Checklist

Name _____ Date _____ Class _____

PERFORMANCE STANDARDS:
Level 4–Performs skill without supervision and adapts to problem situations.
Level 3–Performs skill satisfactorily without assistance or supervision.
Level 2–Performs skill satisfactorily, but requires assistance/supervision.
Level 1–Performs parts of skill satisfactorily, but requires considerable assistance/supervision.

Attempt (circle one): **1 2 3 4**

Comments:

_____

_____

_____

_____

PERFORMANCE LEVEL ACHIEVED: _____

_____ 1. Safety rules and practices were followed at all times regarding this job.
_____ 2. Tools and equipment were used properly and stored upon completion of this job.
_____ 3. This completed job met the standards set and was done within the allotted time.
_____ 4. No injury or damage to property occurred during this job.
_____ 5. Upon completion of this job, the work area was cleaned correctly.

Instructor's Signature _____ Date_____

PERFORMING LOCK-UP CONVERTER SYSTEM TESTS

NATEF Standard(s) for Automatic Transmission & Transaxle:
A8 Perform lock-up converter system tests; determine necessary action.
C6 Diagnose electronic transmission control systems using a scan tool; determine necessary action.

Safety First

- [ ] Wear safety glasses at all times.
- [ ] Always connect a vehicle's exhaust to a vent hose before you run an engine in a closed shop. Unvented exhaust fumes in a closed shop can cause death.
- [ ] Keep jewelry, loose clothing, and hair away from moving parts while an engine is running.
- [ ] Be careful of hot engine parts.
- [ ] Follow all safety rules when using a lift.
- [ ] Be extremely careful if you have to be under the vehicle during this test. Contact with rotating drive wheels and shafts can cause serious injury.

Tools & Equipment:
- Jumper wires
- Lift
- Scan tool
- Vehicle service information

PROCEDURES Refer to the vehicle service information for specifications and special procedures. Then perform the lock-up converter tests on the transmission provided by your instructor.

_____ 1. Write up a repair order.

_____ 2. Make sure you follow all test procedures in the vehicle service information.

_____ 3. Start the engine and attach the scan tool (if required).

_____ 4. Safely raise the vehicle on a lift.

_____ 5. Accelerate the vehicle until the top gear is reached.

_____ 6. Read the scan tool to ensure that the converter locks. **Note:** The converter will unlock when the driver taps the brake pedal. This can be felt as a slight driveline bump.

_____ 7. Test the solenoids, if applicable, by removing the electrical connector and checking for specified resistance between the terminals. **Note:** This also can be done by applying voltage and ground to the valve using jumper wires.

_____ 8. Safely lower the vehicle to the ground.

(continued)

Automatic Transmission & Transaxle

Performance ✓ Checklist

Name _____ Date _____ Class _____

PERFORMANCE STANDARDS:
Level 4–Performs skill without supervision and adapts to problem situations.
Level 3–Performs skill satisfactorily without assistance or supervision.
Level 2–Performs skill satisfactorily, but requires assistance/supervision.
Level 1–Performs parts of skill satisfactorily, but requires considerable assistance/supervision.

Attempt (circle one): 1 2 3 4

Comments:

_____
_____
_____
_____

PERFORMANCE LEVEL ACHIEVED: _____

_____ 1. Safety rules and practices were followed at all times regarding this job.
_____ 2. Tools and equipment were used properly and stored upon completion of this job.
_____ 3. This completed job met the standards set and was done within the allotted time.
_____ 4. No injury or damage to property occurred during this job.
_____ 5. Upon completion of this job, the work area was cleaned correctly.

Instructor's Signature _____ Date_____

DIAGNOSING NOISE AND
VIBRATION CONCERNS

NATEF Standard(s) for Automatic Transmission & Transaxle:
A10 Diagnose noise and vibration concerns; determine necessary action.

Safety First

- ☐ Wear safety glasses at all times.
- ☐ Follow all safety rules when using common hand tools.
- ☐ Follow all safety rules when using a lift or jack and jack stands.
- ☐ When operating a vehicle equipped with traction control on a lift, it may be necessary to turn the traction control off so the transmission can shift normally.
- ☐ Use exhaust vent if running engine in closed shop. Unvented exhaust fumes can cause death.
- ☐ Be careful of hot or moving parts.

Tools & Equipment:
- Vehicle service information
- Automotive stethoscope
- Common hand tools
- Lift or jack and jack stands
- Vibration analyzer

PROCEDURES Refer to the vehicle service information for specifications and special procedures. Then diagnose the noise and vibration concerns on the vehicle provided by your instructor.

_____ 1. Write up a repair order.

_____ 2. Make sure you follow all procedures in the vehicle service information.

_____ 3. Inspect the fluid level to ensure it is correct. Follow manufacturer's procedure when checking automatic fluid level.

_____ 4. Raise the vehicle on a lift and place the transmission in DRIVE. Listen for noise or vibrations coming from the transmission/transaxle.

_____ 5. Place transmission in NEUTRAL.

_____ 6. Raise engine speed and listen for any noises or feel for vibration in the transmission.

_____ 7. If noise changes with engine speed, you may have to replace the internal bearings or pump.

_____ 8. If the noise does not change with engine speed, you may have to replace the internal bearing or pinion gear.

Automatic Transmission & Transaxle

(continued)

Performance ✓ Checklist

Name _____ Date _____ Class _____

PERFORMANCE STANDARDS:
Level 4–Performs skill without supervision and adapts to problem situations.
Level 3–Performs skill satisfactorily without assistance or supervision.
Level 2–Performs skill satisfactorily, but requires assistance/supervision.
Level 1–Performs parts of skill satisfactorily, but requires considerable assistance/supervision.

Attempt (circle one): **1 2 3 4**

Comments:

_____

_____

_____

_____

PERFORMANCE LEVEL ACHIEVED: _____

_____ **1.** Safety rules and practices were followed at all times regarding this job.
_____ **2.** Tools and equipment were used properly and stored upon completion of this job.
_____ **3.** This completed job met the standards set and was done within the allotted time.
_____ **4.** No injury or damage to property occurred during this job.
_____ **5.** Upon completion of this job, the work area was cleaned correctly.

Instructor's Signature _____ Date_____

INSPECTING COOLER AND LINES

NATEF Standard(s) for Automatic Transmission & Transaxle:
D1-8 Inspect, leak test, and flush cooler, lines, and fittings.

Safety First
- ☐ Wear safety glasses at all times.
- ☐ Follow all safety rules when using common hand tools.
- ☐ Follow all safety rules when using a lift.

Tools & Equipment:
- Lift
- Common hand tools
- Vehicle service information

PROCEDURES Refer to the vehicle service information for specifications and special procedures. Then inspect the cooler and lines in the transmission provided by your instructor.

_____ 1. Write up a repair order.

_____ 2. Make sure you follow all procedures in the vehicle service information.

_____ 3. Inspect the engine coolant in the radiator. If there is transmission fluid in the coolant, the fluid cooler is leaking.

_____ 4. Check the transmission dipstick for signs of engine coolant. If the fluid is milky pink, coolant is leaking into the transmission.

_____ 5. Safely raise the vehicle on a lift.

_____ 6. Check for leaks in the fluid lines and the fittings between the transmission and the cooler.

_____ 7. Safely lower the vehicle to the ground.

Automatic Transmission & Transaxle

(continued)

Performance ✓ Checklist

Name _____ Date _____ Class _____

PERFORMANCE STANDARDS:
Level 4–Performs skill without supervision and adapts to problem situations.
Level 3–Performs skill satisfactorily without assistance or supervision.
Level 2–Performs skill satisfactorily, but requires assistance/supervision.
Level 1–Performs parts of skill satisfactorily, but requires considerable assistance/supervision.

Attempt (circle one): **1 2 3 4**

Comments:

_____

_____

_____

_____

PERFORMANCE LEVEL ACHIEVED: _____

_____ 1. Safety rules and practices were followed at all times regarding this job.

_____ 2. Tools and equipment were used properly and stored upon completion of this job.

_____ 3. This completed job met the standards set and was done within the allotted time.

_____ 4. No injury or damage to property occurred during this job.

_____ 5. Upon completion of this job, the work area was cleaned correctly.

Instructor's Signature _____ Date_____

INSPECTING, LEAK TESTING & FLUSHING COOLER, LINES, AND FITTINGS

NATEF Standard(s) for Automatic Transmission & Transaxle:
D1-8 Inspect, leak test, and flush cooler, lines, and fittings.

Safety First

☐ Wear safety glasses at all times.
☐ Follow all safety rules when using common hand tools.

Tools & Equipment:
• Vehicle service information   • Common hand tools   • Transmission flushing machine

PROCEDURES Refer to the vehicle service information for specifications and special procedures. Then inspect, leak test, and flush cooler, lines, and fittings on the vehicle provided by your instructor.

_____ 1. Write up a repair order.

_____ 2. Make sure you follow all procedures in the vehicle service information.

_____ 3. Inspect transmission lines, cooler, and fittings for signs of deterioration and leaking transmission fluid.

_____ 4. Repair or replace any lines, fittings, or cooler that has deteriorated or is leaking.

_____ 5. Following recommended procedures, attach a flushing machine to the transmission lines. Flush according to procedure.

_____ 6. Note any leaks while using the machine to verify the fittings and lines are able to hold pressure.

_____ 7. If a leak was detected, replace the faulty component.

_____ 8. Remove the flushing machine and reattach the lines to the vehicle.

_____ 9. Check to make sure the transmission level meets specifications.

Automatic Transmission & Transaxle

(continued)

Performance ✓ Checklist

Name _____ Date _____ Class _____

PERFORMANCE STANDARDS:
Level 4–Performs skill without supervision and adapts to problem situations.
Level 3–Performs skill satisfactorily without assistance or supervision.
Level 2–Performs skill satisfactorily, but requires assistance/supervision.
Level 1–Performs parts of skill satisfactorily, but requires considerable assistance/supervision.

Attempt (circle one): **1 2 3 4**

Comments:

_____

_____

_____

_____

PERFORMANCE LEVEL ACHIEVED: _____

_____ **1.** Safety rules and practices were followed at all times regarding this job.

_____ **2.** Tools and equipment were used properly and stored upon completion of this job.

_____ **3.** This completed job met the standards set and was done within the allotted time.

_____ **4.** No injury or damage to property occurred during this job.

_____ **5.** Upon completion of this job, the work area was cleaned correctly.

Instructor's Signature _____ Date _____

SERVICING COOLER LINES AND HOSES

NATEF Standard(s) for Automatic Transmission & Transaxle:
D1-8 Inspect, leak test, and flush cooler, lines, and fittings.

Safety First

☐ Wear safety glasses at all times.
☐ Follow all safety rules when using common hand tools.
☐ Follow all safety rules when using a lift.
☐ Always connect a vehicle's exhaust to a vent hose before you run an engine in a closed shop. Unvented exhaust fumes in a closed shop can cause death.
☐ Be careful of hot engine parts.

Tools & Equipment:

- Lift
- Rule
- Tubing reamer
- Replacement-hose clamps
- Seamless replacement tubing

- High-grade flexible hose
- New hose clamps
- Tubing bender
- Common hand tools

- String
- Tubing cutter
- Flaring tool
- Vehicle service information

PROCEDURES Refer to the vehicle service information for specifications and special procedures. Then service the cooler lines and hoses in the transmission provided by your instructor.

_____ 1. Write up a repair order.

_____ 2. Make sure you follow all procedures in the vehicle service information.

_____ 3. Safely raise the vehicle on a lift. **CAUTION:** Do not attempt to repair a damaged cooler line. Instead, you should replace the entire line.

Make and Replace Metal Cooler Lines

_____ 1. Measure the damaged length of tubing as accurately as possible. *Note:* Use a length of string and then measure the string.

_____ 2. Use a tubing cutter to cut a section of new tubing to the length measured in Step 2.

_____ 3. Ream one cut end of the new tubing with a tubing reamer.

_____ 4. Use a tubing bender to shape the new tubing to the same shape as the old tubing.

_____ 5. Recheck the length and cut to size as necessary. *Note:* When formed to the shape of the damaged line, the new tubing may be slightly longer than the old.

_____ 6. Ream the other end of the tubing.

_____ 7. Install the necessary fitting on each end of the tubing.

(continued)

_____ 8. Use a flaring tool to double-flare one end of the tubing.

_____ 9. Double-flare the other end of the tubing.

_____ 10. Position the new line loosely under the existing clamps on the engine or chassis.

_____ 11. Connect each end of the line and torque fitting to specifications.

_____ 12. Tighten mounting clamps.

_____ 13. Run engine at idle and check for leaks.

Replace Damaged Section of Flexible Hose

_____ 1. Measure the old length of flexible hose and cut a new hose to the measured length.

_____ 2. Install the new hose so that it fits flush against the radiator at one end and extends at least 2 inches over the flared end of tubing at the other end.

_____ 3. Secure the hose with new clamps and tighten to meet manufacturer's specifications.

_____ 4. Safely lower the vehicle to the ground.

Performance ✓ Checklist

Name _____ Date _____ Class _____

PERFORMANCE STANDARDS:
Level 4–Performs skill without supervision and adapts to problem situations.
Level 3–Performs skill satisfactorily without assistance or supervision.
Level 2–Performs skill satisfactorily, but requires assistance/supervision.
Level 1–Performs parts of skill satisfactorily, but requires considerable assistance/supervision.

Attempt (circle one): 1 2 3 4

Comments:

_____

_____

_____

_____

PERFORMANCE LEVEL ACHIEVED: _____

_____ 1. Safety rules and practices were followed at all times regarding this job.

_____ 2. Tools and equipment were used properly and stored upon completion of this job.

_____ 3. This completed job met the standards set and was done within the allotted time.

_____ 4. No injury or damage to property occurred during this job.

_____ 5. Upon completion of this job, the work area was cleaned correctly.

Instructor's Signature _____ Date_____

INSPECTING THE PLANETARY GEAR ASSEMBLY

NATEF Standard(s) for Automatic Transmission & Transaxle:
A11 Diagnose transmission/transaxle gear reduction/multiplication concerns using driving, driven, and held member (power flow) principles.
D3-5 Inspect and measure planetary gear assembly (includes sun, ring gear, thrust washers, planetary gears, and carrier assembly); determine necessary action.

Safety First
- ☐ Wear safety glasses at all times.
- ☐ Follow all safety rules when using common hand tools.

Tools & Equipment:
- Petroleum jelly
- Vehicle service information
- Feeler gauge
- Common hand tools

PROCEDURES Refer to the vehicle service information for specifications and special procedures. Then inspect the planetary gear assembly in the transmission provided by your instructor.

_____ 1. Write up a repair order.

_____ 2. Make sure you follow all procedures in the vehicle service information.

_____ 3. Inspect the thrust washer for wear.

_____ 4. Remove the planetary carrier and examine the splines for wear.

_____ 5. Inspect the planetary gears for damage.

_____ 6. Examine the planetary center spline for stripping or distortion.

_____ 7. Inspect the gear teeth for damage.

_____ 8. Measure the gear bushings for wear using a feeler gauge.

_____ 9. If any abnormalities are found, the planetary gear must be replaced as an assembly.

_____ 10. Reassemble the planetary drive train and install the thrust washers and snap rings.
Use petroleum jelly to hold the thrust washers in place during reassembly.

_____ 11. After reassembly, make a final check of gear train endplay.

Automatic Transmission & Transaxle

(_continued_)

Performance ✓ Checklist

Name _____ Date _____ Class _____

PERFORMANCE STANDARDS:
Level 4–Performs skill without supervision and adapts to problem situations.
Level 3–Performs skill satisfactorily without assistance or supervision.
Level 2–Performs skill satisfactorily, but requires assistance/supervision.
Level 1–Performs parts of skill satisfactorily, but requires considerable assistance/supervision.

Attempt (circle one): **1 2 3 4**

Comments:

_____

_____

_____

_____

PERFORMANCE LEVEL ACHIEVED: _____

_____ **1.** Safety rules and practices were followed at all times regarding this job.
_____ **2.** Tools and equipment were used properly and stored upon completion of this job.
_____ **3.** This completed job met the standards set and was done within the allotted time.
_____ **4.** No injury or damage to property occurred during this job.
_____ **5.** Upon completion of this job, the work area was cleaned correctly.

Instructor's Signature _____ Date_____

DIAGNOSING PRESSURE CONCERNS USING HYDRAULIC PRINCIPLES

NATEF Standard(s) for Automatic Transmission & Transaxle:
A12 Diagnose pressure concerns in the transmission using hydraulic principles (Pascal's law).

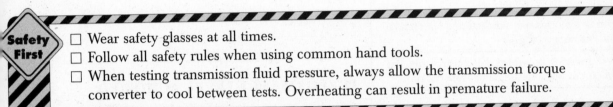

Safety First

☐ Wear safety glasses at all times.
☐ Follow all safety rules when using common hand tools.
☐ When testing transmission fluid pressure, always allow the transmission torque converter to cool between tests. Overheating can result in premature failure.

Tools & Equipment:
• Vehicle service information   • Common hand tools   • Pressure gauges

PROCEDURES Refer to the vehicle service information for specifications and special procedures. Then diagnose pressure concerns on the vehicle provided by your instructor.

_____ 1. Write up a repair order.

_____ 2. Make sure you follow all procedures in the vehicle service information.

_____ 3. Verify that the fluid level is within manufacturer's specifications before beginning your test.

_____ 4. Identify the location of pressure taps on the transmission.

_____ 5. Install pressure gauges to the transmission using the pressure taps.

_____ 6. Following the manufacturer's procedure, note pressure readings at specific rpm and gear.

Automatic Transmission & Transaxle

(continued)

Performance ✓ Checklist

Name _____ Date _____ Class _____

PERFORMANCE STANDARDS:
Level 4–Performs skill without supervision and adapts to problem situations.
Level 3–Performs skill satisfactorily without assistance or supervision.
Level 2–Performs skill satisfactorily, but requires assistance/supervision.
Level 1–Performs parts of skill satisfactorily, but requires considerable assistance/supervision.

Attempt (circle one): **1 2 3 4**

Comments:

_____

_____

_____

_____

PERFORMANCE LEVEL ACHIEVED: _____

_____ 1. Safety rules and practices were followed at all times regarding this job.

_____ 2. Tools and equipment were used properly and stored upon completion of this job.

_____ 3. This completed job met the standards set and was done within the allotted time.

_____ 4. No injury or damage to property occurred during this job.

_____ 5. Upon completion of this job, the work area was cleaned correctly.

Instructor's Signature _____ Date _____

INSPECTING AN OIL PUMP ASSEMBLY

NATEF Standard(s) for Automatic Transmission & Transaxle:
D2-3 Inspect, measure, and reseal oil pump assembly and components.

Safety First
- ☐ Wear safety glasses at all times.
- ☐ Follow all safety rules when using common hand tools.
- ☐ Check Material Safety Data Sheets for chemical safety information.

Tools & Equipment:
- Feeler gauge
- RTV sealant
- Lathe-cut replacement seal
- Spring-loaded center punch
- Special bushing-removing tool
- Straightedge
- Sharp knife
- Automatic transmission fluid
- Common hand tools
- Seal-installing tool
- Pump removal tools
- Torque wrench
- Vehicle service information

PROCEDURES Refer to the vehicle service information for specifications and special procedures. Then inspect the oil pump assembly in the transmission provided by your instructor.

_____ 1. Write up a repair order.

_____ 2. Make sure you follow all procedures in the vehicle service information.

_____ 3. Remove the pump from the case.

_____ 4. Remove the outer lathe-cut seal ring and pump seal.

_____ 5. Following the manufacturer's specifications, disassemble the pump.

  Source: _____ *Page:* _____

_____ 6. Mark the gears with a punch to ensure proper replacement.

_____ 7. Wash the individual parts of the pump assembly thoroughly. Then dry the parts and spread them out on a workbench.

_____ 8. Closely examine the housing for cracks, nicks, burrs, or damage to the sealing surfaces.

_____ 9. Inspect the gears for burrs, nicks, or scratches.

_____ 10. Examine the reaction-shaft support gears for cracks and pump gear wear pattern.

_____ 11. Use a feeler gauge to measure the clearance between the outer pump gear and the pump pocket in the housing. Compare your findings with the vehicle service information's specifications.

_____ 12. Check the clearance between the outer pump gear teeth and the crescent with a feeler gauge. Compare your findings with the vehicle service information's specifications.

(continued)

_____ 13. Using a feeler gauge, check the clearance between the inner pump gear teeth and the crescent. Compare your findings with the vehicle service information's specifications.

_____ 14. Check the pump-gear side clearance with a feeler gauge and straightedge. Compare your findings with the vehicle service information's specifications. *Note:* Replace the pump if it exceeds any vehicle service information clearance specifications.

_____ 15. Use the special bushing-removal tool to remove the pump-housing bushing.

_____ 16. Install the pump seal with the appropriate seal-installing tool.

_____ 17. Examine the seal rings on the reaction-shaft support. If necessary, replace them.

_____ 18. Check to make sure that the seal rings rotate in their grooves.

_____ 19. Place some transmission fluid in the pocket of the pump housing to provide initial lubrication and sealing.

_____ 20. Install the pump gears into the pump housing. *Note:* Observe markings to ensure that individual gears are not inverted.

_____ 21. Align and install the reaction-shaft support. *Note:* Be sure to properly torque the shaft support bolts.

_____ 22. Lubricate and install a new lathe-cut seal in the groove around the pump body.

_____ 23. Install the pump in the case and torque to specifications.

Performance ✓ Checklist

Name _____   Date _____   Class _____

PERFORMANCE STANDARDS:
Level 4–Performs skill without supervision and adapts to problem situations.
Level 3–Performs skill satisfactorily without assistance or supervision.
Level 2–Performs skill satisfactorily, but requires assistance/supervision.
Level 1–Performs parts of skill satisfactorily, but requires considerable assistance/supervision.

Attempt (circle one): **1 2 3 4**

Comments:

_____

_____

_____

_____

PERFORMANCE LEVEL ACHIEVED: _____

_____ 1. Safety rules and practices were followed at all times regarding this job.

_____ 2. Tools and equipment were used properly and stored upon completion of this job.

_____ 3. This completed job met the standards set and was done within the allotted time.

_____ 4. No injury or damage to property occurred during this job.

_____ 5. Upon completion of this job, the work area was cleaned correctly.

Instructor's Signature _____   Date_____

PERFORMING PRESSURE TESTS

NATEF Standard(s) for Automatic Transmission & Transaxle:
A2 Identify and interpret transmission/transaxle concern; assure proper engine operation; determine necessary action.
A6 Perform pressure tests; determine necessary action.

Safety First

- ☐ Wear safety glasses at all times.
- ☐ Always connect a vehicle's exhaust to a vent hose before you run an engine in a closed shop. Unvented exhaust fumes in a closed shop can cause death.
- ☐ Keep jewelry, loose clothing, and hair away from moving parts while an engine is running.
- ☐ Be careful of hot engine parts.
- ☐ Follow all safety rules when using a lift.
- ☐ Be extremely careful if you have to be under the vehicle during this test. Contact with the spinning driveline or wheels can cause serious injury.

Tools & Equipment:
- Manufacturer's pressure test charts
- Lift
- Tachometer
- T-fitting and vacuum tubing (if required)
- Pressure gauges
- Droplight
- Manufacturer's pressure test troubleshooting charts
- Vacuum gauge (if required)
- Vehicle service information

PROCEDURES Refer to the vehicle service information for specifications and special procedures. Then perform the pressure tests on the transmission provided by your instructor.

_____ 1. Write up a repair order.

_____ 2. Make sure you follow all procedures in the vehicle service information.

_____ 3. Start the engine and warm it to normal operating temperature. Then shut it off.

_____ 4. Connect a tachometer to the engine. Make sure that all gauges and instruments are properly connected before restarting the engine.

_____ 5. Work with a helper inside the vehicle to start the engine and operate the transmission, accelerator, and brake controls.

_____ 6. Safely raise the vehicle on a lift to a comfortable working height.

_____ 7. Connect pressure gauges to specified test ports.

_____ 8. If applicable, use a T-fitting and vacuum tubing to connect the vacuum gauge with a shutoff or bleed-down valve to the vacuum modulator line.

(continued)

Automatic Transmission & Transaxle

_____ 9. Have your helper restart the engine and move the gear selector to the first test position.

_____ 10. Have your helper run the engine at the specified rpm.

_____ 11. Record readings on gauges. _____

_____ 12. Have your helper move the transmission controls as specified.

_____ 13. Record pressures or changes in pressures. _____

_____ 14. Have your helper slow the engine and press gently on the service brakes to stop the spinning wheels and driveline.

_____ 15. Have your helper shut off the engine.

_____ 16. Attach the pressure and/or vacuum gauges to the proper pressure ports for the next test.

_____ 17. Repeat the steps above for each test procedure. Make notes on a separate sheet of paper of all pressures and pressure changes for each test.

_____ 18. Refer to vehicle service information pressure test specification charts to interpret pressure test readings, but you may have to refer to a separate pressure test troubleshooting chart.

_____ 19. Safely lower the vehicle to the ground.

Performance ✓ Checklist

Name _____ Date _____ Class _____

PERFORMANCE STANDARDS:
Level 4–Performs skill without supervision and adapts to problem situations.
Level 3–Performs skill satisfactorily without assistance or supervision.
Level 2–Performs skill satisfactorily, but requires assistance/supervision.
Level 1–Performs parts of skill satisfactorily, but requires considerable assistance/supervision.

Attempt (circle one): **1 2 3 4**

Comments:

_____

_____

_____

_____

PERFORMANCE LEVEL ACHIEVED: _____

_____ 1. Safety rules and practices were followed at all times regarding this job.

_____ 2. Tools and equipment were used properly and stored upon completion of this job.

_____ 3. This completed job met the standards set and was done within the allotted time.

_____ 4. No injury or damage to property occurred during this job.

_____ 5. Upon completion of this job, the work area was cleaned correctly.

Instructor's Signature _____ Date _____

REMOVING & REPAIRING OR REPLACING THE VALVE BODY

NATEF Standard(s) for Automatic Transmission & Transaxle:
D1-4 Inspect, measure, clean, and replace valve body (includes surfaces and bores, springs, valves, sleeves, retainers, brackets, check-balls, screens, spacers, and gaskets).

Safety First
- ☐ Wear safety glasses at all times.
- ☐ Follow all safety rules when using common hand tools.
- ☐ Follow all safety rules when using a lift.
- ☐ Check Material Safety Data Sheets for chemical safety information.

Tools & Equipment:
- Lift
- ¼-drive torque wrench
- Automatic transmission fluid
- Common hand tools
- Assembly lubricant
- Vehicle service information

PROCEDURES Refer to the vehicle service information for specifications and special procedures. Then remove and repair or replace the valve body in the transmission provided by your instructor.

_____ 1. Write up a repair order.

_____ 2. Make sure you follow all procedures in the vehicle service information.

_____ 3. Safely raise the vehicle on a lift.

_____ 4. Drain the fluid from the transmission.

_____ 5. Remove the oil pan or valve body cover.

_____ 6. Remove the valve body attaching bolts and valve body. During removal of the valve body, be sure to release it slowly and to catch any springs or parts that may fall free.

_____ 7. Take each valve body part out one at a time. Clean, inspect, lightly oil, and replace each part before going to the next part. After reinstalling each valve, make sure that it moves freely in its bore. **_Note:_** Always keep subassembly parts grouped together in a logical sequence. This will greatly aid reassembly of parts.

_____ 8. Reassemble the valve body and subassemblies.

_____ 9. Reinstall the valve body. When doing this, use assembly lubricant to retain the check-balls. **CAUTION:** Do not use gasket cement when reinstalling the valve body.

(continued)

_____ **10.** Tighten the valve body attaching bolts to the manufacturer's specifications.

   Source: _____ _Page:_ _____

_____ **11.** Safely lower the vehicle to the ground.

_____ **12.** Fill the vehicle with the proper amount and type of automatic transmission fluid.

Performance ✓ Checklist

Name _____ Date _____ Class _____

PERFORMANCE STANDARDS:
Level 4–Performs skill without supervision and adapts to problem situations.
Level 3–Performs skill satisfactorily without assistance or supervision.
Level 2–Performs skill satisfactorily, but requires assistance/supervision.
Level 1–Performs parts of skill satisfactorily, but requires considerable assistance/supervision.

Attempt (circle one): **1 2 3 4**

Comments:

_____

_____

_____

_____

PERFORMANCE LEVEL ACHIEVED: _____

_____ **1.** Safety rules and practices were followed at all times regarding this job.

_____ **2.** Tools and equipment were used properly and stored upon completion of this job.

_____ **3.** This completed job met the standards set and was done within the allotted time.

_____ **4.** No injury or damage to property occurred during this job.

_____ **5.** Upon completion of this job, the work area was cleaned correctly.

Instructor's Signature _____ Date_____

INSPECTING CASE BORES, PASSAGES, BUSHINGS, VENTS, AND MATING SURFACES

NATEF Standard(s) for Automatic Transmission & Transaxle:
D3-6 Inspect case bores, passages, bushings, vents, and mating surfaces; determine necessary action.

Safety First
- ☐ Wear safety glasses at all times.
- ☐ Follow all safety rules when using common hand tools.

Tools & Equipment:
- Vehicle service information
- Feeler gauge
- Common hand tools
- Dial bore gauge

PROCEDURES Refer to the vehicle service information for specifications and special procedures. Then inspect case bores, passages, bushings, vents, and mating surfaces on the vehicle provided by your instructor.

_____ 1. Write up a repair order.

_____ 2. Make sure you follow all procedures in the vehicle service information.

_____ 3. Inspect the transmission case bores, passages, bushings, and vents for signs of metal burrs, cracks, or foreign material.

_____ 4. Inspect mating surfaces for signs of distortion, warping, cracking, or metal burrs.

_____ 5. Dress any metal burrs according to the manufacturer's specifications. Replace any parts that are cracked, distorted, or warped.

_____ 6. After inspection, clean the parts with a suitable cleaning agent, such as a parts cleaning solution.

_____ 7. Make sure all cleaning agents are wiped from the part using paper rags only. Check that the parts are free of any dirt or cloth particles. Allow the part to dry before reassembly.

Automatic Transmission & Transaxle

(continued)

Performance ✓ Checklist

Name _____ Date _____ Class _____

PERFORMANCE STANDARDS:
Level 4–Performs skill without supervision and adapts to problem situations.
Level 3–Performs skill satisfactorily without assistance or supervision.
Level 2–Performs skill satisfactorily, but requires assistance/supervision.
Level 1–Performs parts of skill satisfactorily, but requires considerable assistance/supervision.

Attempt (circle one): **1 2 3 4**

Comments:

_____

_____

_____

_____

PERFORMANCE LEVEL ACHIEVED: _____

_____ 1. Safety rules and practices were followed at all times regarding this job.

_____ 2. Tools and equipment were used properly and stored upon completion of this job.

_____ 3. This completed job met the standards set and was done within the allotted time.

_____ 4. No injury or damage to property occurred during this job.

_____ 5. Upon completion of this job, the work area was cleaned correctly.

Instructor's Signature _____ Date_____

INSPECTING, ADJUSTING, OR REPLACING
THROTTLE (TV) LINKAGES

NATEF Standard(s) for Automatic Transmission & Transaxle:

B1 Inspect, adjust, or replace throttle valve (TV) linkages or cables; manual shift linkages or cables; transmission range sensor; check gear select indicator (as applicable).

Safety First

☐ Wear safety glasses at all times.

☐ Follow all safety rules when using common hand tools.

☐ Always connect a vehicle's exhaust to a vent hose before you run an engine in a closed shop. Unvented exhaust fumes in a closed shop can cause death.

☐ Keep jewelry, loose clothing, and hair away from moving parts while an engine is running.

☐ Be careful of hot engine parts.

Tools & Equipment:
• Common hand tools
• Vehicle service information
• Pressure gauge

PROCEDURES Refer to the vehicle service information for specifications and special procedures. Then inspect, adjust, or replace the throttle (TV) linkages in the transmission provided by your instructor.

_____ 1. Write up a repair order.

_____ 2. Make sure you follow all procedures in the vehicle service information.

_____ 3. With the engine off, check to ensure that the throttle is closed. *Note:* TV linkage inspections are normally made with the engine off. However, if there are indications that the linkage binds only with the engine running, some of the checks should be repeated with the engine running.

_____ 4. Check the cable or rod for any slack or free movement. There should be none.

_____ 5. Fully open the throttle and check the cable or rod for free or sloppy movement, binding, or sticking. The TV linkage should move to wide open and release freely and smoothly.

_____ 6. Try moving the TV linkage farther with the throttle fully open. It should move only a very small amount before contacting its stop.

_____ 7. When inspecting TV linkages with the engine running, always set the parking brake and shift the gear selector to NEUTRAL.

_____ 8. Pull the end of the TV cable or move the TV rod so that it moves through its travel, and then release it. During the linkage movement, check for free or sloppy movement, sticking, or binding. There should be none.

(continued)

Automatic Transmission & Transaxle

_____ 9. Many times it is possible to drop the pan and note the position of the TV plunger relative to the throttle position. The TV plunger should move directly with the throttle.

_____ 10. Move the throttle lever while watching the kickdown rod or cable, if used. If the rod or cable is locked in the full-throttle position, the transmission will be slow to upshift.

_____ 11. The most exact check for proper TV linkage adjustment is made using a pressure gauge installed in the transmission. *Note:* All TV linkage has an adjustment built into it. However, there is little standardization among manufacturers. The adjustment mechanism may be a sliding swivel, snap lock, or self-adjusting slider with lock tabs.

_____ 12. After adjustment, always repeat the linkage check to ensure correct adjustment.

_____ 13. When the TV linkage is rusty, worn, or frayed, replace it.

Performance ✓ Checklist

Name _____ Date _____ Class _____

PERFORMANCE STANDARDS:
Level 4–Performs skill without supervision and adapts to problem situations.
Level 3–Performs skill satisfactorily without assistance or supervision.
Level 2–Performs skill satisfactorily, but requires assistance/supervision.
Level 1–Performs parts of skill satisfactorily, but requires considerable assistance/supervision.

Attempt (circle one): **1 2 3 4**

Comments:

_____

_____

_____

_____

PERFORMANCE LEVEL ACHIEVED: _____

_____ 1. Safety rules and practices were followed at all times regarding this job.

_____ 2. Tools and equipment were used properly and stored upon completion of this job.

_____ 3. This completed job met the standards set and was done within the allotted time.

_____ 4. No injury or damage to property occurred during this job.

_____ 5. Upon completion of this job, the work area was cleaned correctly.

Instructor's Signature _____ Date_____

DIAGNOSING ELECTRONIC, MECHANICAL, HYDRAULIC, AND VACUUM CONTROL SYSTEM CONCERNS

NATEF Standard(s) for Automatic Transmission & Transaxle:
A9 Diagnose electronic, mechanical, hydraulic, vacuum control system concerns; determine necessary action.

Safety First

- ☐ Wear safety glasses at all times.
- ☐ Follow all safety rules when using common hand tools.
- ☐ Use exhaust vent if running engine in closed shop. Unvented exhaust fumes can cause death.
- ☐ Watch out for hot or moving parts.
- ☐ Follow all safety rules when using an automatic lift.

Tools & Equipment:
- Vehicle service information
- Automotive lift
- Common hand tools
- Drop light
- Scan tool
- DVOM

PROCEDURES Refer to the vehicle service information for specifications and special procedures. Then diagnose the causes of electronic, mechanical, hydraulic, and vacuum control system concerns in the vehicle provided by your instructor.

_____ 1. Write up a repair order.

_____ 2. Make sure you follow all procedures in the vehicle service information.

_____ 3. Use a scan tool to identify control system problems as necessary.

_____ 4. Check for opens, shorts, or unintentional grounds in wiring circuits as necessary.

_____ 5. Check for proper operation of sensors as necessary.

_____ 6. Check for proper operation of solenoid valves as necessary.

_____ 7. Check for proper operation of shift linkage as necessary.

_____ 8. Check for proper operation of throttle valve linkage as necessary.

_____ 9. Check for proper operation of vacuum modulators and vacuum circuits as necessary.

_____ 10. Check for proper operation of governor as necessary.

Automatic Transmission & Transaxle

(continued)

Performance ✓ Checklist

Name _____   Date _____   Class _____

PERFORMANCE STANDARDS:
Level 4–Performs skill without supervision and adapts to problem situations.
Level 3–Performs skill satisfactorily without assistance or supervision.
Level 2–Performs skill satisfactorily, but requires assistance/supervision.
Level 1–Performs parts of skill satisfactorily, but requires considerable assistance/supervision.

Attempt (circle one): **1 2 3 4**

Comments:

_____

_____

_____

_____

PERFORMANCE LEVEL ACHIEVED: _____

_____ 1. Safety rules and practices were followed at all times regarding this job.

_____ 2. Tools and equipment were used properly and stored upon completion of this job.

_____ 3. This completed job met the standards set and was done within the allotted time.

_____ 4. No injury or damage to property occurred during this job.

_____ 5. Upon completion of this job, the work area was cleaned correctly.

Instructor's Signature _____   Date_____

REMOVING & REPLACING SCREW-IN AND PUSH-IN VACUUM MODULATORS

NATEF Standard(s) for Automatic Transmission & Transaxle:
C1 Inspect, adjust, or replace (as applicable) vacuum modulator; inspect and repair or replace lines and hoses.

Safety First
- ☐ Wear safety glasses at all times.
- ☐ Follow all safety rules when using common hand tools.
- ☐ Follow all safety rules when using a lift.

Tools & Equipment:
- Lift
- Vehicle service information
- Special torquing modulator wrench
- Other tools and equipment as required
- Common hand tools

PROCEDURES Refer to the vehicle service information for specifications and special procedures. Then remove and replace the vacuum modulator in the transmission provided by your instructor.

_____ 1. Write up a repair order.

_____ 2. Make sure you follow all procedures in the vehicle service information.

_____ 3. Safely raise the vehicle on a lift.

Remove and Replace a Screw-in Vacuum Modulator

_____ 1. To remove a screw-in vacuum modulator, disconnect the lines from the modulator vacuum port(s).

_____ 2. Using the tool recommended in the vehicle service information for breaking the seal without damaging the modulator, unscrew the modulartor unit.

_____ 3. Carefully withdraw the control rod or valve from the transmission case.

_____ 4. To replace a screw-in vacuum modulator, install the control rod or valve in the transmission case.

_____ 5. Thread the modulator into the case. Then use a modulator wrench that's recommended by the manufacturer to tighten the modulator to proper torque specifications.

   Source: _____ *Page:* _____

_____ 6. Connect the vacuum line(s) to the modulator port(s).

(continued)

Automatic Transmission & Transaxle

Remove and Replace a Push-in Vacuum Modulator

_____ 1. To remove a push-in vacuum modulator, disconnect the lines from the modulator vacuum port(s).

_____ 2. Remove the retaining bolt and bracket. **CAUTION:** Do not pry or bend the bracket. This can distort the modulator unit and damage the diaphragm inside.

_____ 3. Remove the modulator from the transmission case by pulling straight out.

_____ 4. Carefully withdraw the control rod or valve from the transmission case.

_____ 5. To replace a push-in vacuum modulator, install the control rod or valve in the transmission case.

_____ 6. Insert the modulator sleeve into the case, carefully working the O-ring into the bore and pushing the unit into the case as far as it will go.

_____ 7. Position the retainer bracket and install the attaching bolt. Torque the attaching bolt to manufacturer's specifications.

Source: _____ _Page:_ _____

_____ 8. Connect the vacuum line(s) to the modulator vacuum port(s).

_____ 9. Safely lower the vehicle to the ground.

Performance ✓ Checklist

Name _____ Date _____ Class _____

PERFORMANCE STANDARDS:
Level 4–Performs skill without supervision and adapts to problem situations.
Level 3–Performs skill satisfactorily without assistance or supervision.
Level 2–Performs skill satisfactorily, but requires assistance/supervision.
Level 1–Performs parts of skill satisfactorily, but requires considerable assistance/supervision.

Attempt (circle one): 1 2 3 4

Comments:

_____

_____

_____

_____

PERFORMANCE LEVEL ACHIEVED: _____

_____ 1. Safety rules and practices were followed at all times regarding this job.
_____ 2. Tools and equipment were used properly and stored upon completion of this job.
_____ 3. This completed job met the standards set and was done within the allotted time.
_____ 4. No injury or damage to property occurred during this job.
_____ 5. Upon completion of this job, the work area was cleaned correctly.

Instructor's Signature _____ Date_____

TESTING FOR MANIFOLD VACUUM, DIAPHRAGM LEAKAGE, AND MODULATOR SPRING PRESSURE

NATEF Standard(s) for Automatic Transmission & Transaxle:
C1 Inspect, adjust, or replace (as applicable) vacuum modulator; inspect and repair or replace lines and hoses.

Safety First

☐ Wear safety glasses at all times.
☐ Follow all safety rules when using common hand tools.
☐ Follow all safety rules when using a lift.
☐ Always connect a vehicle's exhaust to a vent hose before you run an engine in a closed shop. Unvented exhaust fumes in a closed shop can cause death.
☐ Be careful of hot engine parts.

Tools & Equipment:
- Lift
- Other tools and equipment as required
- Vacuum gauge
- Hand-operated vacuum pump
- Common hand tools
- Vehicle service information

PROCEDURES Refer to the vehicle service information for specifications and special procedures. Then test the manifold vacuum, diaphragm, and modulator spring pressure in the transmission provided by your instructor.

_____ 1. Write up a repair order.

_____ 2. Make sure you follow all procedures in the vehicle service information.

_____ 3. Safely raise the vehicle on a lift.

Test for Manifold Vacuum

_____ 1. Disconnect the vacuum line from the modulator port and connect a vacuum gauge to the vacuum line.

_____ 2. Start the engine and check vacuum-gauge readings against the vehicle service information for acceptable vacuum limits.

Source: _____ Page: _____

_____ 3. Accelerate the engine momentarily. The vacuum gauge should show a rapid drop.

_____ 4. Release the accelerator. The vacuum gauge should recover to its idle reading.

(continued)

Automatic Transmission & Transaxle

_____ 5. If the vacuum gauge reading does not change or if it changes very slowly during Steps 3 and 4, check for plugged, restricted, leaking, or incorrectly connected vacuum line or plugged manifold or restricted exhaust system.

Test for Diaphragm Leakage (Hand-Pump Method)

_____ 1. Disconnect the vacuum line at the modulator port. *Note:* If transmission fluid is present in the hose or modulator port, replace the unit.

_____ 2. Connect the hand-operated vacuum pump to the modulator port.

_____ 3. Apply approximately 17 to 20 inches of vacuum to the diaphragm and close the shutoff on the pump.

_____ 4. Replace the modulator if it does not hold vacuum for at least 30 to 60 seconds.

_____ 5. Safely lower the vehicle to the ground.

Performance ✓ Checklist

Name _____ Date _____ Class _____

PERFORMANCE STANDARDS:
Level 4–Performs skill without supervision and adapts to problem situations.
Level 3–Performs skill satisfactorily without assistance or supervision.
Level 2–Performs skill satisfactorily, but requires assistance/supervision.
Level 1–Performs parts of skill satisfactorily, but requires considerable assistance/supervision.

Attempt (circle one): **1 2 3 4**

Comments:

_____

_____

_____

_____

PERFORMANCE LEVEL ACHIEVED: _____

_____ 1. Safety rules and practices were followed at all times regarding this job.
_____ 2. Tools and equipment were used properly and stored upon completion of this job.
_____ 3. This completed job met the standards set and was done within the allotted time.
_____ 4. No injury or damage to property occurred during this job.
_____ 5. Upon completion of this job, the work area was cleaned correctly.

Instructor's Signature _____ Date_____

INSPECTING & REPLACING THE GOVERNOR ASSEMBLY

NATEF Standard(s) for Automatic Transmission & Transaxle:
C2 Inspect, repair, and replace governor assembly.

Safety First

- ☐ Wear safety glasses at all times.
- ☐ Follow all safety rules when using common hand tools.
- ☐ Check Material Safety Data Sheets for chemical safety information.
- ☐ Follow all safety rules when using compressed air.
- ☐ Follow all safety rules when using a lift.

Tools & Equipment:
- Lift
- Other tools and equipment as required
- Cleaning solution
- Common hand tools
- Air compressor
- Vehicle service information

PROCEDURES Refer to the vehicle service information for specifications and special procedures. Then inspect and replace the governor assembly in the transmission provided by your instructor.

_____ 1. Write up a repair order.

_____ 2. Make sure you follow all procedures in the vehicle service information.

_____ 3. Safely raise the vehicle on a lift.

_____ 4. Following the manufacturer's specifications, inspect the governor for the following defects:

    _____ **a.** Scored or rusted valve bore; a scored, nicked, or rusted spool valve

    _____ **b.** Damaged or plugged governor screen (the screen may be located in the case or in the main valve body)

    _____ **c.** Damaged or distorted mating surfaces

    _____ **d.** Binding, sticking, or damaged weights (the weights and valve should move freely up and down)

    _____ **e.** Broken or distorted springs

    _____ **f.** Cut, distorted, or loose seals

    _____ **g.** Broken snap rings and damaged snap-ring grooves

    _____ **h.** Plugged or restricted fluid passages in the case or output shaft

    _____ **i.** Damaged or excessively loose driven gear

(continued)

Automatic Transmission & Transaxle

_____ **5.** If any defects are found, repair or replace the governor as an assembly.

_____ **6.** If the governor has a drive ball, remove it with a magnet so that the ball will not fall free.

_____ **7.** If necessary, tap the governor assembly with a plastic-tipped hammer to loosen it from the shaft.

_____ **8.** Wash the governor in a parts-cleaning solution and blow out all the passages.

_____ **9.** Replace all seals and O-rings.

_____ **10.** Put the governor (repaired or new) back in the vehicle.

_____ **11.** Safely lower the vehicle to the ground.

_____ **12.** Check vehicle shift speeds.

_____ **13.** Return transmission fluid level to the full mark.

Performance ✓ Checklist

Name _____ Date _____ Class _____

PERFORMANCE STANDARDS:
Level 4–Performs skill without supervision and adapts to problem situations.
Level 3–Performs skill satisfactorily without assistance or supervision.
Level 2–Performs skill satisfactorily, but requires assistance/supervision.
Level 1–Performs parts of skill satisfactorily, but requires considerable assistance/supervision.

Attempt (circle one): **1 2 3 4**

Comments:

_____

_____

_____

_____

PERFORMANCE LEVEL ACHIEVED: _____

_____ **1.** Safety rules and practices were followed at all times regarding this job.

_____ **2.** Tools and equipment were used properly and stored upon completion of this job.

_____ **3.** This completed job met the standards set and was done within the allotted time.

_____ **4.** No injury or damage to property occurred during this job.

_____ **5.** Upon completion of this job, the work area was cleaned correctly.

Instructor's Signature _____ Date_____

INSPECTING & REPLACING SPRAG AND ROLLER CLUTCHES

NATEF Standard(s) for Automatic Transmission & Transaxle:
D4-4 Inspect roller and sprag clutches, races, rollers, sprags, springs, cages, and retainers; replace as needed.

Safety First
- ☐ Wear safety glasses at all times.
- ☐ Follow all safety rules when using common hand tools.

Tools & Equipment:
- Overhauled sprag and roller clutches
- Common hand tools
- Vehicle service information

PROCEDURES Refer to the vehicle service information for specifications and special procedures. Then inspect and replace the sprag and roller clutches in the transmission provided by your instructor.

_____ 1. Write up a repair order.

_____ 2. Make sure you follow all procedures in the vehicle service information.

_____ 3. Inspect one-way clutches for the following defects:

_____ a. Excessively worn or damaged rollers or sprags and springs

_____ b. Bent or damaged spring retainers in sprag clutch assemblies

_____ c. Flat spots or chipped edges on clutch rollers

_____ d. Dents or grooves (brinelling) on roller contact surfaces of clutch races

_____ e. Distorted snap rings

_____ f. Plugged or restricted lubrication holes in the inner clutch races

_____ 4. If any part of the clutch is damaged, the entire clutch assembly should be replaced with a new one by following the vehicle service information's instructions and cautions.

Source: _____ _Page:_ _____

Automatic Transmission & Transaxle

(continued)

Performance ✓ Checklist

Name _____ Date _____ Class _____

PERFORMANCE STANDARDS:
Level 4–Performs skill without supervision and adapts to problem situations.
Level 3–Performs skill satisfactorily without assistance or supervision.
Level 2–Performs skill satisfactorily, but requires assistance/supervision.
Level 1–Performs parts of skill satisfactorily, but requires considerable assistance/supervision.

Attempt (circle one): **1 2 3 4**

Comments:

_____

_____

_____

_____

PERFORMANCE LEVEL ACHIEVED: _____

_____ **1.** Safety rules and practices were followed at all times regarding this job.

_____ **2.** Tools and equipment were used properly and stored upon completion of this job.

_____ **3.** This completed job met the standards set and was done within the allotted time.

_____ **4.** No injury or damage to property occurred during this job.

_____ **5.** Upon completion of this job, the work area was cleaned correctly.

Instructor's Signature _____ Date_____

INSPECTING CLUTCH ASSEMBLIES

NATEF Standard(s) for Automatic Transmission & Transaxle:
D4-1 Inspect clutch drum, piston, check-balls, springs, retainers, seals, and friction and pressure plates; determine necessary action.

Safety First

- ☐ Wear safety glasses at all times.
- ☐ Follow all safety rules when using common hand tools.
- ☐ Use caution when using compressed air.
- ☐ Use caution when using solvents.
- ☐ Follow all OSHA and EPA guidelines when using solvents.

Tools & Equipment:

- Vehicle service information
- Compressed air supply and blow gun
- Common hand tools
- Lint-free towels
- Feeler gauge
- Cleaning solvent
- Straightedge
- Clutch piston spring compressor

PROCEDURES Refer to the vehicle service information for specifications and special procedures. Then inspect the clutch assembly on the vehicle provided by your instructor.

_____ 1. Write up a repair order.

_____ 2. Make sure you follow all procedures in the vehicle service information.

_____ 3. Inspect condition of steel plates and friction discs.

_____ 4. Use a straightedge and feeler gauge to check proper clutch pack clearances.

_____ 5. Check for worn splines and bushings.

_____ 6. Check for a porous or cracked piston.

_____ 7. Check for a missing check-ball.

_____ 8. Check for weak springs and/or bent retainers.

_____ 9. Check condition of all seals.

(continued)

Automatic Transmission & Transaxle

Performance ✓ Checklist

Name _____ Date _____ Class _____

PERFORMANCE STANDARDS:
Level 4–Performs skill without supervision and adapts to problem situations.
Level 3–Performs skill satisfactorily without assistance or supervision.
Level 2–Performs skill satisfactorily, but requires assistance/supervision.
Level 1–Performs parts of skill satisfactorily, but requires considerable assistance/supervision.

Attempt (circle one): **1 2 3 4**

Comments:

_____

_____

_____

_____

PERFORMANCE LEVEL ACHIEVED: _____

_____ **1.** Safety rules and practices were followed at all times regarding this job.

_____ **2.** Tools and equipment were used properly and stored upon completion of this job.

_____ **3.** This completed job met the standards set and was done within the allotted time.

_____ **4.** No injury or damage to property occurred during this job.

_____ **5.** Upon completion of this job, the work area was cleaned correctly.

Instructor's Signature _____ Date _____

REMOVING & REPAIRING OR REPLACING A SERVO ASSEMBLY

NATEF Standard(s) for Automatic Transmission & Transaxle:
D1-5 Inspect servo bore, piston, seals, pin, spring, and retainers; determine necessary action.

Safety First
- ☐ Wear safety glasses at all times.
- ☐ Follow all safety rules when using common hand tools.
- ☐ Follow all safety rules when using a lift.
- ☐ Check Material Safety Data Sheets for chemical safety information.

Tools & Equipment:
- Lift
- Vehicle service information
- Cleaning fluid
- Common hand tools

PROCEDURES Refer to the vehicle service information for specifications and special procedures. Then remove and repair or replace the servo assembly in the transmission provided by your instructor.

_____ 1. Write up a repair order.

_____ 2. Make sure you follow all procedures in the vehicle service information.

_____ 3. Safely raise the vehicle on a lift.

_____ 4. Safely remove the servo assembly. **CAUTION:** The servo assembly is under spring tension, so be careful that the cover and servo parts do not fly loose when removed.

_____ 5. Keep the subassembly parts grouped together in a logical sequence. This will greatly aid in reassembly.

_____ 6. Inspect the servo piston bore for scoring or damage. If the bore wall is scored, replace the transmission case.

_____ 7. Check the piston and piston rod for nicks, scoring, and burrs. Repair or replace either of these parts if you find these problems.

_____ 8. Replace any springs that are damaged, collapsed, or distorted.

_____ 9. Inspect the mating surfaces between the piston and bore wall for nicks, scoring, or damage. Make sure that the piston moves freely in the bore.

_____ 10. Check fluid passages for any restrictions. Remove any debris and clean passages thoroughly with cleaning fluid.

_____ 11. Inspect the fluid passages for leaks.

(continued)

Automatic Transmission & Transaxle

_____ 12. Remove the seals and replace them with new ones. *Note:* Some piston and seal assemblies must be replaced as units.

_____ 13. Check the length of the apply pin against the vehicle service information's specifications.

Source: _____ *Page:* _____

_____ 14. Clean the servo parts with the appropriate cleaning fluid and let them air dry.

_____ 15. Reassemble the servo assembly by reversing the order of removal.

_____ 16. Safely lower the vehicle to the ground.

Performance ✓ Checklist

Name _____ Date _____ Class _____

PERFORMANCE STANDARDS:
Level 4–Performs skill without supervision and adapts to problem situations.
Level 3–Performs skill satisfactorily without assistance or supervision.
Level 2–Performs skill satisfactorily, but requires assistance/supervision.
Level 1–Performs parts of skill satisfactorily, but requires considerable assistance/supervision.

Attempt (circle one): **1 2 3 4**

Comments:

_____

_____

_____

_____

PERFORMANCE LEVEL ACHIEVED: _____

_____ 1. Safety rules and practices were followed at all times regarding this job.
_____ 2. Tools and equipment were used properly and stored upon completion of this job.
_____ 3. This completed job met the standards set and was done within the allotted time.
_____ 4. No injury or damage to property occurred during this job.
_____ 5. Upon completion of this job, the work area was cleaned correctly.

Instructor's Signature _____ Date_____

INSPECTING AN ACCUMULATOR

NATEF Standard(s) for Automatic Transmission & Transaxle:

D1-6 Inspect accumulator bore, piston, seals, spring, and retainer; determine necessary action.

Safety First
- ☐ Wear safety glasses at all times.
- ☐ Follow all safety rules when using common hand tools.
- ☐ Follow all safety rules when using a lift.
- ☐ Check Material Safety Data Sheets for chemical safety information.

Tools & Equipment:
- Lift
- Common hand tools
- Cleaning fluid
- Vehicle service information
- Replacement parts as appropriate

PROCEDURES Refer to the vehicle service information for specifications and special procedures. Then inspect the accumulator in the transmission provided by your instructor.

_____ 1. Write up a repair order.

_____ 2. Make sure you follow all procedures in the vehicle service information.

_____ 3. Safely raise the vehicle on a lift. *Note:* Accumulators may not be accessible without partially disassembling the transmission/transaxle.

_____ 4. Inspect the transmission case bore for scoring and damage. *Note:* Replace the transmission case if the transmission bore walls are scored.

_____ 5. Check the piston and piston rod for nicks, scoring, burrs, or worn areas. Replace them if they are damaged.

_____ 6. Test the spring for strong, positive action. Replace the spring if it is weak.

_____ 7. Inspect the spring for damage, collapse, or distortion. Replace it if any of these conditions exists.

_____ 8. Check the mating surfaces between the piston and bore wall. The surfaces should be smooth and not restrict piston movement.

_____ 9. Inspect the fluid passages for restrictions.

_____ 10. Remove any debris from the fluid passages and clean them thoroughly with cleaning fluid.

_____ 11. Safely lower the vehicle to the ground.

Automatic Transmission & Transaxle

(continued)

Performance ✓ Checklist

Name _____ Date _____ Class _____

PERFORMANCE STANDARDS:
Level 4–Performs skill without supervision and adapts to problem situations.
Level 3–Performs skill satisfactorily without assistance or supervision.
Level 2–Performs skill satisfactorily, but requires assistance/supervision.
Level 1–Performs parts of skill satisfactorily, but requires considerable assistance/supervision.

Attempt (circle one): **1 2 3 4**

Comments:

_____

_____

_____

_____

PERFORMANCE LEVEL ACHIEVED: _____

_____ 1. Safety rules and practices were followed at all times regarding this job.

_____ 2. Tools and equipment were used properly and stored upon completion of this job.

_____ 3. This completed job met the standards set and was done within the allotted time.

_____ 4. No injury or damage to property occurred during this job.

_____ 5. Upon completion of this job, the work area was cleaned correctly.

Instructor's Signature _____ Date _____

INSPECTING & ADJUSTING BANDS AND REPLACING DRUMS

NATEF Standard(s) for Automatic Transmission & Transaxle:
D4-5 Inspect bands and drums; determine necessary action.

Safety First

☐ Wear safety glasses at all times.
☐ Follow all safety rules when using common hand tools.
☐ Check Material Safety Data Sheets for chemical safety information.

Tools & Equipment:
- Common hand tools
- Torque wrench (¼" drive)
- Automatic transmission fluid (ATF)
- Automatic transmission bands
- Vehicle service information

PROCEDURES Refer to the vehicle service information for specifications and special procedures. Then inspect and adjust the bands and replace the drums in the transmission provided by your instructor.

_____ 1. Write up a repair order.

_____ 2. Make sure you follow all procedures in the vehicle service information.

Inspecting Bands

_____ 1. Wipe each band clean with a dry, lint-free cloth. **CAUTION:** Do not force open thin flex-type bands any farther than their normal relaxed position. Opening them beyond that point could cause the lining friction material to crack.

_____ 2. Inspect each band for the following defects, paying particular attention to the ends (they tend to wear more there because they apply there first). Look for distortion; discolored or checked lining; cracked ends; excessive or uneven wear; burn marks, charring, or glazed surfaces; poor lining bond to band; and flaking and pitting.

_____ 3. Scratch the lining surface with your thumbnail. Replace the band if lining material builds up on top of your nail.

_____ 4. Replace any defective band with a new one.

_____ 5. Before installing new bands, soak them in ATF for the length of time recommended in the vehicle service information. _Note:_ Many automatic transmission specialists recommend soaking the clutch pack for 30 minutes in a tray of ATF.

Source: _____ _Page:_____

(_continued_)

Automatic Transmission & Transaxle

_____ 6. Follow the vehicle service information's instructions for the installation of new bands.

Source: _____ *Page:* _____

Adjusting Bands and Replacing Drums

_____ 1. Adjust the bands as part of in-vehicle service or as part of a transmission overhaul. Most bands can only be adjusted during overhaul or by replacing the servo pins.

_____ 2. If the surface of a drum is grooved or scored, replace it by following the procedures in the vehicle service information.

Source: _____ *Page:* _____

_____ 3. If a drum's surface is glazed, replace it by following the manufacturer's specifications.

Source: _____ *Page:* _____

Performance ✓ Checklist

Name _____ Date _____ Class _____

PERFORMANCE STANDARDS:
Level 4–Performs skill without supervision and adapts to problem situations.
Level 3–Performs skill satisfactorily without assistance or supervision.
Level 2–Performs skill satisfactorily, but requires assistance/supervision.
Level 1–Performs parts of skill satisfactorily, but requires considerable assistance/supervision.

Attempt (circle one): **1 2 3 4**

Comments:

_____

_____

_____

_____

PERFORMANCE LEVEL ACHIEVED: _____

_____ 1. Safety rules and practices were followed at all times regarding this job.
_____ 2. Tools and equipment were used properly and stored upon completion of this job.
_____ 3. This completed job met the standards set and was done within the allotted time.
_____ 4. No injury or damage to property occurred during this job.
_____ 5. Upon completion of this job, the work area was cleaned correctly.

Instructor's Signature _____ Date_____

MEASURING & ADJUSTING CLUTCH PACK CLEARANCE

NATEF Standard(s) for Automatic Transmission & Transaxle:
D4-2 Measure clutch pack clearance; determine necessary action.

Safety First

☐ Wear safety glasses at all times.
☐ Follow all safety rules when using common hand tools.

Tools & Equipment:
- Feeler gauge
- Common hand tools
- Automatic transmission overhaul kit
- Vehicle service information
- Multiple-disc clutch assembly

PROCEDURES Refer to the vehicle service information for specifications and special procedures. Then measure and adjust the clutch pack clearance in the transmission provided by your instructor.

_____ 1. Write up a repair order.

_____ 2. Make sure you follow all procedures in the vehicle service information.

Adjust Clearance (Outer Snap-Ring Method)

_____ 1. Install the clutch pack and pressure plate.

_____ 2. Use a feeler gauge to check the distance between the pressure plate and the outer snap ring. Compare this reading to the manufacturer's specifications.

   Source: _____ *Page:* _____

_____ 3. If the clearance is greater than specified, install a thicker snap ring to take up this extra clearance.

_____ 4. If the clearance is not great enough, install a thinner snap ring to increase this clearance. ***Note:*** On vehicles that have a wave-type outer snap ring, place the feeler gauge between the flat pressure plate and the wave in the snap ring farthest from the pressure plate.

(continued)

Automatic Transmission & Transaxle

Adjust Clearance (Varying Clutch Pressure Plate Method)

_____ 1. Install the clutch pack and pressure plate.

_____ 2. Use a feeler gauge to check the distance between the pressure plate and the outer snap ring. Compare this reading to the manufacturer's specifications.

Source: _____ Page:_____

_____ 3. If the clearance is greater than specified, install a thicker pressure plate to take up this extra clearance.

_____ 4. If the clearance is not great enough, install a thinner pressure plate to increase this clearance.

Performance ✓ Checklist

Name _____ Date _____ Class _____

PERFORMANCE STANDARDS:
Level 4–Performs skill without supervision and adapts to problem situations.
Level 3–Performs skill satisfactorily without assistance or supervision.
Level 2–Performs skill satisfactorily, but requires assistance/supervision.
Level 1–Performs parts of skill satisfactorily, but requires considerable assistance/supervision.

Attempt (circle one): 1 2 3 4

Comments:

_____

_____

_____

_____

PERFORMANCE LEVEL ACHIEVED: _____

_____ 1. Safety rules and practices were followed at all times regarding this job.

_____ 2. Tools and equipment were used properly and stored upon completion of this job.

_____ 3. This completed job met the standards set and was done within the allotted time.

_____ 4. No injury or damage to property occurred during this job.

_____ 5. Upon completion of this job, the work area was cleaned correctly.

Instructor's Signature _____ Date_____

AIR TESTING THE CLUTCH PACK AND INTERNAL SERVO ASSEMBLIES

NATEF Standard(s) for Automatic Transmission & Transaxle:
D4-3 Air test operation of clutch and servo assemblies.

Safety First
- ☐ Wear safety glasses at all times.
- ☐ Follow all safety rules when using common hand tools.
- ☐ Follow all safety rules when using compressed air.

Tools & Equipment:
- Transmission or transaxle
- Vehicle service information
- Bench vise
- Common hand tools
- Compressed air gun with conical nozzle

PROCEDURES Refer to the vehicle service information for specifications and special procedures. Then air test the clutch pack and internal servo assemblies in the transmission provided by your instructor.

Note: Some manufacturers require that an air test plate be installed onto the transmission case to perform clutch and servo tests. These plates are special tools. Check the vehicle service information for applications requiring air test plates.

_____ 1. Write up a repair order.

_____ 2. Make sure you follow all procedures in the vehicle service information.

_____ 3. Assemble the clutch drums.

_____ 4. Place the pump assembly in a bench vise with the stator support facing down.

_____ 5. Assemble the clutch drums on the pump hub. *Note:* Slowly rotate the clutch drums back and forth to allow each friction disc to align with the splines on the clutch hub. The drums will drop as the inside teeth align with the clutch hub splines.

_____ 6. Adjust the compressed air supply to 25 psi.

_____ 7. Using a rubber-tipped blow gun, apply air into each hole in the pump housing. You should hear a solid "thud" with little air seepage. The clutch also can be felt as it applies. **CAUTION:** The clutch drums may separate when air is applied. This may damage the Teflon® seal rings. To prevent such separation, place a hand on top of the clutch drums while you are air checking them.

(continued)

Automatic Transmission & Transaxle

_____ 8. Servos can be tested in a similar manner. Assemble the clutch drums and bands in the transmission case. The drum should rotate freely.

_____ 9. Apply air pressure to the servo apply hole. The servo should extend, causing the band to tighten against the drum.

_____ 10. When the pressure is released, the band should loosen.

Performance ✓ Checklist

Name _____ Date _____ Class _____

PERFORMANCE STANDARDS:
Level 4–Performs skill without supervision and adapts to problem situations.
Level 3–Performs skill satisfactorily without assistance or supervision.
Level 2–Performs skill satisfactorily, but requires assistance/supervision.
Level 1–Performs parts of skill satisfactorily, but requires considerable assistance/supervision.

Attempt (circle one): 1 2 3 4

Comments:

_____

_____

_____

_____

PERFORMANCE LEVEL ACHIEVED: _____

_____ 1. Safety rules and practices were followed at all times regarding this job.

_____ 2. Tools and equipment were used properly and stored upon completion of this job.

_____ 3. This completed job met the standards set and was done within the allotted time.

_____ 4. No injury or damage to property occurred during this job.

_____ 5. Upon completion of this job, the work area was cleaned correctly.

Instructor's Signature _____ Date_____

DIAGNOSING ELECTRICAL/ELECTRONIC CONCERNS USING PRINCIPLES OF ELECTRICITY

NATEF Standard(s) for Automatic Transmission & Transaxle:
A13 Diagnose electrical/electronic concerns using principles of electricity (Ohm's law).

Safety First
- ☐ Wear safety glasses at all times.
- ☐ Follow all safety rules when using common hand tools.
- ☐ Follow all safety rules for working with electrical equipment.

Tools & Equipment:
- Vehicle service information
- Oscilloscope or equivalent
- Common hand tools
- Graphing multimeter (GMM) or equivalent
- Digital volt-ohm-meter (DVOM)

PROCEDURES Refer to the vehicle service information for specifications and special procedures. Then using the principles of electricity, diagnose the electrical/electronic concerns in the vehicle provided by your instructor.

_____ 1. Write up a repair order.

_____ 2. Make sure you follow all procedures in the vehicle service information.

_____ 3. Identify the faulty circuit by measuring for proper resistance and voltage drop and comparing to manufacturer's specifications.

_____ 4. Determine if the wiring or a sensor is faulty by either comparing sensor waveform to a known good waveform or reading incorrect voltage to or from sensor.

_____ 5. Verify proper voltage supply and ground to sensor.

_____ 6. Repair or replace faulty components.

Automatic Transmission & Transaxle

(continued)

Performance ✓ Checklist

Name _____ Date _____ Class _____

PERFORMANCE STANDARDS:
Level 4–Performs skill without supervision and adapts to problem situations.
Level 3–Performs skill satisfactorily without assistance or supervision.
Level 2–Performs skill satisfactorily, but requires assistance/supervision.
Level 1–Performs parts of skill satisfactorily, but requires considerable assistance/supervision.

Attempt (circle one): **1 2 3 4**

Comments:

_____

_____

_____

_____

PERFORMANCE LEVEL ACHIEVED: _____

_____ 1. Safety rules and practices were followed at all times regarding this job.

_____ 2. Tools and equipment were used properly and stored upon completion of this job.

_____ 3. This completed job met the standards set and was done within the allotted time.

_____ 4. No injury or damage to property occurred during this job.

_____ 5. Upon completion of this job, the work area was cleaned correctly.

Instructor's Signature _____ Date_____

TESTING TRANSMISSION/TRANSAXLE-RELATED ELECTRICAL/ELECTRONIC COMPONENTS

NATEF Standard(s) for Automatic Transmission & Transaxle:
C6 Diagnose electronic transmission control systems using a scan tool; determine necessary action.

Safety First

☐ Wear safety glasses at all times.
☐ Follow all safety rules when using common hand tools.
☐ Follow all safety rules when using a lift.
☐ Check Material Safety Data Sheets for chemical safety information.

Tools & Equipment:
- Lift
- Ohmmeter
- Vehicle service information
- Digital multimeter
- Electrical parts cleaner
- Scan tool
- Common hand tools

PROCEDURES Refer to the vehicle service information for specifications and special procedures. Then test the transmission/transaxle-related electrical/electronic components in the vehicle provided by your instructor.

_____ 1. Write up a repair order.

_____ 2. Make sure you follow all procedures in the vehicle service information.

_____ 3. Safely raise the vehicle on a lift.

_____ 4. Visually inspect the wiring for loose, damaged, or corroded connections.

_____ 5. Clean the connectors with electrical parts cleaner.

_____ 6. Check for diagnostic trouble codes (DTCs) using a scan tool. **CAUTION:** Consult the vehicle service information before performing these tests. They vary with manufacturer and model of transmission/transaxle.

_____ 7. Use a digital multimeter, if required, to check sensor operation.

_____ 8. When testing the wiring harness and/or connections, wiggle the wires to help pinpoint intermittent problems.

_____ 9. When testing for circuit continuity, check the wiring using an ohmmeter. Disconnect the wires at both ends of the circuit and attach an ohmmeter lead to both ends. If the circuit is complete, a continuity reading will show. An open circuit will result in an infinity reading.

Automatic Transmission & Transaxle

(continued)

_____ **10.** When testing for a short circuit, connect one ohmmeter lead to one end of the wire in question and the other lead to ground. A continuity reading suggests a short to ground. The circuit is not grounded if the ohmmeter shows infinity.

_____ **11.** Safely lower the vehicle to the ground.

Performance ✓ Checklist

Name _____ Date _____ Class _____

PERFORMANCE STANDARDS:
Level 4–Performs skill without supervision and adapts to problem situations.
Level 3–Performs skill satisfactorily without assistance or supervision.
Level 2–Performs skill satisfactorily, but requires assistance/supervision.
Level 1–Performs parts of skill satisfactorily, but requires considerable assistance/supervision.

Attempt (circle one): **1 2 3 4**

Comments:

_____

_____

_____

_____

PERFORMANCE LEVEL ACHIEVED: _____

_____ **1.** Safety rules and practices were followed at all times regarding this job.

_____ **2.** Tools and equipment were used properly and stored upon completion of this job.

_____ **3.** This completed job met the standards set and was done within the allotted time.

_____ **4.** No injury or damage to property occurred during this job.

_____ **5.** Upon completion of this job, the work area was cleaned correctly.

Instructor's Signature _____ Date_____

REMOVING & REPLACING SPEEDOMETER DRIVEN GEAR AND RETAINERS

NATEF Standard(s) for Automatic Transmission & Transaxle:
C5 Inspect and replace speedometer drive gear, driven gear, vehicle speed sensor (VSS), and retainers.

Safety First

☐ Wear safety glasses at all times.
☐ Follow all safety rules when using common hand tools.
☐ Follow all safety rules when using a lift.

Tools & Equipment:
- Common hand tools
- Lift
- Other tools and equipment as required
- Vehicle service information

PROCEDURES Refer to the vehicle service information for specifications and special procedures. Then remove and replace the speedometer driven gear and retainers in the transmission provided by your instructor.

_____ 1. Write up a repair order.

_____ 2. Make sure you follow all procedures in the vehicle service information.

_____ 3. Safely raise the vehicle on a lift.

_____ 4. Index the speedometer gear housing, if appropriate.

_____ 5. Remove the snap ring or bolt and lockplate holding the speedometer driven gear housing in the transmission.

_____ 6. Remove the speedometer gear housing from its hole in the side of the transmission.

_____ 7. Remove the gear from its housing and inspect the gear teeth for wear and damage. Also count the number of teeth on the gear.

_____ 8. Replace the old gear with one of the same size and having the exact number of teeth as the original.

_____ 9. Reinstall the speedometer housing in the transmission. Make sure you properly index the housing, if appropriate.

_____ 10. Secure the speedometer gear housing with the snap ring or bolt and lockplate.

_____ 11. Safely lower the vehicle to the ground.

Automatic Transmission & Transaxle

(continued)

Performance ✓ Checklist

Name _____ Date _____ Class _____

PERFORMANCE STANDARDS:
Level 4–Performs skill without supervision and adapts to problem situations.
Level 3–Performs skill satisfactorily without assistance or supervision.
Level 2–Performs skill satisfactorily, but requires assistance/supervision.
Level 1–Performs parts of skill satisfactorily, but requires considerable assistance/supervision.

Attempt (circle one): **1 2 3 4**

Comments:

_____

_____

_____

_____

PERFORMANCE LEVEL ACHIEVED: _____

_____ 1. Safety rules and practices were followed at all times regarding this job.

_____ 2. Tools and equipment were used properly and stored upon completion of this job.

_____ 3. This completed job met the standards set and was done within the allotted time.

_____ 4. No injury or damage to property occurred during this job.

_____ 5. Upon completion of this job, the work area was cleaned correctly.

Instructor's Signature _____ Date _____

REMOVING A TRANSMISSION

NATEF Standard(s) for Automatic Transmission & Transaxle:
D1-1 Remove and reinstall transmission and torque converter (rear-wheel drive).

Tools & Equipment:
- Lift
- Safety strap or chain
- Vehicle service information
- Large drain pan
- Air compressor
- Hot-solution parts-cleaning tank
- C-clamp
- Common hand tools
- Tilting-base transmission jack

PROCEDURES Refer to the vehicle service information for specifications and special procedures. Then remove the transmission from the vehicle provided by your instructor.

_____ 1. Write up a repair order.

_____ 2. Make sure you follow all procedures in the vehicle service information.

_____ 3. Open the hood and remove the distributor cap from engines having a rear-mounted distributor.

_____ 4. Disconnect the battery ground cable.

_____ 5. Raise the vehicle on a lift.

_____ 6. Remove exhaust as necessary for clearance.

_____ 7. Remove the transmission oil pan and drain fluid.

_____ 8. Remove the starter motor.

_____ 9. Remove the drive shaft(s). Use tape to secure the universal joint bearing caps.

_____ 10. Disconnect the vacuum hoses at the modulator and/or transfer case (if equipped).

_____ 11. Disconnect linkages, taking note of their location for reassembly.

_____ 12. Disconnect wiring at transmission.

_____ 13. Disconnect speedometer cable (if equipped).

_____ 14. Remove the transmission filler tube. **CAUTION:** Be careful not to damage it.

(continued)

Automatic Transmission & Transaxle

_____ 15. Using a safety chain, attach the transmission jack and raise the transmission slightly.

_____ 16. If required, remove rear transmission mount and crossmember.

_____ 17. Remove the torque converter splash shield. *Note:* Some vehicles require oil filter removal.

_____ 18. Using spray paint, mark the torque converter location.

_____ 19. Remove torque converter bolts.

_____ 20. Remove and plug the cooler lines.

_____ 21. Remove engine-to-transmission bolts using swivel sockets and a long extension.

_____ 22. Pull the transmission rearward and slowly lower it.

_____ 23. Transfer the transmission to a clean workbench. **CAUTION:** Be careful not to tip the jack on the floor. Since these components are heavy, this may require an assistant.

Performance ✓ Checklist

Name _____ Date _____ Class _____

PERFORMANCE STANDARDS:
Level 4–Performs skill without supervision and adapts to problem situations.
Level 3–Performs skill satisfactorily without assistance or supervision.
Level 2–Performs skill satisfactorily, but requires assistance/supervision.
Level 1–Performs parts of skill satisfactorily, but requires considerable assistance/supervision.

Attempt (circle one): **1 2 3 4**

Comments:

_____

_____

_____

_____

PERFORMANCE LEVEL ACHIEVED: _____

_____ 1. Safety rules and practices were followed at all times regarding this job.

_____ 2. Tools and equipment were used properly and stored upon completion of this job.

_____ 3. This completed job met the standards set and was done within the allotted time.

_____ 4. No injury or damage to property occurred during this job.

_____ 5. Upon completion of this job, the work area was cleaned correctly.

Instructor's Signature _____ Date_____

REINSTALLING & ADJUSTING A TRANSMISSION

NATEF Standard(s) for Automatic Transmission & Transaxle:

D1-1 Remove and reinstall transmission and torque converter (rear-wheel drive).

Safety First

☐ Wear safety glasses at all times.
☐ Follow all safety rules when using common hand tools.
☐ Follow all safety rules when using a lift.
☐ Follow all safety rules when using a transmission jack.
☐ Follow all safety rules when using compressed air.

Tools & Equipment:

- Lift
- Torque wrench
- Safety strap or chain
- Transmission fluid
- Large drain pan
- Flywheel turner
- Vehicle service information
- Pressure gauge
- Air compressor
- Tilting-base transmission jack
- Common hand tools
- Replacement dipstick tube O-ring seal or gasket

PROCEDURES Refer to the vehicle service information for specifications and special procedures. Then reinstall and adjust the transmission provided by your instructor.

_____ 1. Write up a repair order.

_____ 2. Make sure you follow all procedures in the vehicle service information.

_____ 3. Safely raise the vehicle on a lift.

_____ 4. Install the transmission on the transmission jack and chain it in place.

_____ 5. Install the torque converter in the transmission, sliding it over the stator support onto the front pump. Be sure to feel for two distinct clunks as both sets of splines engage.

_____ 6. Raise the transmission on the jack as necessary to align the bellhousing pilot holes with the dowels on the engine.

_____ 7. Install the bellhousing bolts and tighten them using a long extension and swivel socket.

_____ 8. If necessary, replace the crossmember at the rear of the transmission.

_____ 9. Lower the transmission and bolt the transmission mount to the crossmember.

_____ 10. Remove the safety chain from the transmission and remove the jack.

_____ 11. Align the torque converter to the flexplate and tighten bolts.

_____ 12. Install the converter splash shield.

(continued)

_____ 13. Reinstall the starter motor.

_____ 14. Install the dipstick tube using a new O-ring.

_____ 15. Using compressed air, reverse flush the transmission cooler.

_____ 16. Reconnect the cooler lines.

_____ 17. Reconnect the speedometer cable.

_____ 18. Reconnect the linkages.

_____ 19. Reconnect the vacuum line to the modulator, if required.

_____ 20. Reconnect all electrical connectors.

_____ 21. If you lowered the exhaust system for clearance, reconnect it.

_____ 22. Install the drive shafts.

_____ 23. Safely lower the vehicle to the ground.

_____ 24. Add the proper amount of fluid.

_____ 25. Reconnect any parts removed or disconnected under the hood.

_____ 26. Start the engine and inspect for leaks.

_____ 27. Run the engine until it is at normal operating temperature.

_____ 28. Cycle the transmission through all gears and check the fluid level. *Note:* Be sure to check the fluid level in the correct gear.

_____ 29. Using the vehicle service information, make any necessary adjustments to the linkages.

Performance ✓ Checklist

Name _____ Date _____ Class _____

PERFORMANCE STANDARDS:
Level 4–Performs skill without supervision and adapts to problem situations.
Level 3–Performs skill satisfactorily without assistance or supervision.
Level 2–Performs skill satisfactorily, but requires assistance/supervision.
Level 1–Performs parts of skill satisfactorily, but requires considerable assistance/supervision.

Attempt (circle one): **1 2 3 4**

Comments:

_____

_____

_____

_____

PERFORMANCE LEVEL ACHIEVED: _____

_____ 1. Safety rules and practices were followed at all times regarding this job.

_____ 2. Tools and equipment were used properly and stored upon completion of this job.

_____ 3. This completed job met the standards set and was done within the allotted time.

_____ 4. No injury or damage to property occurred during this job.

_____ 5. Upon completion of this job, the work area was cleaned correctly.

Instructor's Signature _____ Date_____

REMOVING A TRANSAXLE

NATEF Standard(s) for Automatic Transmission & Transaxle:
D1-2 Remove and reinstall transaxle and torque converter assembly.

Safety First

- ☐ Wear safety glasses at all times.
- ☐ Follow all safety rules when using common hand tools.
- ☐ Follow all safety rules when using a lift.
- ☐ Follow all safety rules when using a transmission jack.
- ☐ Check Material Safety Data Sheets for chemical safety information.

Tools & Equipment:

- Lift
- Safety strap or chain
- Tilting-base transmission jack suitable for FWD vehicles
- Large drain pan
- Common hand tools
- Engine support bracket
- Vehicle service information

PROCEDURES Refer to the vehicle service information for specifications and special procedures. Then remove the transaxle from the vehicle provided by your instructor.

_____ 1. Write up a repair order.

_____ 2. Make sure you follow all procedures in the vehicle service information.

_____ 3. Open the hood and disconnect the battery ground cable.

_____ 4. Unbolt and remove the starter motor.

_____ 5. Support the engine with a bracket.

_____ 6. Disconnect any electrical connections to the engine or transaxle that would interfere with transaxle removal.

_____ 7. Safely raise the vehicle to the proper working height on a lift.

_____ 8. Disconnect the speedometer and cruise control (if equipped).

_____ 9. Remove the dipstick tube.

_____ 10. Drain the transaxle of fluid.

_____ 11. If required, separate the ball joint from the control arm.

_____ 12. Remove the axle shafts from the transaxle.

_____ 13. Remove the crossmember or cradle, if required.

_____ 14. If required for clearance, remove the exhaust.

_____ 15. Remove the splash shield from the converter housing.

(continued)

_____ 16. Using paint, mark the torque converter and remove bolts.

_____ 17. Disconnect the transaxle cooler lines and plug them to prevent leakage.

_____ 18. Using a safety chain, attach a transmission jack securely to the transaxle.

_____ 19. If required, remove the transaxle mounts and engine mount.

_____ 20. Remove the engine-to-transaxle bolts.

_____ 21. Pry between the engine and transmission to separate them.

_____ 22. Slowly lower the transaxle. **CAUTION:** Be careful not to get the transaxle caught on the engine, the frame, or the body.

_____ 23. Transfer the transaxle to a clean workbench. **CAUTION:** Be careful not to tip the jack on the floor. Since these components are heavy, this may require an assistant.

Performance ✓ Checklist

Name _____ Date _____ Class _____

PERFORMANCE STANDARDS:
Level 4–Performs skill without supervision and adapts to problem situations.
Level 3–Performs skill satisfactorily without assistance or supervision.
Level 2–Performs skill satisfactorily, but requires assistance/supervision.
Level 1–Performs parts of skill satisfactorily, but requires considerable assistance/supervision.

Attempt (circle one): **1 2 3 4**

Comments:

_____

_____

_____

_____

PERFORMANCE LEVEL ACHIEVED: _____

_____ 1. Safety rules and practices were followed at all times regarding this job.

_____ 2. Tools and equipment were used properly and stored upon completion of this job.

_____ 3. This completed job met the standards set and was done within the allotted time.

_____ 4. No injury or damage to property occurred during this job.

_____ 5. Upon completion of this job, the work area was cleaned correctly.

Instructor's Signature _____ Date_____

DISASSEMBLING & CLEANING THE TRANSMISSION

NATEF Standard(s) for Automatic Transmission & Transaxle:
D1-3 Disassemble, clean, and inspect transmission/transaxle.

Safety First
☐ Wear safety glasses at all times.
☐ Follow all safety rules when using common hand tools.
☐ Follow all safety rules when using compressed air.
☐ Check Material Safety Data Sheets for chemical safety information.

Tools & Equipment:
- Air compressor
- Vehicle service information
- Parts-cleaning tank
- Common hand tools

PROCEDURES Refer to the vehicle service information for specifications and special procedures. Then disassemble and clean the transmission in the vehicle provided by your instructor.

_____ 1. Write up a repair order.

_____ 2. Make sure you follow all procedures in the vehicle service information.

_____ 3. Remove the torque converter.

_____ 4. Remove any external components that could be damaged during cleaning and handling.

_____ 5. Disassemble the transmission in an orderly and sequential manner. Lay out all parts in the order of removal. **CAUTION:** Avoid using force during disassembly.

_____ 6. Clean all hard parts in parts-cleaning solvent and air dry.

_____ 7. Clean the case thoroughly in parts-cleaning solvent and blow out all passages with compressed air.

_____ 8. Inspect all parts as instructed in the vehicle service information. **CAUTION:** Do not discard old seals, rings, and gaskets. They will be needed later for comparison with the replacement parts.

(continued)

Automatic Transmission & Transaxle

Performance ✓ Checklist

Name _____ Date _____ Class _____

PERFORMANCE STANDARDS:
Level 4–Performs skill without supervision and adapts to problem situations.
Level 3–Performs skill satisfactorily without assistance or supervision.
Level 2–Performs skill satisfactorily, but requires assistance/supervision.
Level 1–Performs parts of skill satisfactorily, but requires considerable assistance/supervision.

Attempt (circle one): **1 2 3 4**

Comments:

_____

_____

_____

_____

PERFORMANCE LEVEL ACHIEVED: _____

_____ 1. Safety rules and practices were followed at all times regarding this job.

_____ 2. Tools and equipment were used properly and stored upon completion of this job.

_____ 3. This completed job met the standards set and was done within the allotted time.

_____ 4. No injury or damage to property occurred during this job.

_____ 5. Upon completion of this job, the work area was cleaned correctly.

Instructor's Signature _____ Date_____

DISASSEMBLING & CLEANING THE TRANSAXLE

NATEF Standard(s) for Automatic Transmission & Transaxle:
D1-3 Disassemble, clean, and inspect transmission/transaxle.

Safety First
- ☐ Wear safety glasses at all times.
- ☐ Follow all safety rules when using common hand tools.
- ☐ Follow all safety rules when using compressed air.
- ☐ Check Material Safety Data Sheets for chemical safety information.

Tools & Equipment:
- Air compressor
- Vehicle service information
- Parts-cleaning tank
- Common hand tools

PROCEDURES Refer to the vehicle service information for specifications and special procedures. Then disassemble and clean the transaxle in the vehicle provided by your instructor.

_____ 1. Write up a repair order.

_____ 2. Make sure you follow all procedures in the vehicle service information.

_____ 3. Remove the torque converter.

_____ 4. Remove any external components that could be damaged during cleaning and handling.

_____ 5. Disassemble the transaxle in an orderly and sequential manner. Lay out all parts in the order of removal. **CAUTION:** Avoid using force during disassembly.

_____ 6. Clean all hard parts in the parts-cleaning solvent and air dry.

_____ 7. Clean the case thoroughly in parts-cleaning solvent and blow out all passages with compressed air.

_____ 8. Inspect all parts as instructed in the vehicle service information. **CAUTION:** Do not discard old seals, rings, and gaskets. They will be needed later for comparison with the replacement parts.

Automatic Transmission & Transaxle

(continued)

Performance ✓ Checklist

Name _____ Date _____ Class _____

PERFORMANCE STANDARDS:
Level 4–Performs skill without supervision and adapts to problem situations.
Level 3–Performs skill satisfactorily without assistance or supervision.
Level 2–Performs skill satisfactorily, but requires assistance/supervision.
Level 1–Performs parts of skill satisfactorily, but requires considerable assistance/supervision.

Attempt (circle one): **1 2 3 4**

Comments:

_____

_____

_____

_____

PERFORMANCE LEVEL ACHIEVED: _____

_____ **1.** Safety rules and practices were followed at all times regarding this job.

_____ **2.** Tools and equipment were used properly and stored upon completion of this job.

_____ **3.** This completed job met the standards set and was done within the allotted time.

_____ **4.** No injury or damage to property occurred during this job.

_____ **5.** Upon completion of this job, the work area was cleaned correctly.

Instructor's Signature _____ Date _____

INSPECTING THE TRANSAXLE DRIVE LINK ASSEMBLY

NATEF Standard(s) for Automatic Transmission & Transaxle:
D3-7 Inspect transaxle drive, link chains, sprockets, gears, bearings, and bushings; perform necessary action.

Safety First
- ☐ Wear safety glasses at all times.
- ☐ Follow all safety rules when using common hand tools.
- ☐ Follow all safety rules when using a lift.
- ☐ Follow all safety rules when using a transmission jack.

Tools & Equipment:
- Lift
- Engine support bracket
- Vehicle service information
- Large drain pan
- Tilting-base transmission jack (for FWD)
- Safety strap or chain
- Common hand tools

PROCEDURES Refer to the vehicle service information for specifications and special procedures. Then inspect the transaxle drive link assembly in the vehicle provided by your instructor.

_____ 1. Write up a repair order.

_____ 2. Make sure you follow all procedures in the vehicle service information.

_____ 3. Safely raise the vehicle on a lift.

_____ 4. Remove the transaxle (see Job Sheet AT-42).

_____ 5. Following the vehicle service information's instructions, remove the chain cover.

Source: _____ *Page:* _____

_____ 6. Using the manufacturer's specifications, check the drive chain for excessive stretch.

Source: _____ *Page:* _____

_____ 7. Rotate the drive link assembly to check whether the drive link chain binds on the sprockets. If it does, inspect all the teeth of the drive sprocket (all should be identical). Replace the sprocket if any teeth are different or damaged.

_____ 8. Inspect drive and driven sprocket teeth. If any teeth are different or damaged, replace the suspected parts.

(continued)

Automatic Transmission & Transaxle

_____ 9. Check areas where sprockets ride against the bearings or thrust washers. If necessary, replace the sprockets.

_____ 10. Inspect drive and driven sprocket support roller bearings for damage or missing needle(s). Replace these bearings if they are damaged or missing needle(s). *Note:* Plastic thrust washers should also be replaced if necessary.

_____ 11. Safely lower the vehicle to the ground.

Performance ✓ Checklist

Name _____ Date _____ Class _____

PERFORMANCE STANDARDS:
Level 4–Performs skill without supervision and adapts to problem situations.
Level 3–Performs skill satisfactorily without assistance or supervision.
Level 2–Performs skill satisfactorily, but requires assistance/supervision.
Level 1–Performs parts of skill satisfactorily, but requires considerable assistance/supervision.

Attempt (circle one): **1 2 3 4**

Comments:

_____

_____

_____

_____

PERFORMANCE LEVEL ACHIEVED: _____

_____ 1. Safety rules and practices were followed at all times regarding this job.
_____ 2. Tools and equipment were used properly and stored upon completion of this job.
_____ 3. This completed job met the standards set and was done within the allotted time.
_____ 4. No injury or damage to property occurred during this job.
_____ 5. Upon completion of this job, the work area was cleaned correctly.

Instructor's Signature _____ Date_____

INSPECTING TRANSAXLE FINAL DRIVE COMPONENTS

NATEF Standard(s) for Automatic Transmission & Transaxle:
D3-8 Inspect, measure, repair, adjust or replace transaxle final drive components.

Safety First
- ☐ Wear safety glasses at all times.
- ☐ Follow all safety rules when using common hand tools.

Tools & Equipment:
- Feeler gauge
- Vehicle service information
- Pin punch
- Common hand tools
- Replacement washers and bushings

PROCEDURES Refer to the vehicle service information for specifications and special procedures. Then inspect the transaxle final drive components in the vehicle provided by your instructor.

_____ 1. Write up a repair order.

_____ 2. Make sure you follow all procedures in the vehicle service information.

_____ 3. Inspect the internal gear for damaged teeth, worn bearing surfaces, a broken parking pawl spring, or damage to the parking pawl (if applicable).

_____ 4. Check the sun gear for wear or damage.

_____ 5. Inspect the thrust bearings for wear or damage.

_____ 6. Check the parking gear and governor drive gear for wear or damage.

_____ 7. Inspect the final drive carrier pinion gears for excessive endplay. To do this, measure the endplay between the carrier and the pinion gear with a feeler gauge. Then compare this measurement to the vehicle service information's specifications.

Source: _____ _Page:_____

_____ 8. Drive the differential pinion shaft retainer pin from the carrier with a pin punch.

_____ 9. Remove the pinion shaft, pinion and side gears, and thrust washers. Inspect these parts for damage or abnormal wear.

_____ 10. Check the final drive sun gear shaft for damaged splines or journals.

_____ 11. Follow the vehicle service information for type and material of replacement washers.

Source: _____ _Page:_____

(continued)

Performance ✓ Checklist

Name _____ Date _____ Class _____

PERFORMANCE STANDARDS:
Level 4–Performs skill without supervision and adapts to problem situations.
Level 3–Performs skill satisfactorily without assistance or supervision.
Level 2–Performs skill satisfactorily, but requires assistance/supervision.
Level 1–Performs parts of skill satisfactorily, but requires considerable assistance/supervision.

Attempt (circle one): **1 2 3 4**

Comments:

_____

_____

_____

_____

PERFORMANCE LEVEL ACHIEVED: _____

_____ 1. Safety rules and practices were followed at all times regarding this job.

_____ 2. Tools and equipment were used properly and stored upon completion of this job.

_____ 3. This completed job met the standards set and was done within the allotted time.

_____ 4. No injury or damage to property occurred during this job.

_____ 5. Upon completion of this job, the work area was cleaned correctly.

Instructor's Signature _____ Date_____

REMOVING, INSPECTING & REPLACING BUSHINGS

NATEF Standard(s) for Automatic Transmission & Transaxle:
D3-4 Inspect bushings; determine necessary action.

Safety First
- ☐ Wear safety glasses at all times.
- ☐ Follow all safety rules when using common hand tools.

Tools & Equipment:
- Bushings
- Vehicle service information
- Feeler gauge
- Removal and replacement tools
- Common hand tools
- Inside and outside micrometers

PROCEDURES Refer to the vehicle service information for specifications and special procedures. Then remove, inspect, and replace the bushings in the transmission provided by your instructor.

_____ 1. Write up a repair order.

_____ 2. Make sure you follow all procedures in the vehicle service information.

_____ 3. Inspect bushing condition by appearance and feel. If you find any galling, scoring, flaking, excess wear, or observe rough operation, replace the bushing.

_____ 4. If you want to measure bushing clearance for practice, anything over 0.006 inch is excessive.

_____ 5. Several methods can be used to remove a bushing, depending on the shape of the bore and the size of the bushing. Use the vehicle service information's recommended methods and tools.

Source: _____ Page:_____

_____ 6. If a bushing cutter or sharp chisel is recommended for removing bushings from stepped bores, locate them in the bushing and work next to it, folding the bushing inward to collapse it for removal.

_____ 7. Install new bushings. To prevent damaging the bushing or its bore, always use a bushing installer to push the new bushing into its bore.

_____ 8. During installation, be sure to press or drive the bushing into its bore to the correct depth. *Note:* Some bushing installers (drivers) have steps that bottom out and stop at the correct depth.

(continued)

Automatic Transmission & Transaxle

Performance ✓ Checklist

Name _____ Date _____ Class _____

PERFORMANCE STANDARDS:
Level 4–Performs skill without supervision and adapts to problem situations.
Level 3–Performs skill satisfactorily without assistance or supervision.
Level 2–Performs skill satisfactorily, but requires assistance/supervision.
Level 1–Performs parts of skill satisfactorily, but requires considerable assistance/supervision.

Attempt (circle one): **1 2 3 4**

Comments:

_____

_____

_____

_____

PERFORMANCE LEVEL ACHIEVED: _____

_____ **1.** Safety rules and practices were followed at all times regarding this job.
_____ **2.** Tools and equipment were used properly and stored upon completion of this job.
_____ **3.** This completed job met the standards set and was done within the allotted time.
_____ **4.** No injury or damage to property occurred during this job.
_____ **5.** Upon completion of this job, the work area was cleaned correctly.

Instructor's Signature _____ Date_____

INSPECTING, MEASURING & REPLACING THRUST WASHERS AND BEARINGS

NATEF Standard(s) for Automatic Transmission & Transaxle:

D3-2 Inspect, measure, and replace thrust washers and bearings.

Safety First
- ☐ Wear safety glasses at all times.
- ☐ Follow all safety rules when using common hand tools.
- ☐ Never use compressed air to dry a bearing. The high pressure can cause the bearing to come apart and cause injury.

Tools & Equipment:
- Vehicle service information
- Feeler gauge
- Common hand tools
- Dial indicator

PROCEDURES Refer to the vehicle service information for specifications and special procedures. Then inspect, measure, and replace the thrust washers and bearings on the vehicle provided by your instructor.

_____ 1. Write up a repair order.

_____ 2. Make sure you follow all procedures in the vehicle service information.

_____ 3. Inspect thrust washers for signs of deterioration due to either lack of lubrication or because of friction.

_____ 4. Inspect bearings for any signs of damage.

_____ 5. Replace any component that exhibits signs of wear.

_____ 6. Following the manufacturer's procedure, measure endplay, gap, or thickness to determine proper clearance. If clearance is not within specifications and is adjustable, adjust to specifications.

(continued)

Automatic Transmission & Transaxle

Performance ✓ Checklist

Name _____ Date _____ Class _____

PERFORMANCE STANDARDS:
Level 4–Performs skill without supervision and adapts to problem situations.
Level 3–Performs skill satisfactorily without assistance or supervision.
Level 2–Performs skill satisfactorily, but requires assistance/supervision.
Level 1–Performs parts of skill satisfactorily, but requires considerable assistance/supervision.

Attempt (circle one): **1 2 3 4**

Comments:

_____

_____

_____

_____

PERFORMANCE LEVEL ACHIEVED: _____

_____ 1. Safety rules and practices were followed at all times regarding this job.
_____ 2. Tools and equipment were used properly and stored upon completion of this job.
_____ 3. This completed job met the standards set and was done within the allotted time.
_____ 4. No injury or damage to property occurred during this job.
_____ 5. Upon completion of this job, the work area was cleaned correctly.

Instructor's Signature _____ Date_____

REASSEMBLING THE TRANSMISSION

NATEF Standard(s) for Automatic Transmission & Transaxle:
D1-7 Assemble transmission/transaxle.

Safety First
☐ Wear safety glasses at all times.
☐ Follow all safety rules when using common hand tools.

Tools & Equipment:
- Transmission overhaul kit
- Common hand tools
- Petroleum jelly
- Rubber-tipped blow gun
- Appropriate measuring equipment
- Vehicle service information

PROCEDURES Refer to the vehicle service information for specifications and special procedures. Then reassemble the transmission provided by your instructor.

_____ 1. Write up a repair order.

_____ 2. Make sure you follow all procedures in the vehicle service information.

_____ 3. Replace all clutch piston seals. Make sure that the seal lips point toward the pressure source. Lube the seals using petroleum jelly. If required, assemble the piston using special tools.

_____ 4. Assemble the clutch discs in the drum. Make sure that clearance is set to specifications.

_____ 5. Air test clutches on the pump for operation and leaks.

_____ 6. Assemble the clutch drums in the case. Allow gravity to work for you by supporting the case vertically with the bellhousing facing up.

_____ 7. Align the drum over the hub and slowly rotate the clutch drum back and forth to allow each friction disc to align with the splines on the hub. The assembly will drop as the inside teeth align with the hub splines.

_____ 8. Avoid using force when reassembling parts into the transmission case. If you have to use force to reassemble the part, you have made an error in assembly, used an incorrect snap ring or thrust washer, or installed parts in the wrong order.

_____ 9. Make sure all snap rings are seated in their grooves.

_____ 10. Adjust bands using struts or pins as required.

_____ 11. Using pump alignment procedures that are in the vehicle service information, install the front pump.

_____ 12. Check endplay and adjust as necessary using thrust washers.

_____ 13. Air check clutches and bands.

(*continued*)

_____ **14.** Install the valve body and accumulators. Make sure you torque them to specifications.

_____ **15.** Install the pan.

_____ **16.** Install the torque converter over the stator support and input shaft. Make sure that the torque converter is seated on the oil pump drive lugs.

Performance ✓ Checklist

Name _____ Date _____ Class _____

PERFORMANCE STANDARDS:
Level 4–Performs skill without supervision and adapts to problem situations.
Level 3–Performs skill satisfactorily without assistance or supervision.
Level 2–Performs skill satisfactorily, but requires assistance/supervision.
Level 1–Performs parts of skill satisfactorily, but requires considerable assistance/supervision.

Attempt (circle one): **1 2 3 4**

Comments:

_____

_____

_____

_____

PERFORMANCE LEVEL ACHIEVED: _____

_____ **1.** Safety rules and practices were followed at all times regarding this job.

_____ **2.** Tools and equipment were used properly and stored upon completion of this job.

_____ **3.** This completed job met the standards set and was done within the allotted time.

_____ **4.** No injury or damage to property occurred during this job.

_____ **5.** Upon completion of this job, the work area was cleaned correctly.

Instructor's Signature _____ Date_____

INSPECTING, REPLACING & ALIGNING POWERTRAIN MOUNTS

NATEF Standard(s) for Automatic Transmission & Transaxle:
C7 Inspect, replace, and align powertrain mounts.

Safety First
☐ Wear safety glasses at all times.
☐ Follow all safety rules when using common hand tools.

Tools & Equipment:
• Vehicle service information
• Common hand tools

PROCEDURES Refer to the vehicle service information for specifications and special procedures. Then inspect, replace, and align powertrain mounts on the vehicle provided by your instructor.

_____ 1. Write up a repair order.

_____ 2. Make sure you follow all procedures in the vehicle service information.

_____ 3. Visually inspect the engine and transmission mounts for cracks, deterioration, swelling, or signs of misalignment.

_____ 4. If any problems are found, follow the manufacturer's procedure for replacing a mount.

_____ 5. Check for proper alignment after installing a new mount to ensure components will operate normally.

Automatic Transmission & Transaxle

(continued)

Performance ✓ Checklist

Name _____ Date _____ Class _____

PERFORMANCE STANDARDS:
Level 4–Performs skill without supervision and adapts to problem situations.
Level 3–Performs skill satisfactorily without assistance or supervision.
Level 2–Performs skill satisfactorily, but requires assistance/supervision.
Level 1–Performs parts of skill satisfactorily, but requires considerable assistance/supervision.

Attempt (circle one): **1 2 3 4**

Comments:

_____

_____

_____

_____

PERFORMANCE LEVEL ACHIEVED: _____

_____ **1.** Safety rules and practices were followed at all times regarding this job.

_____ **2.** Tools and equipment were used properly and stored upon completion of this job.

_____ **3.** This completed job met the standards set and was done within the allotted time.

_____ **4.** No injury or damage to property occurred during this job.

_____ **5.** Upon completion of this job, the work area was cleaned correctly.

Instructor's Signature _____ Date_____

MEASURING ENDPLAY OR PRELOAD

NATEF Standard(s) for Automatic Transmission & Transaxle:
D3-1 Measure endplay or preload; determine necessary action.

Safety First
- ☐ Wear safety glasses at all times.
- ☐ Follow all safety rules when using common hand tools.
- ☐ Be careful when working with automatic transmission parts. Unknowingly removing springs or other components under pressure can cause bodily injury.

Tools & Equipment:
- Vehicle service information
- Feeler gauges
- Common hand tools
- Dial indicator

PROCEDURES Refer to the vehicle service information for specifications and special procedures. Then measure the endplay or preload on the vehicle provided by your instructor.

_____ 1. Write up a repair order.

_____ 2. Make sure you follow all procedures in the vehicle service information.

_____ 3. Locate the measurement specifications in the vehicle service information for the specific part to be measured.

_____ 4. Following the proper procedure, measure the endplay/preload with either the dial indicator or feeler gauge as instructed.

_____ 5. If the measurement is not within specifications, use the vehicle service information to determine if the component has different thickness snap rings, shims, or thrust washers to adjust endplay/preload.

_____ 6. Adjust if allowable and necessary.

_____ 7. Remeasure endplay/preload.

Automatic Transmission & Transaxle

(continued)

Performance ✓ Checklist

Name _____ Date _____ Class _____

PERFORMANCE STANDARDS:
Level 4–Performs skill without supervision and adapts to problem situations.
Level 3–Performs skill satisfactorily without assistance or supervision.
Level 2–Performs skill satisfactorily, but requires assistance/supervision.
Level 1–Performs parts of skill satisfactorily, but requires considerable assistance/supervision.

Attempt (circle one): **1 2 3 4**

Comments:

_____
_____
_____
_____

PERFORMANCE LEVEL ACHIEVED: _____

_____ **1.** Safety rules and practices were followed at all times regarding this job.

_____ **2.** Tools and equipment were used properly and stored upon completion of this job.

_____ **3.** This completed job met the standards set and was done within the allotted time.

_____ **4.** No injury or damage to property occurred during this job.

_____ **5.** Upon completion of this job, the work area was cleaned correctly.

Instructor's Signature _____ Date_____

Name _____ Date _____ Class _____

DIAGNOSING THE CLUTCH

NATEF Standard(s) for Manual Drive Train & Axles:
B1 Diagnose clutch noise, binding, slippage, pulsation, and chatter; determine necessary action.

DIRECTIONS: Fill in the blanks below by identifying (1) Safety First Practices that must be followed, (2) Tools and Equipment required, (3) three possible causes of clutch problems, and (4) corrective actions that should be taken.

Safety First

- _____
- _____
- _____
- _____

Tools and Equipment Required:

- _____   - _____
- _____   - _____
- _____   - _____

QUICK ✔ Diagnostic for Cause(s)

List at least three common causes of clutch problems:

1. _____
2. _____
3. _____
4. _____
5. _____

QUICK ✔ Diagnostic for Corrective Action(s)

List possible corrective action(s):

1. _____
2. _____
3. _____
4. _____
5. _____

Manual Drive Train & Axles

DIAGNOSING MANUAL DRIVE TRAINS

NATEF Standard(s) for Manual Drive Train & Axles:
C4 Diagnose noise, hard shifting, jumping out of gear, and fluid leakage concerns; determine necessary action.

DIRECTIONS: Fill in the blanks below by identifying (1) Safety First Practices that must be followed, (2) Tools and Equipment required, (3) three possible causes of manual drive train problems, and (4) corrective actions that should be taken.

Safety First

- _____
- _____
- _____
- _____

Tools and Equipment Required:

- _____ • _____
- _____ • _____
- _____ • _____

QUICK ✔ Diagnostic for Cause(s)

List at least three common causes of manual drive train problems:

1. _____
2. _____
3. _____
4. _____
5. _____

QUICK ✔ Diagnostic for Corrective Action(s)

List possible corrective action(s):

1. _____
2. _____
3. _____
4. _____
5. _____

Name _____   Date _____   Class _____

DIAGNOSING DIFFERENTIAL
NOISE AND VIBRATION

NATEF Standard(s) for Manual Drive Train & Axles:
E1-1 Diagnose noise and vibration concerns; determine necessary action.

DIRECTIONS: Fill in the blanks below by identifying (1) Safety First Practices that must be followed, (2) Tools and Equipment required, (3) three possible causes of differential noise and vibration problems, and (4) corrective actions that should be taken.

Safety First

- _____
- _____
- _____
- _____

Tools and Equipment Required:

- _____   - _____
- _____   - _____
- _____   - _____

QUICK ✔ Diagnostic for Cause(s)

List at least three common causes of differential noise and vibration problems:

1. _____
2. _____
3. _____
4. _____

QUICK ✔ Diagnostic for Corrective Action(s)

List possible corrective action(s):

1. _____
2. _____
3. _____
4. _____

Manual Drive Train
& Axles

DIAGNOSING LIMITED SLIP DIFFERENTIALS

NATEF Standard(s) for Manual Drive Train & Axles:
E2-1 Diagnose noise, slippage, and chatter concerns; determine necessary action.

DIRECTIONS: Fill in the blanks below by identifying (1) Safety First Practices that must be followed, (2) Tools and Equipment required, (3) three possible causes of differential noise, slippage, and chatter problems, and (4) corrective actions that should be taken.

Safety First

- _____
- _____
- _____
- _____

Tools and Equipment Required:

- _____   • _____
- _____   • _____
- _____   • _____

QUICK ✔ Diagnostic for Cause(s)

List at least three common causes of differential noise, slippage, and chatter problems:

1. _____
2. _____
3. _____
4. _____
5. _____

QUICK ✔ Diagnostic for Corrective Action(s)

List possible corrective action(s):

1. _____
2. _____
3. _____
4. _____
5. _____

Name _____ Date _____ Class _____

DIAGNOSING DIFFERENTIAL CASE LEAKS

NATEF Standard(s) for Manual Drive Train & Axles:
E1-2 Diagnose fluid leakage concerns; determine necessary action.

DIRECTIONS: Fill in the blanks below by identifying (1) Safety First Practices that must be followed, (2) Tools and Equipment required, (3) three possible causes of differential case fluid leakage problems, and (4) corrective actions that should be taken.

Safety First

- _____
- _____
- _____
- _____

Tools and Equipment Required:

- _____   • _____
- _____   • _____
- _____   • _____

QUICK ✔ Diagnostic for Cause(s)

List at least three common causes of differential case fluid leakage problems:

1. _____
2. _____
3. _____
4. _____
5. _____

QUICK ✔ Diagnostic for Corrective Action(s)

List possible corrective action(s):

1. _____
2. _____
3. _____
4. _____
5. _____

Manual Drive Train & Axles

DIAGNOSING AXLE SHAFT, BEARINGS, AND SEALS

NATEF Standard(s) for Manual Drive Train & Axles:

E3-1 Diagnose drive axle shafts, bearings, and seals for noise, vibration, and fluid leakage concerns; determine necessary action.

DIRECTIONS: Fill in the blanks below by identifying (1) Safety First Practices that must be followed, (2) Tools and Equipment required, (3) three possible causes of drive axle shaft, bearing, and lubricant seal problems, and (4) corrective actions that should be taken.

Safety First
- _____
- _____
- _____
- _____

Tools and Equipment Required:
- _____   • _____
- _____   • _____
- _____   • _____

QUICK ✔ Diagnostic for Cause(s)

List at least three common causes of drive axle shaft, bearing, and lubricant seal problems:

1. _____
2. _____
3. _____
4. _____

QUICK ✔ Diagnostic for Corrective Action(s)

List possible corrective action(s):

1. _____
2. _____
3. _____
4. _____

Name _____   Date _____   Class _____

DIAGNOSING THE TRANSAXLE
FINAL DRIVE ASSEMBLY

NATEF Standard(s) for Manual Drive Train & Axles:
C13 Diagnose transaxle final drive assembly noise and vibration concerns; determine necessary action.

DIRECTIONS: Fill in the blanks below by identifying (1) Safety First Practices that must be followed, (2) Tools and Equipment required, (3) three possible causes of transaxle final drive assembly problems, and (4) corrective actions that should be taken.

Safety First
- _____
- _____
- _____
- _____

Tools and Equipment Required:

- _____   - _____
- _____   - _____
- _____   - _____

QUICK ✔ Diagnostic for Cause(s)

List at least three common causes of transaxle final drive assembly problems:

1. _____
2. _____
3. _____
4. _____

QUICK ✔ Diagnostic for Corrective Action(s)

List possible corrective action(s):

1. _____
2. _____
3. _____
4. _____

Manual Drive Train & Axles

Automotive Excellence Technical Applications Volume 2   **367**

DIAGNOSING CV JOINTS

NATEF Standard(s) for Manual Drive Train & Axles:
D1 Diagnose constant-velocity (CV) joint noise and vibration concerns; determine necessary action.

DIRECTIONS: Fill in the blanks below by identifying (1) Safety First Practices that must be followed, (2) Tools and Equipment required, (3) three possible causes of CV joint problems, and (4) corrective actions that should be taken.

Safety First

- _____
- _____
- _____
- _____

Tools and Equipment Required:

- _____   • _____
- _____   • _____
- _____   • _____

QUICK ✔ Diagnostic for Cause(s)

List at least three common causes of CV joint problems:

1. _____
2. _____
3. _____
4. _____
5. _____

QUICK ✔ Diagnostic for Corrective Action(s)

List possible corrective action(s):

1. _____
2. _____
3. _____
4. _____
5. _____

DIAGNOSING FOUR-WHEEL-DRIVE SYSTEM'S ELECTRICAL/ELECTRONIC COMPONENTS

NATEF Standard(s) for Manual Drive Train & Axles:
F7 Diagnose, test, adjust, and replace electrical/electronic components of four-wheel-drive systems.

DIRECTIONS: Fill in the blanks below by identifying (1) Safety First Practices that must be followed, (2) Tools and Equipment required, (3) three possible causes of electrical/electronic component problems in a four-wheel-drive system, and (4) corrective actions that should be taken.

Safety First
- _____
- _____
- _____
- _____

Tools and Equipment Required:
- _____ • _____
- _____ • _____
- _____ • _____

QUICK ✔ Diagnostic for Cause(s)

List at least three common causes of electrical/electronic component problems in a four-wheel-drive system:

1. _____
2. _____
3. _____
4. _____

QUICK ✔ Diagnostic for Corrective Action(s)

List possible corrective action(s):

1. _____
2. _____
3. _____
4. _____

Manual Drive Train & Axles

DIAGNOSING NOISE, VIBRATION, AND STEERING CONCERNS

NATEF Standard(s) for Manual Drive Train & Axles:
F1 Diagnose noise, vibration, and unusual steering concerns; determine necessary action.

DIRECTIONS: Fill in the blanks below by identifying (1) Safety First Practices that must be followed, (2) Tools and Equipment required, (3) three possible causes of noise, vibration, and steering problems, and (4) corrective actions that should be taken.

Safety First

- _____
- _____
- _____
- _____

Tools and Equipment Required:

- _____   • _____
- _____   • _____
- _____   • _____

QUICK ✔ Diagnostic for Cause(s)

List at least three common causes of noise, vibration, and steering problems:

1. _____
2. _____
3. _____
4. _____
5. _____

QUICK ✔ Diagnostic for Corrective Action(s)

List possible corrective action(s):

1. _____
2. _____
3. _____
4. _____
5. _____

COMPLETING A VEHICLE REPAIR ORDER FOR A MANUAL DRIVE TRAIN AND AXLES CONCERN

NATEF Standard(s) for Manual Drive Train & Axles:

A1 Complete work order to include customer information, vehicle identifying information, customer concern, related service history, cause, and correction.

SAMPLE VEHICLE REPAIR ORDER   Vehicle Repair Order # _____

Date ____/____/____

Customer Name & Phone #: _____   Vehicle Make/Type: _____   VIN: _____   Mileage: _____

Service History: _____

Customer Concern: _____

Cause of Concern: _____

Suggested Repairs/Maintenance: _____

Services Performed: _____

Parts				Labor	Time In:
Item	Description	Price		Diagnosis Time:	Time Complete:
1				Repair Time:	Total Hours:
2				I hereby authorize the above repair work to be done using the necessary material, and hereby grant you and/or your employees permission to operate the vehicle herein described on streets, highways, or elsewhere for the purpose of testing and/or inspection. An express mechanic's lien is hereby acknowledged on above vehicle to secure the amount of repairs thereof.	
3					
4					
5			X _____	
6					

PROCEDURES Refer to the vehicle service information for specifications and special procedures. Then prepare a vehicle repair order for the vehicle provided by your instructor.

_____ 1. **Write legibly.** Others will be reading what you have written.

_____ 2. **Make sure all information is accurate.** Inaccurate information will slow the repair process.

_____ 3. **Complete every part of the Vehicle Repair Order.** Every part must be completed.

_____ 4. **Number the Vehicle Repair Order.** This will help others track the repair.

_____ 5. **Date the Vehicle Repair Order.** This will help document the service history.

_____ 6. **Enter the Customer Name and Phone Number.** Make sure you have spelled the Customer Name correctly. Double-check the Phone Number.

_____ 7. **Enter the Vehicle Make/Type.** This information is essential.

(continued)

Manual Drive Train & Axles

_____ 8. **Enter the VIN (vehicle identification number).** This is a string of coded data that is unique to the vehicle. The location of the VIN depends on the manufacturer. It is usually found on the dashboard next to the windshield on the driver's side. The VIN is a rich source of information. It is needed to properly use a scan tool to read diagnostic trouble codes. Double-check the VIN to ensure accuracy.

_____ 9. **Enter the Mileage of the vehicle.** This information is part of the service history.

_____ 10. **Complete the Service History.** The service history is a history of all the service operations performed on a vehicle. The service history alerts the technician to previous problems with the vehicle. In the case of recurring problems, it helps the technician identify solutions that were ineffective.

   • A detailed service history is usually kept by the service facility where the vehicle is regularly serviced.

   • Information on service performed on the vehicle at other service centers is not available unless the customer makes it available. For this reason, ask the customer about service performed outside of the present service center.

_____ 11. **Identify the Customer Concern.** This should be a reasonably detailed and accurate description of the problem that the customer is having with the vehicle. The customer is usually the best source of information regarding the problem. This information can be used to perform the initial diagnosis. The customer has a passenger car. She says, "My car shakes violently when I engage the clutch. Once the clutch is engaged, the shaking stops." Enter her concern on the Customer Concern line.

_____ 12. **Identify the Cause of Concern.** This will identify the problem. In this case, there may be several possible causes. Enter the possible causes on the Cause of Concern line.

_____ 13. Ask the customer to read the text at the bottom of the Labor box. By signing on the line at the bottom of this box, the customer authorizes repair work on the vehicle according to the terms specified.

_____ 14. **Identify Suggested Repairs/Maintenance.** This will identify what needs to be done to correct the problem.

_____ 15. **Identify Services Performed.** This will identify the specific maintenance and repair procedures that were performed to correct the problem.

_____ 16. **Provide Parts information.** This includes a numbered list of items used to complete the repair. It includes a description of each part with the price of the part.

_____ 17. **Provide Labor information.** The Diagnosis Time and the Repair Time are totaled to give the Total Hours.

Performance ✓ Checklist

Name _____ Date _____ Class _____

PERFORMANCE STANDARDS:
Level 4–Performs skill without supervision and adapts to problem situations.
Level 3–Performs skill satisfactorily without assistance or supervision.
Level 2–Performs skill satisfactorily, but requires assistance/supervision.
Level 1–Performs parts of skill satisfactorily, but requires considerable assistance/supervision.

Attempt (circle one): **1 2 3 4**

Comments:

_____
_____
_____
_____

PERFORMANCE LEVEL ACHIEVED: _____

_____ 1. Safety rules and practices were followed at all times regarding this job.
_____ 2. Tools and equipment were used properly and stored upon completion of this job.
_____ 3. This completed job met the standards set and was done within the allotted time.
_____ 4. No injury or damage to property occurred during this job.
_____ 5. Upon completion of this job, the work area was cleaned correctly.

Instructor's Signature _____ Date_____

Manual Drive Train & Axles

DISASSEMBLING & REPAIRING CLUTCH LINKAGE AND RELEASE BEARING

NATEF Standard(s) for Manual Drive Train & Axles:

B4 Inspect release (throw-out) bearing, lever, and pivot; determine necessary action.

Safety First

☐ Wear safety glasses at all times.
☐ Follow all safety rules when using common hand tools.
☐ Check Material Safety Data Sheets for chemical safety information.
☐ Follow all safety rules when using a transmission jack.

Tools & Equipment:

• Common hand tools • Transmission jack • Vehicle service information

PROCEDURES Refer to the vehicle service information for specifications and special procedures. Then inspect the linkage components and release bearing in the vehicle provided by your instructor.

_____ 1. Write up a repair order.

_____ 2. Look up the procedure for clutch linkage disassembly, inspection, and repair procedures in the vehicle service information.

   Source: _____ *Page:* _____

_____ 3. Remove the transmission from the vehicle.

_____ 4. Disconnect the release bearing and collar from the release fork.

_____ 5. Inspect for worn or damaged parts.

_____ 6. Replace the release bearing and lube the bearing retainer with high-temperature grease.

_____ 7. Check that all parts operate smoothly.

_____ 8. Replace the transmission.

_____ 9. Following vehicle service information instructions, adjust the clutch linkage.

(continued)

Performance ✓ Checklist

Name _____ Date _____ Class _____

PERFORMANCE STANDARDS:
Level 4–Performs skill without supervision and adapts to problem situations.
Level 3–Performs skill satisfactorily without assistance or supervision.
Level 2–Performs skill satisfactorily, but requires assistance/supervision.
Level 1–Performs parts of skill satisfactorily, but requires considerable assistance/supervision.

Attempt (circle one): **1 2 3 4**

Comments:

_____

_____

_____

_____

PERFORMANCE LEVEL ACHIEVED: _____

_____ **1.** Safety rules and practices were followed at all times regarding this job.

_____ **2.** Tools and equipment were used properly and stored upon completion of this job.

_____ **3.** This completed job met the standards set and was done within the allotted time.

_____ **4.** No injury or damage to property occurred during this job.

_____ **5.** Upon completion of this job, the work area was cleaned correctly.

Instructor's Signature _____ Date_____

SERVICING CRANKSHAFT PILOT BEARINGS AND BUSHINGS

NATEF Standard(s) for Manual Drive Train & Axles:
B7 Inspect, remove or replace crankshaft pilot bearing or bushing (as applicable).

Safety First
- ☐ Wear safety glasses at all times.
- ☐ Follow all safety rules when using common hand tools.
- ☐ Check Material Safety Data Sheets for chemical safety information.
- ☐ Follow all safety rules when using jack stands, ramps, or lifts.

Tools & Equipment:
- Pilot bearing puller/installer
- Soft-faced mallet
- Cleaning solvent
- Vehicle service information
- Bristle brush
- Torque wrench
- Lubricant
- Clutch alignment tool
- Jack stands, ramps, or lifts
- Common hand tools

PROCEDURES Refer to the vehicle service information for specifications and special procedures. Then remove and replace the pilot bushing or bearing in the vehicle provided by your instructor.

_____ 1. Write up a repair order.

_____ 2. Make sure you follow all procedures in the vehicle service information.

_____ 3. Look up the vehicle's pilot bushing/bearing removal, inspection, and replacement procedures in the vehicle service information.

Source: _____ *Page:* _____

_____ 4. Safely raise the vehicle.

_____ 5. Remove the transmission and clutch housing if they have not already been removed.

_____ 6. Remove any excess grease from the pilot bearing or bushing and then remove it from the crankshaft with the bearing puller.

_____ 7. Examine the bushing/bearing for damage or wear that might indicate other transmission damage. *Note:* Always replace bearings or bushings you have removed regardless of whether or not they appear to be damaged.

_____ 8. Clean the crankshaft by using the bristle brush and an appropriate cleaning solvent.

_____ 9. Lubricate the new bushing/bearing according to the manufacturer's specifications.

_____ 10. Center the bushing/bearing with the crankshaft bore on the bearing installer (or clutch alignment tool) and gently tap it into place with the soft-faced mallet.

(continued)

Manual Drive Train & Axles

_____ 11. Remove the bearing installer or clutch alignment tool from the crankshaft.

_____ 12. Coat the bushing/bearing with an approved high-temperature lubricant. **CAUTION:** Avoid over-lubricating, which can contaminate the clutch disc and cause it to slip.

_____ 13. Re-install the transmission.

_____ 14. Safely lower the vehicle to the ground.

Performance ✓ Checklist

Name _____ Date _____ Class _____

PERFORMANCE STANDARDS:
Level 4–Performs skill without supervision and adapts to problem situations.
Level 3–Performs skill satisfactorily without assistance or supervision.
Level 2–Performs skill satisfactorily, but requires assistance/supervision.
Level 1–Performs parts of skill satisfactorily, but requires considerable assistance/supervision.

Attempt (circle one): **1 2 3 4**

Comments:

_____

_____

_____

_____

PERFORMANCE LEVEL ACHIEVED: _____

_____ 1. Safety rules and practices were followed at all times regarding this job.

_____ 2. Tools and equipment were used properly and stored upon completion of this job.

_____ 3. This completed job met the standards set and was done within the allotted time.

_____ 4. No injury or damage to property occurred during this job.

_____ 5. Upon completion of this job, the work area was cleaned correctly.

Instructor's Signature _____ Date_____

REPAIRING CLUTCH PEDALS

NATEF Standard(s) for Manual Drive Train & Axles:

B2 Inspect clutch pedal linkage, cables, automatic adjuster mechanisms, brackets, bushings, pivots, and springs; perform necessary action.

Safety First
- ☐ Wear safety glasses at all times.
- ☐ Follow all safety rules when using common hand tools.

Tools & Equipment:
- Common hand tools
- Vehicle service information

PROCEDURES Refer to the vehicle service information for specifications and special procedures. Then adjust or repair the manual clutch in the vehicle provided by your instructor.

_____ 1. Write up a repair order.

_____ 2. Make sure you follow all procedures in the vehicle service information.

_____ 3. Look up the vehicle's clutch adjustment procedure in the vehicle service information.

Source: _____ *Page:* _____

_____ 4. Inspect the cable for kinks, twists, and bends, and remove any that you find. **CAUTION:** Yanking on the cable can damage linkage components.

_____ 5. Inspect the cable for frays. *Note:* The cable is made out of braided or twisted wires that fray when it is damaged. Damaged cables should always be replaced.

_____ 6. Replace and adjust the clutch cable according to the manufacturer's instructions.

(continued)

Manual Drive Train & Axles

Performance ✓ Checklist

Name _____ Date _____ Class _____

PERFORMANCE STANDARDS:
Level 4–Performs skill without supervision and adapts to problem situations.
Level 3–Performs skill satisfactorily without assistance or supervision.
Level 2–Performs skill satisfactorily, but requires assistance/supervision.
Level 1–Performs parts of skill satisfactorily, but requires considerable assistance/supervision.

Attempt (circle one): **1 2 3 4**

Comments:

_____

_____

_____

_____

PERFORMANCE LEVEL ACHIEVED: _____

_____ **1.** Safety rules and practices were followed at all times regarding this job.

_____ **2.** Tools and equipment were used properly and stored upon completion of this job.

_____ **3.** This completed job met the standards set and was done within the allotted time.

_____ **4.** No injury or damage to property occurred during this job.

_____ **5.** Upon completion of this job, the work area was cleaned correctly.

Instructor's Signature _____ Date _____

DIAGNOSING & REPAIRING CLUTCHES

NATEF Standard(s) for Manual Drive Train & Axles:

B3 Inspect hydraulic clutch slave and master cylinders, lines, and hoses; determine necessary action.

B6 Bleed clutch hydraulic system.

Safety First

- ☐ Wear safety glasses at all times.
- ☐ Follow all safety rules when using common hand tools.
- ☐ Check Material Safety Data Sheets for chemical safety information.

Tools & Equipment:

- Common hand tools
- Vehicle service information

PROCEDURES Refer to the vehicle service information for specifications and special procedures. Then adjust or repair the hydraulic clutch in the vehicle provided by your instructor.

_____ 1. Write up a repair order.

_____ 2. Make sure you follow all procedures in the vehicle service information.

_____ 3. Inspect and fill, if necessary, the hydraulic fluid level in the reservoir.

_____ 4. Inspect the master and slave cylinders for leaks, worn seals, corrosion, and air in the system. **CAUTION:** The master and slave cylinders are usually made out of plastic. Make sure the area in which they are located is not too hot. If this area is too hot, the cylinders will melt. Also check the oil temperature. If it is too high, the oil will melt the plastic.

_____ 5. Check all the hydraulic lines for leaks, breaks, kinks, and wear.

_____ 6. Look up the vehicle's procedure for bleeding the hydraulic clutch linkage in the vehicle service information. *Note:* Most hydraulic systems have special screws for bleeding the clutch.

Source: _____ *Page:* _____

_____ 7. Bleed the hydraulic clutch system according to the manufacturer's instructions.

Manual Drive Train & Axles

(continued)

Performance ✓ Checklist

Name _____ Date _____ Class _____

PERFORMANCE STANDARDS:
Level 4–Performs skill without supervision and adapts to problem situations.
Level 3–Performs skill satisfactorily without assistance or supervision.
Level 2–Performs skill satisfactorily, but requires assistance/supervision.
Level 1–Performs parts of skill satisfactorily, but requires considerable assistance/supervision.

Attempt (circle one): **1 2 3 4**

Comments:

_____

_____

_____

_____

PERFORMANCE LEVEL ACHIEVED: _____

_____ 1. Safety rules and practices were followed at all times regarding this job.

_____ 2. Tools and equipment were used properly and stored upon completion of this job.

_____ 3. This completed job met the standards set and was done within the allotted time.

_____ 4. No injury or damage to property occurred during this job.

_____ 5. Upon completion of this job, the work area was cleaned correctly.

Instructor's Signature _____ Date_____

DIAGNOSING CLUTCH NOISES AND CONCERNS

NATEF Standard(s) for Manual Drive Train & Axles:

B1 Diagnose clutch noise, binding, slippage, pulsation, and chatter; determine necessary action.

Safety First

- ☐ Wear safety glasses at all times.
- ☐ Follow all safety rules when using common hand tools.
- ☐ Use exhaust vent if running engine in a closed shop. Unvented exhaust fumes can cause death.
- ☐ Watch out for hot or moving parts.
- ☐ Follow all safety rules when using a lift or jack and jack stands.

Tools & Equipment:
- Vehicle service information
- Lift or jack and jack stands
- Common hand tools
- Trouble light
- Automotive stethoscope
- Pry bar

PROCEDURES Refer to the vehicle service information for specifications and special procedures. Then diagnose the clutch noise and slippage in the vehicle provided by your instructor.

_____ 1. Write up a repair order.

_____ 2. Make sure you follow all procedures in the vehicle service information.

_____ 3. With the engine running, check for a bearing noise when the clutch is disengaged and the transmission/transaxle is in NEUTRAL. This would likely indicate a defective release bearing. On a vehicle with a transmission, shift into low gear and listen for a squealing noise. This could indicate a dry (no lube) pilot bushing or a defective pilot bearing.

_____ 4. A bearing noise when the clutch is engaged may indicate insufficient clutch pedal free play and a worn release bearing.

_____ 5. A clutch that seems to be binding or not releasing may be the result of excessive clutch pedal free play or a low fluid level in a hydraulic clutch master cylinder. This can also be caused by a defective clutch disc.

_____ 6. Clutch slippage can be caused by improper linkage adjustment, resulting in the clutch not being fully engaged. A defective hydraulic clutch master cylinder or oil on the clutch disc can also cause this.

_____ 7. Clutch chatter may be caused by loose or defective engine mounts; oil on the clutch disc; or a worn-out, burned, or glazed clutch disc.

(continued)

Performance ✓ Checklist

Name _____ Date _____ Class _____

PERFORMANCE STANDARDS:
Level 4–Performs skill without supervision and adapts to problem situations.
Level 3–Performs skill satisfactorily without assistance or supervision.
Level 2–Performs skill satisfactorily, but requires assistance/supervision.
Level 1–Performs parts of skill satisfactorily, but requires considerable assistance/supervision.

Attempt (circle one): **1 2 3 4**

Comments:

_____

_____

_____

_____

PERFORMANCE LEVEL ACHIEVED: _____

_____ **1.** Safety rules and practices were followed at all times regarding this job.

_____ **2.** Tools and equipment were used properly and stored upon completion of this job.

_____ **3.** This completed job met the standards set and was done within the allotted time.

_____ **4.** No injury or damage to property occurred during this job.

_____ **5.** Upon completion of this job, the work area was cleaned correctly.

Instructor's Signature _____ Date_____

REPAIRING CLUTCH DISCS AND PRESSURE PLATES

NATEF Standard(s) for Manual Drive Train & Axles:

B5 Inspect and replace clutch pressure plate assembly and clutch disc.

Safety First

- ☐ Wear safety glasses at all times.
- ☐ Follow all safety rules when using common hand tools.
- ☐ Treat every friction material as if it is asbestos.
- ☐ Check Material Safety Data Sheets for chemical safety information.
- ☐ Follow all safety rules when using a transmission jack.
- ☐ Follow all lifting rules when installing heavy components.

Tools & Equipment:

- Scribe
- Transmission jack
- Cleaning solvent
- Clutch alignment tool
- Steel ruler
- Common hand tools
- Flywheel turner
- Torque wrench
- Vehicle service information

PROCEDURES Refer to the vehicle service information for specifications and special procedures. Then inspect and replace, as necessary, the clutch disc or pressure plate in the vehicle provided by your instructor.

_____ 1. Write up a repair order.

_____ 2. Make sure you follow all procedures in the vehicle service information.

_____ 3. Remove the transmission from the vehicle.

_____ 4. Inspect the clutch bellhousing for oil, metal, or pieces of clutch facing. **CAUTION:** Treat every friction component as if it contains asbestos.

_____ 5. Clean the clutch pressure plate with a non-aerosol solvent.

_____ 6. Mark the flywheel and pressure plate with a scribe.

_____ 7. Remove the pressure plate and clutch disc.

_____ 8. Replace the clutch disc and pressure plate using a clutch alignment tool. _Note:_ Tighten pressure plate bolts in a crisscross pattern.

_____ 9. Torque all bolts to specification.

_____ 10. Re-install the transmission.

(continued)

Manual Drive Train & Axles

Performance ✓ Checklist

Name _____ Date _____ Class _____

PERFORMANCE STANDARDS:
Level 4–Performs skill without supervision and adapts to problem situations.
Level 3–Performs skill satisfactorily without assistance or supervision.
Level 2–Performs skill satisfactorily, but requires assistance/supervision.
Level 1–Performs parts of skill satisfactorily, but requires considerable assistance/supervision.

Attempt (circle one): **1 2 3 4**

Comments:

_____

_____

_____

_____

PERFORMANCE LEVEL ACHIEVED: _____

_____ 1. Safety rules and practices were followed at all times regarding this job.

_____ 2. Tools and equipment were used properly and stored upon completion of this job.

_____ 3. This completed job met the standards set and was done within the allotted time.

_____ 4. No injury or damage to property occurred during this job.

_____ 5. Upon completion of this job, the work area was cleaned correctly.

Instructor's Signature _____ Date_____

INSPECTING, REMOVING & REPLACING FLYWHEELS AND RING GEARS

NATEF Standard(s) for Manual Drive Train & Axles:

B8 Inspect flywheel and ring gear for wear and cracks, measure runout; determine necessary action.

Safety First

- ☐ Wear safety glasses at all times.
- ☐ Follow all safety rules when using common hand tools.
- ☐ Check Material Safety Data Sheets for chemical safety information.
- ☐ Follow all safety rules when using jack stands, ramps, or lifts.
- ☐ Follow all lifting rules when installing heavy components.

Tools & Equipment:

- Drift punch
- Soft-faced mallet
- Torque wrench
- Common hand tools
- Hammer
- Flywheel turner
- Jack stands, ramps, or lifts
- Vehicle service information
- Vise grip pliers
- Sharp awl
- Cleaning solvent
- Crocus (emery) cloth

PROCEDURES Refer to the vehicle service information for specifications and special procedures. Then remove, inspect, and reinstall or replace the flywheel and ring gear in the vehicle provided by your instructor.

_____ 1. Write up a repair order.

_____ 2. Make sure you follow all procedures in the vehicle service information.

_____ 3. Look up the removal procedure for a flywheel in the vehicle service information.

Source: _____ _Page:_ _____

_____ 4. Safely raise the vehicle.

_____ 5. Inspect the flywheel surface for oil leaks that might be coming from the rear main bearing seal, the camshaft plug, the pilot bushing/bearing, the flywheel mounting bolts, the transmission input shaft, or a clogged transmission vent.

_____ 6. Clean the flywheel and ring gear by using an approved solvent and a clean rag.

_____ 7. Look for indications of heat damage on the flywheel—such as any black, brown, or bluish discoloration.

_____ 8. Check the flywheel surface for cracks, grooves, wear, or other damage. _Note:_ A flywheel must be resurfaced by an automotive machinist or be replaced.

_____ 9. Examine any flywheels with dowel pins for damaged, loose, and missing pins. _Note:_ Damaged, loose, and missing pins must be replaced.

(_continued_)

_____ 10. Inspect the flywheel ring gear teeth by turning the flywheel with the flywheel turner.

_____ 11. Remove the flywheel by making alignment marks across it and the crankshaft flange and loosening the flywheel bolts gradually in a crisscross pattern with a socket wrench. **CAUTION:** The flywheel is heavy. Be sure to have help when removing it.

_____ 12. Remove the flywheel and ring gear from the crankshaft flange.

_____ 13. Inspect the flywheel mounting bolts and replace those that are worn or damaged. *Note:* Use replacement bolts that are of the same hardness and quality of the old ones.

_____ 14. Clean and inspect the crankshaft flange and remove any rough spots with a crocus (emery) cloth.

_____ 15. Inspect the replacement flywheel to be sure that the machinist made all the repairs.

_____ 16. Install the flywheel (with the help of an assistant); then tighten the mounting bolts gradually in a crisscross pattern.

_____ 17. Torque the mounting bolts to the manufacturer's specifications.

_____ 18. Look up flywheel runout specifications for the vehicle in the vehicle service information.

Source: _____ *Page:* _____

_____ 19. Check the flywheel runout to make sure it is within the manufacturer's specifications.

_____ 20. Safely lower the vehicle to the ground.

Performance ✓ Checklist

Name _____   Date _____   Class _____

PERFORMANCE STANDARDS:
Level 4–Performs skill without supervision and adapts to problem situations.
Level 3–Performs skill satisfactorily without assistance or supervision.
Level 2–Performs skill satisfactorily, but requires assistance/supervision.
Level 1–Performs parts of skill satisfactorily, but requires considerable assistance/supervision.

Attempt (circle one): **1 2 3 4**

Comments:

_____

_____

_____

_____

PERFORMANCE LEVEL ACHIEVED: _____

_____ 1. Safety rules and practices were followed at all times regarding this job.

_____ 2. Tools and equipment were used properly and stored upon completion of this job.

_____ 3. This completed job met the standards set and was done within the allotted time.

_____ 4. No injury or damage to property occurred during this job.

_____ 5. Upon completion of this job, the work area was cleaned correctly.

Instructor's Signature _____   Date _____

MEASURING CRANKSHAFT ENDPLAY AND FLYWHEEL RUNOUT

NATEF Standard(s) for Manual Drive Train & Axles:

B10 Measure flywheel runout and crankshaft endplay; determine necessary action.

Safety First

- ☐ Wear safety glasses at all times.
- ☐ Follow all safety rules when using common hand tools.
- ☐ Check Material Safety Data Sheets for chemical safety information.

Tools & Equipment:
- Bristle brush
- Flywheel turner
- Vehicle service information
- Torque wrench
- Cleaning solvent
- Dial indicator
- Common hand tools

PROCEDURES Refer to the vehicle service information for specifications and special procedures. Then measure the crankshaft endplay and flywheel runout in the vehicle provided by your instructor.

_____ 1. Write up a repair order.

_____ 2. Make sure you follow all procedures in the vehicle service information.

_____ 3. Clean and dry the crankshaft flange with the bristle brush and an approved cleaner.

_____ 4. Look up the manufacturer's specifications for crankshaft endplay.

   Source: _____ _Page:_ _____

_____ 5. Push the crankshaft as far as you can towards the front of the vehicle by pressing on the crankshaft flange.

_____ 6. Mount the dial indicator on the engine block with its plunger at a 90° angle to the crankshaft flange.

_____ 7. Move the crankshaft as far as you can towards the open clutch area by pushing on the harmonic balancer.

_____ 8. Record the measurement on the dial indicator.

   _____

_____ 9. Compare your measurement to the manufacturer's specifications. _Note:_ If the crankshaft endplay is excessive, the flywheel runout will not be accurate either.

_____ 10. Look up the manufacturer's specifications for flywheel runout.

   Source: _____ _Page:_ _____

(_continued_)

_____ 11. Clean and dry the flywheel friction surface. Make sure the bolts are also clean and dry.

_____ 12. Mount a dial indicator at a 90° angle to the side of the flywheel that contacts the pressure plate.

_____ 13. Measure the flywheel runout on the indicator by turning the flywheel with the flywheel turner.

_____ 14. Compare the runout to the manufacturer's specifications. If it is excessive, refer the flywheel to a machinist for possible resurfacing.

_____ 15. Re-check the runout on the replacement flywheel after it has been reinstalled. **CAUTION:** Be sure to torque all the bolts to the manufacturer's recommendations.

Performance ✓ Checklist

Name _____   Date _____   Class _____

PERFORMANCE STANDARDS:
Level 4–Performs skill without supervision and adapts to problem situations.
Level 3–Performs skill satisfactorily without assistance or supervision.
Level 2–Performs skill satisfactorily, but requires assistance/supervision.
Level 1–Performs parts of skill satisfactorily, but requires considerable assistance/supervision.

Attempt (circle one): **1 2 3 4**

Comments:

_____

_____

_____

_____

PERFORMANCE LEVEL ACHIEVED: _____

_____ 1. Safety rules and practices were followed at all times regarding this job.
_____ 2. Tools and equipment were used properly and stored upon completion of this job.
_____ 3. This completed job met the standards set and was done within the allotted time.
_____ 4. No injury or damage to property occurred during this job.
_____ 5. Upon completion of this job, the work area was cleaned correctly.

Instructor's Signature _____   Date_____

INSPECTING THE ENGINE BLOCK, CLUTCH HOUSING, AND TRANSMISSION CASE

NATEF Standard(s) for Manual Drive Train & Axles:

B9 Inspect engine block, clutch (bell) housing, and transmission/transaxle case mating surfaces, and alignment dowels; determine necessary action.

Safety First

☐ Wear safety glasses at all times.
☐ Follow all safety rules when using common hand tools.
☐ Check Material Safety Data Sheets for chemical safety information.
☐ Follow all safety rules when using a transmission jack.

Tools & Equipment:
- Transmission jack
- Vehicle service information
- Cleaning solvent
- Common hand tools

PROCEDURES Refer to the vehicle service information for specifications and special procedures. Then inspect and clean the mating surfaces of the engine block, clutch housing, and transmission case in the vehicle provided by your instructor.

_____ 1. Write up a repair order.

_____ 2. Make sure you follow all procedures in the vehicle service information.

_____ 3. Remove the transmission.

_____ 4. Inspect the engine block, clutch housing, and transmission case mating surfaces for oil leaks, cracks, or damage.

_____ 5. Install the transmission.

Manual Drive Train & Axles

(continued)

Performance ✓ Checklist

Name _____ Date _____ Class _____

PERFORMANCE STANDARDS:
Level 4–Performs skill without supervision and adapts to problem situations.
Level 3–Performs skill satisfactorily without assistance or supervision.
Level 2–Performs skill satisfactorily, but requires assistance/supervision.
Level 1–Performs parts of skill satisfactorily, but requires considerable assistance/supervision.

Attempt (circle one): **1 2 3 4**

Comments:

_____

_____

_____

_____

PERFORMANCE LEVEL ACHIEVED: _____

_____ **1.** Safety rules and practices were followed at all times regarding this job.

_____ **2.** Tools and equipment were used properly and stored upon completion of this job.

_____ **3.** This completed job met the standards set and was done within the allotted time.

_____ **4.** No injury or damage to property occurred during this job.

_____ **5.** Upon completion of this job, the work area was cleaned correctly.

Instructor's Signature _____ Date_____

INSPECTING, ADJUSTING & REINSTALLING SHIFT LINKAGES

NATEF Standard(s) for Manual Drive Train & Axles:

C5 Inspect, adjust, and reinstall shift linkages, brackets, bushings, cables, pivots, and levers.

Safety First

☐ Wear safety glasses at all times.
☐ Follow all safety rules when using common hand tools.
☐ Check Material Safety Data Sheets for chemical safety information.
☐ Follow all safety rules when using jacks and jack stands, ramps, or lifts.

Tools & Equipment:

- Shifter adjustment pin
- Common hand tools
- Cleaning solvent
- Vehicle service information
- Jacks and jack stands, ramps, or lifts

PROCEDURES Refer to the vehicle service information for specifications and special procedures. Then inspect, adjust, and reinstall the shift linkage in the vehicle provided by your instructor.

_____ 1. Write up a repair order.

_____ 2. Make sure you follow all procedures in the vehicle service information.

_____ 3. Look up the manufacturer's recommended procedure for removing and servicing the shift linkage.

Source: _____ *Page:* _____

_____ 4. Shift the transmission into NEUTRAL and raise the car on jacks or a lift. *Note:* Some external linkages can be reached from under the hood.

_____ 5. Find the shift linkage rods and clean them with a clean rag and an appropriate cleaner.

_____ 6. Inspect the linkage components by looking for missing, sticky, loose, bent, damaged and worn parts. *Note:* Some rods are normally bent.

_____ 7. Replace any worn or damaged parts before adjusting the linkage. *Note:* Be sure to look closely at the many plastic and rubber parts that wear and break easily.

_____ 8. Adjust the linkage by first loosening the linkage rod adjustment screws, locknuts, and other linkage rod hold-downs.

_____ 9. Put the transmission-side shifter arms in NEUTRAL and insert the shifter adjustment pin through both the linkage rod levers and the gear shift lever bracket. *Note:* This holds the rod levers in NEUTRAL.

(continued)

_____ 10. Tighten the adjustment screws and locknuts to hold the linkage rod at the manufacturer's recommended distance between the shifter arm and the linkage rod levers.

_____ 11. Remove the shifter adjustment pin.

_____ 12. Adjust the transmission lock rod linkage. *Note:* It should not be loose and the ignition key should turn the ignition switch easily into and out of the lock position.

_____ 13. Lubricate the linkage with an appropriate lubricant.

_____ 14. Lower the vehicle to the ground and move the transmission through all its gears. *Note:* If the gear shift lever does not shift smoothly, recheck the linkage before considering a transmission overhaul.

Performance ✓ Checklist

Name _____ Date _____ Class _____

PERFORMANCE STANDARDS:
Level 4–Performs skill without supervision and adapts to problem situations.
Level 3–Performs skill satisfactorily without assistance or supervision.
Level 2–Performs skill satisfactorily, but requires assistance/supervision.
Level 1–Performs parts of skill satisfactorily, but requires considerable assistance/supervision.

Attempt (circle one): 1 2 3 4

Comments:

_____

_____

_____

_____

PERFORMANCE LEVEL ACHIEVED: _____

_____ 1. Safety rules and practices were followed at all times regarding this job.

_____ 2. Tools and equipment were used properly and stored upon completion of this job.

_____ 3. This completed job met the standards set and was done within the allotted time.

_____ 4. No injury or damage to property occurred during this job.

_____ 5. Upon completion of this job, the work area was cleaned correctly.

Instructor's Signature _____ Date_____

DRAINING & ADDING FLUID

NATEF Standard(s) for Manual Drive Train & Axles:

A6 Drain and fill manual transmission/transaxle and final drive unit.

Safety First
- ☐ Wear safety glasses at all times.
- ☐ Follow all safety rules when using common hand tools.
- ☐ Follow all safety rules when using an automotive lift.

Tools & Equipment:
- Vehicle service information
- Proper lubricant and pump
- Common hand tools
- Fluid drain container
- Automotive lift
- Suction pump

PROCEDURES Refer to the vehicle service information for specifications and special procedures. Then drain and fill the manual transmission/transaxle in the vehicle provided by your instructor.

_____ 1. Write up a repair order.

_____ 2. Make sure you follow all procedures in the vehicle service information.

Draining and Filling a Manual Transmission/Transaxle

_____ 1. Drain the lubricant into an approved container. Some transmissions/transaxles have a lubricant drain plug. Some drain the lubricant through a lower bolt attaching a housing to the case. For others, the lubricant is removed through the filler hole.

_____ 2. Carefully inspect the plug for clinging metal fragments. Some transmission/transaxle drain plugs are magnetized. Metal fragments can indicate internal problems.

_____ 3. Before adding new lubricant, make sure it is the correct viscosity and type. Some manufacturers specify 90 weight gear oil, while others may specify automatic transmission fluid or engine oil.

_____ 4. Use a pump to fill the transmission/transaxle to the specified fluid level and replace the filler plug. **CAUTION:** Before removing a transmission/transaxle filler or drain plug, be sure to identify the proper plug. It is a common mistake to accidentally remove an external bolt, causing major internal problems.

(continued)

Manual Drive Train & Axles

Draining and Filling a Final Drive Unit

_____ 1. The final drive unit is usually a part of the transaxle and uses the same fluid. However, some final drives may have a separate lubricant. Follow manufacturer's instructions in these rare cases.

_____ 2. Most rear-wheel-drive vehicles have separate transmissions and final drives. The differential or rear axle assembly is serviced separately from the transmission. If these units don't have a drain plug, the gear oil is drained from a lower inspection cover bolt or removed from the filler plug hole using a suction pump. Refill the unit through the filler hole using a pump and the specified gear oil. Be sure to refill the unit to the proper level. **CAUTION:** Some limited slip differentials require a special limited slip lubricant or an additive.

Performance ✓ Checklist

Name _____   Date _____   Class _____

PERFORMANCE STANDARDS:
Level 4–Performs skill without supervision and adapts to problem situations.
Level 3–Performs skill satisfactorily without assistance or supervision.
Level 2–Performs skill satisfactorily, but requires assistance/supervision.
Level 1–Performs parts of skill satisfactorily, but requires considerable assistance/supervision.

Attempt (circle one): **1 2 3 4**

Comments:

_____

_____

_____

_____

PERFORMANCE LEVEL ACHIEVED: _____

_____ 1. Safety rules and practices were followed at all times regarding this job.
_____ 2. Tools and equipment were used properly and stored upon completion of this job.
_____ 3. This completed job met the standards set and was done within the allotted time.
_____ 4. No injury or damage to property occurred during this job.
_____ 5. Upon completion of this job, the work area was cleaned correctly.

Instructor's Signature _____ Date_____

DIAGNOSING FLUID LOSS

NATEF Standard(s) for Manual Drive Train & Axles:
A5 Diagnose fluid loss, level, and condition concerns; determine necessary action.

Safety First

- ☐ Wear safety glasses at all times.
- ☐ Follow all safety rules when using common hand tools.
- ☐ Follow all safety rules when using an automotive lift.

Tools & Equipment:
- Vehicle service information
- Common hand tools
- Automotive lift

PROCEDURES Refer to the vehicle service information for specifications and special procedures. Then diagnose the fluid level concerns in the vehicle provided by your instructor.

_____ 1. Write up a repair order.

_____ 2. Make sure you follow all procedures in the vehicle service information.

_____ 3. Raise the vehicle on a lift and check for transmission lubricant leakage around all rusting surfaces and the area of the front and rear seals.

_____ 4. Check for overfull transmission lubricant. Remove fill plug and let excess lubricant drain to correct level.

_____ 5. With the fill plug removed, use your finger to check the lubricant level and compare to transmission specifications. Add proper lubricant as necessary.

_____ 6. Inspect a sample of the lubricant on your finger for contamination. Metal particles in the lubricant will indicate internal damage. Contaminated lubricant would indicate the need for a lubricant change.

_____ 7. Inspect the transmission vent for restriction and clear blockage as necessary. A plugged vent will cause an internal pressure buildup that may cause leakage.

(continued)

Manual Drive Train & Axles

Performance ✓ Checklist

Name _____ Date _____ Class _____

PERFORMANCE STANDARDS:
Level 4–Performs skill without supervision and adapts to problem situations.
Level 3–Performs skill satisfactorily without assistance or supervision.
Level 2–Performs skill satisfactorily, but requires assistance/supervision.
Level 1–Performs parts of skill satisfactorily, but requires considerable assistance/supervision.

Attempt (circle one): **1 2 3 4**

Comments:

_____

_____

_____

_____

PERFORMANCE LEVEL ACHIEVED: _____

_____ **1.** Safety rules and practices were followed at all times regarding this job.

_____ **2.** Tools and equipment were used properly and stored upon completion of this job.

_____ **3.** This completed job met the standards set and was done within the allotted time.

_____ **4.** No injury or damage to property occurred during this job.

_____ **5.** Upon completion of this job, the work area was cleaned correctly.

Instructor's Signature _____ Date_____

DIAGNOSING NOISE, GEAR PROBLEM, AND FLUID LEAKAGE CONCERNS

NATEF Standard(s) for Manual Drive Train & Axles:

A2 Identify and interpret drive train concern; determine necessary action.

C4 Diagnose noise, hard shifting, jumping out of gear, and fluid leakage concerns; determine necessary action.

Safety First

- ☐ Wear safety glasses at all times.
- ☐ Follow all safety rules when using common hand tools.
- ☐ Use exhaust vent if running engine in a closed shop. Unvented exhaust fumes can cause death.
- ☐ Watch out for hot or moving parts.
- ☐ Follow all safety rules when using a lift or jack and jack stands.

Tools & Equipment:
- Vehicle service information
- Automotive stethoscope
- Engine support fixture (when removing transaxles)
- Common hand tools
- Trouble light
- Lift or jack and jack stands
- Required special tools

PROCEDURES Refer to the vehicle service information for specifications and special procedures. Then diagnose noises, gear problems, and fluid leakage in the vehicle provided by your instructor.

_____ 1. Write up a repair order.

_____ 2. Make sure you follow all procedures in the vehicle service information.

WITH TRANSMISSION/TRANSAXLE IN THE VEHICLE

Diagnosing Noise

_____ 1. Check for proper lubricant fluid level.

_____ 2. Check for loose fasteners or mounts.

_____ 3. Check for sounds of internal gear noise.

_____ 4. Check for sounds of internal bearing noise.

_____ 5. If gears are grinding when shifting into low or reverse, check for proper clutch operation.

Diagnosing Hard Shifting

_____ 1. Check shift linkage for binding and proper adjustment as necessary.

_____ 2. If gears are grinding when shifting into low or reverse, check for proper clutch operation.

(continued)

Manual Drive Train & Axles

Diagnosing Jumping Out of Gear

_____ 1. Check for defective or loose engine or transmission/transaxle mounts.

_____ 2. Check shift linkage for vehicle body or frame contact.

_____ 3. Check for a hardened rubber gear shift boot.

_____ 4. Check for broken shift rail detent ball springs as necessary or accessible.

Diagnosing Fluid Leakage

_____ 1. Inspect for a plugged vent.

_____ 2. Check for excessively high lubricant level.

_____ 3. Inspect around all mating surfaces for loose fasteners or defective gaskets.

_____ 4. Inspect all seals for leakage.

WITH TRANSMISSION/TRANSAXLE OUT
OF VEHICLE AND DISASSEMBLED

Diagnosing Noise

_____ 1. Inspect condition of all gears. Look for abnormal wear or broken teeth.

_____ 2. Inspect for proper gear endplay. Check condition of thrust washers.

_____ 3. Inspect condition of all bearings.

Diagnosing Hard Shifting

_____ 1. Inspect shift linkage for binding.

_____ 2. Inspect shift rails for binding.

_____ 3. Inspect condition of interlock pins or balls.

_____ 4. Inspect for bent or damaged shift forks.

_____ 5. Inspect condition of synchronizers and mating gear teeth.

Diagnosing Jumping Out of Gear

_____ 1. Inspect shift rail detent balls and springs.

_____ 2. Inspect gear/shaft endplay and thrust washers.

_____ 3. Inspect condition of synchronizers and mating gear teeth.

Performance ✓ Checklist

Name _____ Date _____ Class _____

PERFORMANCE STANDARDS:
Level 4–Performs skill without supervision and adapts to problem situations.
Level 3–Performs skill satisfactorily without assistance or supervision.
Level 2–Performs skill satisfactorily, but requires assistance/supervision.
Level 1–Performs parts of skill satisfactorily, but requires considerable assistance/supervision.

Attempt (circle one): **1 2 3 4**

Comments:

_____

_____

_____

_____

PERFORMANCE LEVEL ACHIEVED: _____

_____ 1. Safety rules and practices were followed at all times regarding this job.

_____ 2. Tools and equipment were used properly and stored upon completion of this job.

_____ 3. This completed job met the standards set and was done within the allotted time.

_____ 4. No injury or damage to property occurred during this job.

_____ 5. Upon completion of this job, the work area was cleaned correctly.

Instructor's Signature _____ Date _____

Manual Drive Train
& Axles

REMOVING & REINSTALLING TRANSMISSIONS AND TRANSAXLES

NATEF Standard(s) for Manual Drive Train & Axles:
C1 Remove and reinstall transmission/transaxle.

Safety First
- ☐ Wear safety glasses at all times.
- ☐ Follow all safety rules when using common hand tools.
- ☐ Check Material Safety Data Sheets for chemical safety information.
- ☐ Follow all safety rules when using a transmission jack.
- ☐ Follow all safety rules when using compressed air.

Tools & Equipment:
- Bristle brush
- Transmission jack
- Torque wrench
- Common hand tools
- Cleaning solvent
- Vehicle service information

PROCEDURES Refer to the vehicle service information for specifications and special procedures. Then remove and reinstall or replace the transmission/transaxle in the vehicle provided by your instructor.

_____ 1. Write up a repair order.

_____ 2. Make sure you follow all procedures in the vehicle service information.

_____ 3. Clean and dry the crankshaft flange with the bristle brush and an approved cleaner.

_____ 4. Look up the manufacturer's recommended procedure for removing the transmission/transaxle.
Source: _____ Page:_____

_____ 5. Remove the transmission/transaxle according to the manufacturer's recommendations.
CAUTION: Do not pry on aluminum parts with a screwdriver. Use a soft plastic mallet on hard-to-remove parts.

_____ 6. Reinstall the original or install the replacement transmission/transaxle according to the manufacturer's instructions.

_____ 7. Torque all bolts to the manufacturer's specifications.

(continued)

Performance ✓ Checklist

Name _____   Date _____   Class _____

PERFORMANCE STANDARDS:
Level 4–Performs skill without supervision and adapts to problem situations.
Level 3–Performs skill satisfactorily without assistance or supervision.
Level 2–Performs skill satisfactorily, but requires assistance/supervision.
Level 1–Performs parts of skill satisfactorily, but requires considerable assistance/supervision.

Attempt (circle one): **1 2 3 4**

Comments:

_____

_____

_____

_____

PERFORMANCE LEVEL ACHIEVED: _____

_____ **1.** Safety rules and practices were followed at all times regarding this job.

_____ **2.** Tools and equipment were used properly and stored upon completion of this job.

_____ **3.** This completed job met the standards set and was done within the allotted time.

_____ **4.** No injury or damage to property occurred during this job.

_____ **5.** Upon completion of this job, the work area was cleaned correctly.

Instructor's Signature _____ Date_____

REMOVING & REPLACING TRANSMISSION GASKETS AND SEALANTS

NATEF Standard(s) for Manual Drive Train & Axles:

C7 Inspect and replace gaskets, seals, and sealants; inspect sealing surfaces.

Safety First

☐ Wear safety glasses at all times.
☐ Follow all safety rules when using common hand tools.
☐ Check Material Safety Data Sheets for chemical safety information.
☐ Follow all safety rules when using jacks and jack stands, ramps, or lifts.

Tools & Equipment:

- Bristle brush
- Jacks and jack stands, ramps, or lifts
- Torque wrench
- Vehicle service information
- Cleaning solvent
- Common hand tools

PROCEDURES Refer to the vehicle service information for specifications and special procedures. Then remove and replace the transmission gaskets and sealants in the vehicle provided by your instructor.

_____ 1. Write up a repair order.

_____ 2. Make sure you follow all procedures in the vehicle service information.

_____ 3. Raise the vehicle and identify the transmission installed in the vehicle by the transmission identification tag or number. *Note:* You may have to clean it with the bristle brush to read it.

_____ 4. Look up the manufacturer's recommended procedure for removing and replacing the transmission gaskets and sealants.

Source: _____ Page: _____

_____ 5. Check the drive train and axles for leaks. Record any areas that will need new gaskets.

_____

_____ 6. Shift the transmission into NEUTRAL and then drain the transmission fluid. **CAUTION:** Be sure to dispose of it according to EPA regulations.

_____ 7. If the transmission or transaxle must be removed, follow all the manufacturer's recommendations for removing it. **CAUTION:** Neither an engine support nor a transmission jack can hold both the engine and the transmission safely at once. Be sure to support them adequately before proceeding.

_____ 8. Clean the transmission housings with the bristle brush and an approved solvent.

(continued)

_____ 9. Remove the transmission components according to the manufacturer's recommendations. **CAUTION:** Do not pry on aluminum parts with a screwdriver. Use a soft plastic mallet on hard-to-remove parts.

_____ 10. Inspect the gaskets for cracks, leaks, and other damage.

_____ 11. Check all mating surfaces for scratches, gouges, and left-over gasket material. *Note:* Replace any damaged component.

_____ 12. Remove and discard old gasket material. *Note:* The new gaskets will be installed during the reassembly process.

_____ 13. Clean, inspect, and tag the components as they are removed from the assembly. *Note:* Store the components together in a plastic bag to keep from losing any of them.

_____ 14. Remove and replace the rear engine oil seal before reinstalling the transmission, if necessary.

_____ 15. Install all the components, new gaskets, and sealants according to the manufacturer's recommendations.

_____ 16. Torque the housing bolts to the manufacturer's specifications. *Note:* Check the endplay during the reassembly process.

_____ 17. Shift the transmission through all its gears while turning the input shaft by hand. *Note:* If it does not shift smoothly, the transmission must be disassembled and reinspected.

_____ 18. Safely lower the vehicle to the ground.

Performance ✓ Checklist

Name _____ Date _____ Class _____

PERFORMANCE STANDARDS:
Level 4–Performs skill without supervision and adapts to problem situations.
Level 3–Performs skill satisfactorily without assistance or supervision.
Level 2–Performs skill satisfactorily, but requires assistance/supervision.
Level 1–Performs parts of skill satisfactorily, but requires considerable assistance/supervision.

Attempt (circle one): **1 2 3 4**

Comments:

_____

_____

_____

_____

PERFORMANCE LEVEL ACHIEVED: _____

_____ 1. Safety rules and practices were followed at all times regarding this job.

_____ 2. Tools and equipment were used properly and stored upon completion of this job.

_____ 3. This completed job met the standards set and was done within the allotted time.

_____ 4. No injury or damage to property occurred during this job.

_____ 5. Upon completion of this job, the work area was cleaned correctly.

Instructor's Signature _____ Date_____

INSPECTING & INSTALLING SPEEDOMETER GEARS, SPEED SENSOR, AND RETAINERS

NATEF Standard(s) for Manual Drive Train & Axles:
C12 Inspect and reinstall speedometer drive gear, driven gear, vehicle speed sensor (VSS), and retainers.

Safety First
☐ Wear safety glasses at all times.
☐ Follow all safety rules when using common hand tools.

Tools & Equipment:
- Common hand tools
- Vehicle service information

PROCEDURES Refer to the vehicle service information for specifications and special procedures. Then inspect and install the speedometer gears, speed sensor, and retainers in the vehicle provided by your instructor.

_____ 1. Write up a repair order.

_____ 2. Make sure you follow all procedures in the vehicle service information.

_____ 3. Look up the manufacturer's recommended procedure for inspecting and installing the speedometer gears, speed sensor, and retainers.

   Source: _____ *Page:* _____

_____ 4. Follow all the manufacturer's safety recommendations.

_____ 5. Inspect and reinstall the speedometer gears, speed sensor, and retainers according to the manufacturer's instructions. ***Note:*** In a transmission, the speedometer assembly is often mounted at the rear of the output shaft. In a transaxle, the speedometer assembly may be on the transmission section gear shafts or on the differential case in the differential section of the transmission.

(continued)

Performance ✓ Checklist

Name _____ Date _____ Class _____

PERFORMANCE STANDARDS:
Level 4–Performs skill without supervision and adapts to problem situations.
Level 3–Performs skill satisfactorily without assistance or supervision.
Level 2–Performs skill satisfactorily, but requires assistance/supervision.
Level 1–Performs parts of skill satisfactorily, but requires considerable assistance/supervision.

Attempt (circle one): **1 2 3 4**

Comments:

_____

_____

_____

_____

PERFORMANCE LEVEL ACHIEVED: _____

_____ **1.** Safety rules and practices were followed at all times regarding this job.

_____ **2.** Tools and equipment were used properly and stored upon completion of this job.

_____ **3.** This completed job met the standards set and was done within the allotted time.

_____ **4.** No injury or damage to property occurred during this job.

_____ **5.** Upon completion of this job, the work area was cleaned correctly.

Instructor's Signature _____ Date_____

REMOVING & SERVICING INTERNAL SHIFT ASSEMBLY COMPONENTS

NATEF Standard(s) for Manual Drive Train & Axles:

C9 Inspect, adjust, and reinstall shift cover, forks, levers, grommets, shafts, sleeves, detent mechanism, interlocks, and springs.

Safety First

☐ Wear safety glasses at all times.
☐ Follow all safety rules when using common hand tools.
☐ Check Material Safety Data Sheets for chemical safety information.
☐ Follow all safety rules when using jacks and jack stands, ramps, or lifts.
☐ Follow all safety rules when using compressed air.

Tools & Equipment:

- Bristle brush
- Drift punch
- Cleaning solvent
- Ball peen hammer
- Air compressor
- Common hand tools
- Jacks and jack stands, ramps, or lifts
- Vehicle service information

PROCEDURES Refer to the vehicle service information for specifications and special procedures. Then remove, inspect, and replace or reinstall the internal shift assembly components in the vehicle provided by your instructor.

_____ 1. Write up a repair order.

_____ 2. Make sure you follow all procedures in the vehicle service information.

_____ 3. Identify the transmission installed in the vehicle by the transmission identification tag or number. *Note:* You may have to clean it with the bristle brush to read it.

_____ 4. Look up the manufacturer's recommended procedure for removing and servicing the shift assembly components.

Source: _____ Page: _____

_____ 5. Safely raise the vehicle.

_____ 6. Remove the transmission from the vehicle according to the manufacturer's recommended procedure.

_____ 7. Take out the gear shift lever by removing the retaining bolts and pulling straight out on the lever.

_____ 8. Remove the shift rails. *Note:* Label each rail according to its correct position for reassembly.

_____ 9. Remove the shift forks. *Note:* Keep each fork with its corresponding rail.

_____ 10. Clean all the parts with the bristle brush, an approved solvent, and compressed air.

(continued)

Manual Drive Train & Axles

_____ **11.** Inspect the rails and other parts for bends, cracks, breaks, and other damage.

_____ **12.** Check the rails and their bushings for wear by placing the rails in their bores and pushing them in and out. *Note:* Replace any worn or damaged parts.

_____ **13.** Inspect each shift fork on its rail. *Note:* Replace both of them if they do not slide easily or if they wobble.

_____ **14.** Inspect the shift fork tips, the shift shafts, and the interlock for wear. *Note:* Replace any worn or damaged parts. Always replace every drift pin and expansion plug when the transmission is overhauled.

_____ **15.** Reassemble the internal shift assembly components and shift the transmission through all its gears before installing the transmission into the vehicle. *Note:* If the transmission does not shift smoothly, repeat the inspection and assembly procedure until it shifts consistently.

_____ **16.** Safely lower the vehicle to the ground.

Performance ✓ Checklist

Name _____ Date _____ Class _____

PERFORMANCE STANDARDS:
Level 4–Performs skill without supervision and adapts to problem situations.
Level 3–Performs skill satisfactorily without assistance or supervision.
Level 2–Performs skill satisfactorily, but requires assistance/supervision.
Level 1–Performs parts of skill satisfactorily, but requires considerable assistance/supervision.

Attempt (circle one): **1 2 3 4**

Comments:

_____

_____

_____

_____

PERFORMANCE LEVEL ACHIEVED: _____

_____ **1.** Safety rules and practices were followed at all times regarding this job.

_____ **2.** Tools and equipment were used properly and stored upon completion of this job.

_____ **3.** This completed job met the standards set and was done within the allotted time.

_____ **4.** No injury or damage to property occurred during this job.

_____ **5.** Upon completion of this job, the work area was cleaned correctly.

Instructor's Signature _____ Date_____

REMOVING & REINSTALLING TRANSMISSION COMPONENTS

NATEF Standard(s) for Manual Drive Train & Axles:

C2 Disassemble, clean, and reassemble transmission/transaxle components.

Safety First

- ☐ Wear safety glasses at all times.
- ☐ Follow all safety rules when using common hand tools.
- ☐ Check Material Safety Data Sheets for chemical safety information.
- ☐ Follow all safety rules when using jacks and jack stands, ramps, or lifts.
- ☐ Follow all safety rules when using compressed air.

Tools & Equipment:

- Bristle brush
- Cleaning solvent
- Vehicle service information
- Torque wrench
- Common hand tools
- Jacks and jack stands, ramps, or lifts

PROCEDURES Refer to the vehicle service information for specifications and special procedures. Then remove and reinstall the transmission components in the vehicle provided by your instructor.

_____ 1. Write up a repair order.

_____ 2. Make sure you follow all procedures in the vehicle service information.

_____ 3. Identify the transmission installed in the vehicle by the transmission identification tag or number.

_____ 4. Look up the manufacturer's recommended procedure for removing and servicing the transmission components.

   Source: _____ *Page:* _____

_____ 5. Safely raise the vehicle.

_____ 6. Drain the transmission fluid. *Note:* Be sure to dispose of it according to EPA regulations.

_____ 7. Clean the transmission housings with the bristle brush and an approved solvent.

_____ 8. Inspect the transmission case for cracks and other damage. *Note:* Damage such as stripped threads can be repaired, but damaged housings must be replaced.

_____ 9. Remove the transmission components according to the manufacturer's recommendations. *Note:* Do not pry on aluminum parts with a screwdriver. Use a soft plastic mallet on hard-to-remove parts.

_____ 10. Check all mating surfaces for scratches, gouges, and left-over gasket material. *Note:* Replace any damaged component.

(continued)

Manual Drive Train & Axles

_____ 11. Remove and discard old gasket material.

_____ 12. Clean, inspect, and tag the components as they are removed from the assembly.

_____ 13. Remove and replace the rear engine oil seal before reinstalling the transmission, if necessary.

_____ 14. Replace the bushing or bearing in the transmission extension housing. *Note:* Be sure to align the oil hole in the bushing with the oil slot in the housing.

_____ 15. Lubricate all the clean, inspected components before reinstalling them.

_____ 16. Install all the components, new gaskets, and sealants. Torque the housing bolts to the manufacturer's specifications. *Note:* Check the endplay during the reassembly process.

_____ 17. Shift the transmission through all its gears while turning the input shaft by hand. *Note:* If it does not shift smoothly, the transmission must be disassembled and reinspected.

_____ 18. Safely lower the vehicle to the ground.

Performance ✓ Checklist

Name _____   Date _____   Class _____

PERFORMANCE STANDARDS:
Level 4–Performs skill without supervision and adapts to problem situations.
Level 3–Performs skill satisfactorily without assistance or supervision.
Level 2–Performs skill satisfactorily, but requires assistance/supervision.
Level 1–Performs parts of skill satisfactorily, but requires considerable assistance/supervision.

Attempt (circle one): **1 2 3 4**

Comments:

_____

_____

_____

_____

PERFORMANCE LEVEL ACHIEVED: _____

_____ 1. Safety rules and practices were followed at all times regarding this job.

_____ 2. Tools and equipment were used properly and stored upon completion of this job.

_____ 3. This completed job met the standards set and was done within the allotted time.

_____ 4. No injury or damage to property occurred during this job.

_____ 5. Upon completion of this job, the work area was cleaned correctly.

Instructor's Signature _____   Date_____

INSPECTING TRANSAXLE HOUSINGS, MATING SURFACES, BORES, AND BUSHINGS

NATEF Standard(s) for Manual Drive Train & Axles:

C3 Inspect transmission/transaxle case, extension housing, case mating surfaces, bores, bushings, and vents; perform necessary action.

Safety First
- ☐ Wear safety glasses at all times.
- ☐ Follow all safety rules when using common hand tools.
- ☐ Check Material Safety Data Sheets for chemical safety information.
- ☐ Follow all safety rules when using jacks and jack stands, ramps, or lifts.

Tools & Equipment:
- Bristle brush
- Cleaning solvent
- Vehicle service information
- Torque wrench
- Common hand tools
- Jacks and jack stands, ramps, or lifts

PROCEDURES Refer to the vehicle service information for specifications and special procedures. Then remove, inspect, and reinstall the transaxle components in the vehicle provided by your instructor.

_____ 1. Write up a repair order.

_____ 2. Make sure you follow all procedures in the vehicle service information.

_____ 3. Identify the transmission installed in the vehicle by the transmission identification tag or number. *Note:* You may have to clean it with the bristle brush to read it.

_____ 4. Look up the manufacturer's recommended procedure for removing and servicing the transaxle components.

Source: _____   Page:_____

_____ 5. Safely raise the vehicle.

_____ 6. Shift the transmission into NEUTRAL and then drain the transaxle fluid. **CAUTION:** Be sure to dispose of it according to EPA regulations.

_____ 7. If the engine must be removed, follow all the manufacturer's recommendations for removing the engine. **CAUTION:** Neither an engine support nor a transmission jack can hold both the engine and the transmission safely at once. Be sure to support them adequately before proceeding.

_____ 8. Clean the transaxle housings with the bristle brush and an approved solvent.

_____ 9. Inspect the transmission housing for cracks and other damage. *Note:* Damaged housings must be replaced, but damage such as stripped threads can be repaired. Transaxle housings are subject to more damage and wear than transmission housings because they are usually thinner walled and made out of aluminum.

(continued)

Manual Drive Train & Axles

_____ **10.** Remove the transaxle from the vehicle according to the manufacturer's recommendations.

_____ **11.** Remove the transaxle components according to the manufacturer's recommendations. **CAUTION:** Do not pry on aluminum parts with a screwdriver. Use a soft plastic mallet on hard-to-remove parts.

_____ **12.** Check all mating surfaces for scratches, gouges, and left-over gasket material. *Note:* Replace any damaged component.

_____ **13.** Remove and discard old gasket material. *Note:* New gaskets will be installed during the reassembly process.

_____ **14.** Clean, inspect, and tag the components as they are removed from the assembly. *Note:* Store them together in a plastic bag to keep from losing any of them.

_____ **15.** Remove and replace the rear oil seal before reinstalling the transaxle.

_____ **16.** Replace all the bushings or bearings for all the shafts. *Note:* Be sure to align the oil holes in the bushings or bearings with the oil slots.

_____ **17.** Lubricate all the clean, inspected components before reinstalling them. *Note:* Be sure to use a lubricant appropriate for aluminum parts.

_____ **18.** Install all the components, new gaskets, and sealants. Torque the housing bolts to the manufacturer's specifications. *Note:* Check the endplay during the reassembly process.

_____ **19.** Shift the transmission through all its gears while turning the input shaft by hand. *Note:* If it does not shift smoothly, the transmission must be disassembled and reinspected.

_____ **20.** Safely lower the vehicle to the ground.

Performance ✓ Checklist

Name _____ Date _____ Class _____

PERFORMANCE STANDARDS:
Level 4–Performs skill without supervision and adapts to problem situations.
Level 3–Performs skill satisfactorily without assistance or supervision.
Level 2–Performs skill satisfactorily, but requires assistance/supervision.
Level 1–Performs parts of skill satisfactorily, but requires considerable assistance/supervision.

Attempt (circle one): **1 2 3 4**

Comments:

_____

_____

_____

_____

PERFORMANCE LEVEL ACHIEVED: _____

_____ **1.** Safety rules and practices were followed at all times regarding this job.

_____ **2.** Tools and equipment were used properly and stored upon completion of this job.

_____ **3.** This completed job met the standards set and was done within the allotted time.

_____ **4.** No injury or damage to property occurred during this job.

_____ **5.** Upon completion of this job, the work area was cleaned correctly.

Instructor's Signature _____ Date_____

REMOVING, INSPECTING & REINSTALLING SYNCHRONIZER ASSEMBLY COMPONENTS

NATEF Standard(s) for Manual Drive Train & Axles:

C11 Inspect and reinstall synchronizer hub, sleeve, keys (inserts), springs, and blocking rings.

Safety First

☐ Wear safety glasses at all times.
☐ Follow all safety rules when using common hand tools.
☐ Check Material Safety Data Sheets for chemical safety information.
☐ Follow all safety rules when using a lift.
☐ Follow all safety rules when using compressed air.

Tools & Equipment:

- Feeler gauges
- Common hand tools
- Air compressor
- Scribe
- Vehicle service information
- Lift
- Cleaning solvent
- Bristle brush

PROCEDURES Refer to the vehicle service information for specifications and special procedures. Then remove, inspect, and reinstall the synchronizer assembly components in the vehicle provided by your instructor.

_____ 1. Write up a repair order.

_____ 2. Make sure you follow all procedures in the vehicle service information.

_____ 3. Look up the manufacturer's recommended procedure for servicing the synchronizer assembly.

Source: _____ Page:_____

_____ 4. Raise the vehicle and rotate the synchronizer hub on the shaft, check for any looseness, and replace the entire assembly if there is any looseness.

_____ 5. Mark the locations of the hub, the sleeve, and the synchronizer assembly on the shaft before you remove them for inspection so they can be put back in the same place. *Note:* Keep the rings with their mating gears.

_____ 6. Look up the manufacturer's specification for the distance between the synchronizer rings and the clutch teeth on their speed gears.

Source: _____ Page:_____

_____ 7. Measure the distance between the synchronizer rings and their clutch teeth. *Note:* Replace the synchronizer if the measurement is not within the manufacturer's specifications.

_____ 8. Inspect the assembly by removing the sleeve from the hub, the springs, and the keys.

(continued)

_____ 9. Clean the parts with the bristle brush and an appropriate solvent, then blow them dry with compressed air. **CAUTION:** Protect your eyes when using compressed air.

_____ 10. Inspect the hub and sleeve splines by looking for worn or damaged tips. *Note:* The tips should be pointed.

_____ 11. Check the hub and synchronizer rings for damage or wear. *Note:* Make sure that the teeth on the rings fit into their speed gears correctly, and check the teeth for any damage or wear.

_____ 12. Inspect the springs for bends and breaks, making certain that the small bent ends are not broken.

_____ 13. Lightly oil all the components with the recommended lubricant before reassembling.

_____ 14. Assemble the synchronizer assemblies before installing them on the output shaft, making sure that the springs engage all three keys. *Note:* Remember that the longer ends of the hubs are toward the front of the transmission.

_____ 15. Re-check the ring to gear teeth distance and make sure that the rings rotate freely.

_____ 16. Manually shift the transmission to ensure proper operation of all gears.

_____ 17. Safely lower the vehicle to the ground.

Performance ✓ Checklist

Name _____ Date _____ Class _____

PERFORMANCE STANDARDS:
Level 4–Performs skill without supervision and adapts to problem situations.
Level 3–Performs skill satisfactorily without assistance or supervision.
Level 2–Performs skill satisfactorily, but requires assistance/supervision.
Level 1–Performs parts of skill satisfactorily, but requires considerable assistance/supervision.

Attempt (circle one): **1 2 3 4**

Comments:

_____

_____

_____

_____

PERFORMANCE LEVEL ACHIEVED: _____

_____ 1. Safety rules and practices were followed at all times regarding this job.

_____ 2. Tools and equipment were used properly and stored upon completion of this job.

_____ 3. This completed job met the standards set and was done within the allotted time.

_____ 4. No injury or damage to property occurred during this job.

_____ 5. Upon completion of this job, the work area was cleaned correctly.

Instructor's Signature _____ Date_____

INSPECTING & REINSTALLING POWERTRAIN MOUNTS

NATEF Standard(s) for Manual Drive Train & Axles:

C6 Inspect and reinstall powertrain mounts.

Safety First

☐ Wear safety glasses at all times.
☐ Follow all safety rules when using common hand tools.
☐ Follow all safety rules when using jacks and jack stands, ramps, or lifts.
☐ Be careful of hot engine parts.

Tools & Equipment:

- Alignment bolts
- Common hand tools
- Torque wrench
- Vehicle service information
- Jacks and jack stands, ramps, or lifts

PROCEDURES Refer to the vehicle service information for specifications and special procedures. Then inspect and reinstall powertrain mounts in the vehicle provided by your instructor.

_____ 1. Write up a repair order.

_____ 2. Make sure you follow all procedures in the vehicle service information.

_____ 3. Look up the manufacturer's recommended procedure for removing and installing the powertrain mounts.

Source: _____ Page: _____

_____ 4. Raise the vehicle and inspect the powertrain mounts by prying on the transaxle while you look for loose or damaged mounts. *Note:* If the rubber separates from the metal part of the mount or the transaxle moves up but not down, you will need to replace the mount.

_____ 5. If the engine must be raised to replace a mount, follow recommended procedures.

_____ 6. Install a new mount by first attaching it to the chassis and then tightening the bolts to the manufacturer's specified torque.

_____ 7. Then attach and torque the bolts that connect the mounts to the powertrain and reinstall the engine.

_____ 8. Lower the vehicle if you have not already done so, brace the front wheels, and start the engine.

_____ 9. Shift the transmission into first gear, release the clutch very briefly, depress the clutch again, and then turn off the engine. *Note:* This aligns the powertrain mounts.

_____ 10. Disconnect the negative battery cable, torque any loose bolts to the manufacturer's specifications, and then reconnect the battery.

(continued)

Performance ✓ Checklist

Name _____ Date _____ Class _____

PERFORMANCE STANDARDS:
Level 4–Performs skill without supervision and adapts to problem situations.
Level 3–Performs skill satisfactorily without assistance or supervision.
Level 2–Performs skill satisfactorily, but requires assistance/supervision.
Level 1–Performs parts of skill satisfactorily, but requires considerable assistance/supervision.

Attempt (circle one): **1 2 3 4**

Comments:

_____

_____

_____

_____

PERFORMANCE LEVEL ACHIEVED: _____

_____ 1. Safety rules and practices were followed at all times regarding this job.

_____ 2. Tools and equipment were used properly and stored upon completion of this job.

_____ 3. This completed job met the standards set and was done within the allotted time.

_____ 4. No injury or damage to property occurred during this job.

_____ 5. Upon completion of this job, the work area was cleaned correctly.

Instructor's Signature _____ Date_____

INSPECTING & REPLACING CENTER SUPPORT BEARINGS

NATEF Standard(s) for Manual Drive Train & Axles:

D5 Inspect, service, and replace shaft center support bearings.

Safety First

- ☐ Wear safety glasses at all times.
- ☐ Follow all safety rules when using common hand tools.
- ☐ Follow all safety rules when using a lift.

Tools & Equipment:
- Lift
- Common hand tools
- Vehicle service information

PROCEDURES Refer to the vehicle service information for specifications and special procedures. Then inspect and replace the center support bearing in the vehicle provided by your instructor.

_____ 1. Write up a repair order.

_____ 2. Make sure you follow all procedures in the vehicle service information.

_____ 3. Look up the manufacturer's recommended procedure and specifications for inspecting and replacing center support bearings.

Source: _____ *Page:* _____

_____ 4. Drive the vehicle onto the lift, shift the transmission into NEUTRAL, raise the vehicle on the lift, and support it safely.

_____ 5. Inspect the center support bearing's bracket for looseness, damage, or wear.

_____ 6. Inspect the center support bearing for damage and wear by shaking the drive shaft near the bearing and listening for noise from the bearing.

_____ 7. Examine the rubber bushings and insulation around the bearing for damage or wear.

_____ 8. Remove and replace the center support bearing according to the manufacturer's recommendations.

_____ 9. Safely lower the vehicle to the ground.

(continued)

Performance ✓ Checklist

Name _____ Date _____ Class _____

PERFORMANCE STANDARDS:
Level 4–Performs skill without supervision and adapts to problem situations.
Level 3–Performs skill satisfactorily without assistance or supervision.
Level 2–Performs skill satisfactorily, but requires assistance/supervision.
Level 1–Performs parts of skill satisfactorily, but requires considerable assistance/supervision.

Attempt (circle one): **1 2 3 4**

Comments:

_____

_____

_____

_____

PERFORMANCE LEVEL ACHIEVED: _____

_____ 1. Safety rules and practices were followed at all times regarding this job.

_____ 2. Tools and equipment were used properly and stored upon completion of this job.

_____ 3. This completed job met the standards set and was done within the allotted time.

_____ 4. No injury or damage to property occurred during this job.

_____ 5. Upon completion of this job, the work area was cleaned correctly.

Instructor's Signature _____ Date_____

MEASURING & ADJUSTING DRIVELINE RUNOUT

NATEF Standard(s) for Manual Drive Train & Axles:
D6 Check shaft balance; measure shaft runout; measure and adjust driveline angles.

Safety First

☐ Wear safety glasses at all times.
☐ Follow all safety rules when using common hand tools.
☐ Follow all safety rules when using a lift.
☐ Always connect a vehicle's exhaust to a vent hose before you run an engine in a closed shop. Unvented fumes in a closed shop can cause death.
☐ Keep jewelry, loose clothing, and hair away from moving parts while an engine is running.
☐ Be careful of hot engine parts.
☐ Check Material Safety Data Sheets for chemical safety information.
☐ Follow all lifting rules when installing heavy components.

Tools & Equipment:
- Shop light
- Worm-screw hose clamps
- Common hand tools
- Dial indicator
- Lift
- Vehicle service information
- Torque wrench
- Cleaning solvent

PROCEDURES Refer to the vehicle service information for specifications and special procedures. Then measure and adjust the driveline in the rear-wheel-drive vehicle provided by your instructor.

_____ 1. Write up a repair order.

_____ 2. Make sure you follow all procedures in the vehicle service information.

_____ 3. Look up the manufacturer's recommended procedure and specifications for measuring and adjusting rear-wheel vehicle driveline runout.

Source: _____   *Page:* _____

_____ 4. Drive the vehicle onto the lift, shift the transmission into NEUTRAL, raise the vehicle on the lift, support it safely, and disengage the parking brake.

_____ 5. Measure the driveline runout by mounting the dial indicator on the lift stand, cleaning the driveline with an approved solvent, and turning the driveline with a rear wheel.

_____ 6. Record the runout. *Note:* Three evenly spaced readings should be taken along the drive shaft. Avoid the drive shaft balance weights.

(continued)

Manual Drive Train & Axles

_____ 7. If any of the runout measurements exceed the manufacturer's specifications, disconnect the driveline at the companion flange, rotate it 180 degrees, and re-check the driveline runout in all three positions (the rotation may lessen the runout). Don't let the driveline hang from the universal joint.

_____ 8. If the driveline is still not within specifications, either replace it or have it repaired by a specialty shop.

_____ 9. Safely lower the vehicle to the ground.

Performance ✓ Checklist

Name _____ Date _____ Class _____

PERFORMANCE STANDARDS:
Level 4–Performs skill without supervision and adapts to problem situations.
Level 3–Performs skill satisfactorily without assistance or supervision.
Level 2–Performs skill satisfactorily, but requires assistance/supervision.
Level 1–Performs parts of skill satisfactorily, but requires considerable assistance/supervision.

Attempt (circle one): **1 2 3 4**

Comments:

_____

_____

_____

_____

PERFORMANCE LEVEL ACHIEVED: _____

_____ 1. Safety rules and practices were followed at all times regarding this job.
_____ 2. Tools and equipment were used properly and stored upon completion of this job.
_____ 3. This completed job met the standards set and was done within the allotted time.
_____ 4. No injury or damage to property occurred during this job.
_____ 5. Upon completion of this job, the work area was cleaned correctly.

Instructor's Signature _____ Date_____

DIAGNOSING & ADJUSTING UNIVERSAL JOINTS

NATEF Standard(s) for Manual Drive Train & Axles:

D2 Diagnose universal joint noise and vibration concerns; perform necessary action.

Safety First

☐ Wear safety glasses at all times.
☐ Follow all safety rules when using common hand tools.
☐ Follow all safety rules when using a lift.
☐ Follow all safety rules when using a transmission jack.

Tools & Equipment:
- Ruler
- Lift
- Vehicle service information
- Inclinometer
- Transmission jack
- Shims
- Common hand tools

PROCEDURES Refer to the vehicle service information for specifications and special procedures. Then measure and adjust the universal joints in the vehicle provided by your instructor.

_____ 1. Write up a repair order.

_____ 2. Make sure you follow all procedures in the vehicle service information.

_____ 3. Look up the manufacturer's recommended procedure and specifications for measuring and adjusting the front and rear universal joints. *Note:* Measuring and adjusting the universal joints are the earliest inspections of the driveline. These inspections are made with the driveline on the vehicle.

_____ 4. Measure the rear trim height with the ruler while the vehicle is at curb weight. *Note:* Measure rear trim height from the top of the rear axle to the bottom of the frame. If it is too high, add weight to the vehicle until it is at the specified trim height.

_____ 5. Drive the vehicle onto the lift, shift the transmission into NEUTRAL, release the parking brake, raise the vehicle on the lift, and support it safely.

_____ 6. Rotate the rear universal joint until a bearing cup is positioned vertically.

_____ 7. Clean the bearing cup, put the inclinometer on the cup, center the bubble, and record the measurement.

_____

_____ 8. Rotate the drive train 90 degrees, place the inclinometer on the next bearing cup, and record the measurement.

_____

(*continued*)

_____ 9. Find the difference between the two measurements and compare it to the specified joint angle in the vehicle service information. Repeat the procedure on the front universal joint. **CAUTION:** Weak or broken rear springs can cause incorrect universal joint angles.

_____ 10. If the front universal angle is too high, place a transmission jack under the transmission extension housing, remove the mounting bolts, and raise the transmission extension housing. *Note:* Replace the transmission mount when you adjust the front universal joint angle.

_____ 11. Install the appropriate shims, lower the transmission with the jack, replace the mount bolts, and measure universal joint angle again. Compare it to the specification.

_____ 12. Adjust the rear angle by loosening the rear axle leaf spring nuts and—without removing the nuts—by placing the appropriate shims between the springs and the spring seat. *Note:* Place the thick end of the shim toward the front of the vehicle to decrease the angle, and the thick end toward the rear of the vehicle to increase the angle.

_____ 13. Align the wedge hole with the spring seat, tighten the nuts to specified torque, and measure the universal joint angle.

_____ 14. Repeat the procedure until the angle is correct.

Performance ✓ Checklist

Name _____ Date _____ Class _____

PERFORMANCE STANDARDS:
Level 4–Performs skill without supervision and adapts to problem situations.
Level 3–Performs skill satisfactorily without assistance or supervision.
Level 2–Performs skill satisfactorily, but requires assistance/supervision.
Level 1–Performs parts of skill satisfactorily, but requires considerable assistance/supervision.

Attempt (circle one): **1 2 3 4**

Comments:

_____

_____

_____

_____

PERFORMANCE LEVEL ACHIEVED: _____

_____ 1. Safety rules and practices were followed at all times regarding this job.
_____ 2. Tools and equipment were used properly and stored upon completion of this job.
_____ 3. This completed job met the standards set and was done within the allotted time.
_____ 4. No injury or damage to property occurred during this job.
_____ 5. Upon completion of this job, the work area was cleaned correctly.

Instructor's Signature _____ Date_____

DIAGNOSING NOISE AND
VIBRATION CONCERNS

NATEF Standard(s) for Manual Drive Train & Axles:

E1-1 Diagnose noise and vibration concerns; determine necessary action.

Safety First

- ☐ Wear safety glasses at all times.
- ☐ Follow all safety rules when using common hand tools.
- ☐ Use exhaust vent if running engine in closed shop. Unvented exhaust fumes can cause death.
- ☐ Be careful and avoid moving parts such as drive shafts and wheels.
- ☐ Avoid contact with hot exhaust components.
- ☐ Make sure vehicle is properly supported.

Tools & Equipment:

- Vehicle service information
- Automotive lift
- Common hand tools
- Automotive stethoscope
- Drop light

PROCEDURES Refer to the vehicle service information for specifications and special procedures. Then diagnose noise and vibration concerns in the vehicle provided by your instructor.

_____ 1. Write up a repair order.

_____ 2. Make sure you follow all procedures in the vehicle service information.

_____ 3. The first step of diagnosing a rear axle noise is to check for the proper level of lubricant.

_____ 4. Bearing noise emitted from a rear-drive axle assembly can be caused by a defective axle bearing, pinion bearing, or carrier bearing. To determine which it is, the vehicle can be raised on a lift and run while listening with an automotive stethoscope.

_____ 5. Listen in the area of the pinion shaft, carrier bearings, or axle bearings to determine which bearing(s) are defective. Remember, the pinion bearings are rotating about three times the speed of the carrier or axle bearings.

_____ 6. An axle bearing may be stopped from rotating by pulling on that wheel's parking brake cable to compare noise changes.

_____ 7. A severely failed pinion bearing will likely cause a whine that varies with torque loading and may have noticeable endplay of the pinion shaft.

_____ 8. A differential whine, heard when the vehicle is driven, is caused by improper meshing of the ring gear and pinion gear. This is a result of loose or worn pinion bearings, worn ring and pinion teeth, or improper adjustment.

(continued)

_____ 9. Grinding or vibration of the differential usually indicates broken or damaged teeth on the ring gear or pinion.

_____ 10. A clattering noise made only when turning a corner is usually caused by broken or damaged teeth on the axle side gears or spider pinion gears.

_____ 11. Any of the above problems (other than an axle bearing) will require the complete disassembly and setup of the differential components.

Performance ✓ Checklist

Name _____ Date _____ Class _____

PERFORMANCE STANDARDS:
Level 4–Performs skill without supervision and adapts to problem situations.
Level 3–Performs skill satisfactorily without assistance or supervision.
Level 2–Performs skill satisfactorily, but requires assistance/supervision.
Level 1–Performs parts of skill satisfactorily, but requires considerable assistance/supervision.

Attempt (circle one): **1 2 3 4**

Comments:

_____

_____

_____

_____

PERFORMANCE LEVEL ACHIEVED: _____

_____ 1. Safety rules and practices were followed at all times regarding this job.
_____ 2. Tools and equipment were used properly and stored upon completion of this job.
_____ 3. This completed job met the standards set and was done within the allotted time.
_____ 4. No injury or damage to property occurred during this job.
_____ 5. Upon completion of this job, the work area was cleaned correctly.

Instructor's Signature _____ Date_____

INSPECTING & INSTALLING LIMITED SLIP CLUTCH COMPONENTS

NATEF Standard(s) for Manual Drive Train & Axles:

E2-3 Inspect and reinstall clutch (cone or plate) components.

Safety First

☐ Wear safety glasses at all times.
☐ Follow all safety rules when using common hand tools.
☐ Check Material Safety Data Sheets for chemical safety information.
☐ Use caution when working with the mechanical energy stored in compressed springs.

Tools & Equipment:
• Torque wrench   • Common hand tools   • Vehicle service information

PROCEDURES Refer to the vehicle service information for specifications and special procedures. Then inspect and install limited slip clutch components in the vehicle provided by your instructor.

_____ 1. Write up a repair order.

_____ 2. Make sure you follow all procedures in the vehicle service information.

_____ 3. Look up the manufacturer's recommended procedure and specifications for disassembling and assembling the differential case, assembly, and plate-type clutch components.

Source: _____ *Page:* _____

_____ 4. With the differential case and assembly already disassembled, inspect the clutch disc guides, clutch plates, and discs for worn or damaged parts.

_____ 5. For plate-type clutches, lubricate all the differential parts–except the preload plates–with an approved lubricant and soak the discs and plates in a limited-slip lubricant for thirty minutes. *Note:* For cone-type clutches, go to Step 9.

_____ 6. Assemble the clutch pack for the ring gear side of the differential. *Note:* Put the fiber side of a disc to the side gear hub, and then alternate the rest of the discs by always following a friction disc with a steel one.

_____ 7. Align the clutch disc with the plate ears, and then install the clutch disc in the differential. Repeat the procedure on the other side, and then put the thrust washers on the pinion gears.

_____ 8. Complete the clutch installation procedure according to the manufacturer's instructions.

_____ 9. Look up the manufacturer's recommended procedure for inspecting and installing cone-type clutches.

Source: _____ *Page:* _____

(continued)

Manual Drive Train & Axles

_____ **10.** For cone-type clutches, lubricate all the differential parts except for the thrust plates and any surface that contacts the case bolts.

_____ **11.** Place the correct replacement cone inside the case, seat it in position, and assemble the thrust plate, pinion gears, and thrust washers.

_____ **12.** Put the springs and other thrust plate on the top of the pinion shaft.

_____ **13.** Put the other cone on top of the thrust plate, and then reassemble the differential case according to the manufacturer's instructions.

_____ **14.** Torque the nuts, in a star pattern, to the manufacturer's specifications.

Performance ✓ Checklist

Name _____ Date _____ Class _____

PERFORMANCE STANDARDS:
Level 4–Performs skill without supervision and adapts to problem situations.
Level 3–Performs skill satisfactorily without assistance or supervision.
Level 2–Performs skill satisfactorily, but requires assistance/supervision.
Level 1–Performs parts of skill satisfactorily, but requires considerable assistance/supervision.

Attempt (circle one): **1 2 3 4**

Comments:

_____

_____

_____

_____

PERFORMANCE LEVEL ACHIEVED: _____

_____ **1.** Safety rules and practices were followed at all times regarding this job.
_____ **2.** Tools and equipment were used properly and stored upon completion of this job.
_____ **3.** This completed job met the standards set and was done within the allotted time.
_____ **4.** No injury or damage to property occurred during this job.
_____ **5.** Upon completion of this job, the work area was cleaned correctly.

Instructor's Signature _____ Date_____

DIAGNOSING FLUID LEAKAGE CONCERNS

NATEF Standard(s) for Manual Drive Train & Axles:
E1-2 Diagnose fluid leakage concerns; determine necessary action.

Safety First

☐ Wear safety glasses at all times.
☐ Follow all safety rules when using common hand tools.
☐ Be careful and avoid moving parts such as drive shafts and wheels.
☐ Avoid contact with hot exhaust components.
☐ Follow all safety rules when using a lift or jack and jack stands.

Tools & Equipment:
- Vehicle service information
- Drop light
- Common hand tools
- Lift or jack and jack stands

PROCEDURES Refer to the vehicle service information for specifications and special procedures. Then diagnose fluid leakage concerns in the vehicle provided by your instructor.

_____ 1. Write up a repair order.

_____ 2. Make sure you follow all procedures in the vehicle service information.

_____ 3. Check for defective pinion oil seal. Replace pinion oil seal.

_____ 4. Check for loose differential cover or carrier bolts. Tighten differential cover or carrier bolts.

_____ 5. Check for defective differential cover or carrier gasket. Replace differential cover or carrier gasket.

_____ 6. Check for leaking axle seals. Replace axle seals.

_____ 7. Check for leaking axle housing. Seal or replace axle housing.

(continued)

Performance ✓ Checklist

Name _____ Date _____ Class _____

PERFORMANCE STANDARDS:
Level 4–Performs skill without supervision and adapts to problem situations.
Level 3–Performs skill satisfactorily without assistance or supervision.
Level 2–Performs skill satisfactorily, but requires assistance/supervision.
Level 1–Performs parts of skill satisfactorily, but requires considerable assistance/supervision.

Attempt (circle one): **1 2 3 4**

Comments:

_____

_____

_____

_____

PERFORMANCE LEVEL ACHIEVED: _____

_____ **1.** Safety rules and practices were followed at all times regarding this job.

_____ **2.** Tools and equipment were used properly and stored upon completion of this job.

_____ **3.** This completed job met the standards set and was done within the allotted time.

_____ **4.** No injury or damage to property occurred during this job.

_____ **5.** Upon completion of this job, the work area was cleaned correctly.

Instructor's Signature _____ Date_____

CLEANING & INSPECTING DIFFERENTIAL HOUSING

NATEF Standard(s) for Manual Drive Train & Axles:

E2-2 Clean and inspect differential housing; refill with correct lubricant.

Safety First

☐ Wear safety glasses at all times.
☐ Follow all safety rules when using common hand tools.

Tools & Equipment:
- Vehicle service information
- Common hand tools
- Rear differential lubricant

PROCEDURES Refer to the vehicle service information for specifications and special procedures. Then clean and inspect the differential housing on the vehicle provided by your instructor.

_____ 1. Write up a repair order.

_____ 2. Make sure you follow all procedures in the vehicle service information.

_____ 3. Remove the cover from the rear differential and drain the lubricant into a suitable container.

_____ 4. Look for pieces of metal in the bottom of the differential case. Small fillings of metal are commonly found.

_____ 5. Clean the differential using an approved cleaning agent, such as a parts washer.

_____ 6. Inspect the gears for signs of excessive wear, misalignment, or broken/missing teeth.

_____ 7. If there are no signs of excessive wear or broken/missing gear teeth, clean and reinstall the differential cover using a suitable sealant.

_____ 8. Remove the fill plug and refill using the approved lubricant according to the manufacturer's specifications.

Manual Drive Train & Axles

(continued)

Performance ✓ Checklist

Name _____ Date _____ Class _____

PERFORMANCE STANDARDS:
Level 4–Performs skill without supervision and adapts to problem situations.
Level 3–Performs skill satisfactorily without assistance or supervision.
Level 2–Performs skill satisfactorily, but requires assistance/supervision.
Level 1–Performs parts of skill satisfactorily, but requires considerable assistance/supervision.

Attempt (circle one): **1 2 3 4**

Comments:

_____

_____

_____

_____

PERFORMANCE LEVEL ACHIEVED: _____

_____ 1. Safety rules and practices were followed at all times regarding this job.

_____ 2. Tools and equipment were used properly and stored upon completion of this job.

_____ 3. This completed job met the standards set and was done within the allotted time.

_____ 4. No injury or damage to property occurred during this job.

_____ 5. Upon completion of this job, the work area was cleaned correctly.

Instructor's Signature _____ Date_____

INSPECTING & REPLACING DIFFERENTIAL CASE COMPONENTS AND SIDE BEARINGS

NATEF Standard(s) for Manual Drive Train & Axles:

E1-10 Disassemble, inspect, measure, and adjust or replace differential pinion gears (spiders), shaft, side gears, side bearings, thrust washers, and case.

Safety First
- ☐ Wear safety glasses at all times.
- ☐ Follow all safety rules when using common hand tools.
- ☐ Use care when handling sharp parts.
- ☐ Check Material Safety Data Sheets for chemical safety information.

Tools & Equipment:
- Bearing puller
- Brass drift
- Vehicle service information
- Crocus (emery) cloth
- Cleaning solvent
- Bearing installation tool
- Common hand tools

PROCEDURES Refer to the vehicle service information for specifications and special procedures. Then inspect and replace the differential case components and side bearings in the vehicle provided by your instructor.

_____ 1. Write up a repair order.

_____ 2. Make sure you follow all procedures in the vehicle service information.

_____ 3. Look up the manufacturer's recommended procedure and specifications for inspecting and replacing differential case components and side bearings.

   Source: _____ *Page:* _____

_____ 4. With the differential case removed from the axle housing and in a bench vise, remove a differential side bearing with the bearing puller. Then reverse the case in the vise and remove the other differential side bearing.

_____ 5. Inspect the bearing bores in the case, remove any flaws with the crocus cloth, and apply a thin coat of rear axle lubricant to the bearing cones, hubs, and shoulders in the case.

_____ 6. With the case on the bench, carefully drive a side bearing cone into the case with a hammer and driver. Then drive the other side bearing cone into the case the same way. *Note:* Always use new replacement bearings.

_____ 7. To inspect and replace the other parts of a conventional differential case assembly, place the differential case firmly in a bench vise, mesh the splines of an axle shaft with a side gear, and then rock the differential side gears against the pinion gears. *Note:* If play is excessive, the pinion gears will need to be adjusted during reassembly.

(continued)

_____ **8.** Mark the components in the case so they can be replaced in their original locations.

_____ **9.** Remove the components according to the manufacturer's recommendations. Then clean them and the case with an appropriate solvent and the bristle brush. Allow the components and the case to air dry or dry them with low-pressure compressed air.

_____ **10.** Thoroughly inspect the case and the components for damage and lightly lubricate them.

_____ **11.** Inspect the gear teeth and thrust surfaces for wear and damage such as uneven tooth patterns, nicks, burrs, or scratches on any of the differential parts or the case.

_____ **12.** Smooth away minor nicks, burrs, and scratches with the crocus cloth and replace any damaged parts.

_____ **13.** Reassemble the differential case assembly according to the manufacturer's recommendations.

Performance ✓ Checklist

Name _____ Date _____ Class _____

PERFORMANCE STANDARDS:
Level 4–Performs skill without supervision and adapts to problem situations.
Level 3–Performs skill satisfactorily without assistance or supervision.
Level 2–Performs skill satisfactorily, but requires assistance/supervision.
Level 1–Performs parts of skill satisfactorily, but requires considerable assistance/supervision.

Attempt (circle one): **1 2 3 4**

Comments:

_____

_____

_____

_____

PERFORMANCE LEVEL ACHIEVED: _____

_____ **1.** Safety rules and practices were followed at all times regarding this job.
_____ **2.** Tools and equipment were used properly and stored upon completion of this job.
_____ **3.** This completed job met the standards set and was done within the allotted time.
_____ **4.** No injury or damage to property occurred during this job.
_____ **5.** Upon completion of this job, the work area was cleaned correctly.

Instructor's Signature _____ Date_____

REASSEMBLING & INSTALLING DIFFERENTIAL CASE ASSEMBLIES

NATEF Standard(s) for Manual Drive Train & Axles:

E1-11 Reassemble and reinstall differential case assembly; measure runout; determine necessary action.

Safety First

☐ Wear safety glasses at all times.
☐ Follow all safety rules when using common hand tools.
☐ Check Material Safety Data Sheets for chemical safety information.
☐ Follow all lifting rules when installing heavy components.

Tools & Equipment:
- Dial indicator
- Common hand tools
- V-blocks
- Vehicle service information
- Torque wrench

PROCEDURES Refer to the vehicle service information for specifications and special procedures. Then reassemble and install the differential case assembly and measure the runout in the vehicle provided by your instructor.

_____ 1. Write up a repair order.

_____ 2. Make sure you follow all procedures in the vehicle service information.

_____ 3. Look up the manufacturer's recommended procedure and specifications for measuring runout and reassembling the differential case and assembly.

Source: _____ Page: _____

_____ 4. Make sure the axle housing has been cleaned and inspected. Then mount the differential case in the V-blocks by putting each side bearing in a block. *Note:* See your instructor for special directions concerning vehicles with integral carriers.

_____ 5. Mount the dial indicator on a stationary surface and rest the plunger at a right angle against the differential case flange. Measure and record the runout.

_____

_____ 6. Replace the differential case if the runout is not within specifications. *Note:* If the runout is still out of spec after replacing the case, replace the ring gear also. The pinion gear should always be replaced with the ring gear. If the runout is still too high after replacing the case, the ring gear, and the pinion gear, torque the ring gear bolts again.

_____ 7. Lightly lubricate the differential side bearing cones, bearing races, and adjusting nuts.

(continued)

Manual Drive Train
& Axles

_____ **8.** Place the differential case in the housing or carrier and put the match mark on the drive pinion gear between the two marked teeth on the ring gear.

_____ **9.** Measure and adjust the side bearing preload according to the manufacturer's recommendations. Then measure and adjust the ring-to-pinion-gear backlash.

_____ **10.** Check the ring gear tooth patterns and make any necessary adjustments.

_____ **11.** Reseal or replace the differential cover gasket.

_____ **12.** Torque the nuts, in a star pattern, to the manufacturer's specifications.

_____ **13.** Fill the differential with lubricant.

Performance ✓ Checklist

Name _____ Date _____ Class _____

PERFORMANCE STANDARDS:
Level 4–Performs skill without supervision and adapts to problem situations.
Level 3–Performs skill satisfactorily without assistance or supervision.
Level 2–Performs skill satisfactorily, but requires assistance/supervision.
Level 1–Performs parts of skill satisfactorily, but requires considerable assistance/supervision.

Attempt (circle one): **1 2 3 4**

Comments:

_____

_____

_____

_____

PERFORMANCE LEVEL ACHIEVED: _____

_____ **1.** Safety rules and practices were followed at all times regarding this job.

_____ **2.** Tools and equipment were used properly and stored upon completion of this job.

_____ **3.** This completed job met the standards set and was done within the allotted time.

_____ **4.** No injury or damage to property occurred during this job.

_____ **5.** Upon completion of this job, the work area was cleaned correctly.

Instructor's Signature _____ Date_____

MEASURING & ADJUSTING DRIVE PINION BEARING PRELOAD

NATEF Standard(s) for Manual Drive Train & Axles:
E1-7 Measure and adjust drive pinion bearing preload.

Safety First
- ☐ Wear safety glasses at all times.
- ☐ Follow all safety rules when using common hand tools.
- ☐ Follow all safety rules when using a lift.

Tools & Equipment:
- Beam torque wrench
- Common hand tools
- Flange-holding tool
- Vehicle service information
- Lift

PROCEDURES Refer to the vehicle service information for specifications and special procedures. Then measure and adjust the drive pinion bearing preload in the vehicle provided by your instructor.

_____ 1. Write up a repair order.

_____ 2. Make sure you follow all procedures in the vehicle service information.

_____ 3. Look up the manufacturer's recommended procedure and specifications for measuring and adjusting the drive pinion bearing preload.

Source: _____ _Page:_ _____

_____ 4. Safely raise the vehicle on a lift, clean the axle housing, properly reassemble drive pinion components, install new pinion bearings, and select the correct shims. After you have done all of this, it is important to hold the drive pinion gear absolutely constant in relation to the ring gear. **CAUTION:** This can be done by installing the collapsible spacer between the front and rear pinion bearings, then slowly torquing the pinion nut to the manufacturer's specification.

_____ 5. Reassemble the drive pinion assembly according to the manufacturer's recommendations.

_____ 6. Hold the companion flange with the holding tool and slowly tighten the pinion nut.
Note: Always use a new pinion nut.

_____ 7. While you are tightening the nut, stop and gradually turn the drive pinion by hand to set the roller bearings, and then check the drive pinion preload with a beam type (lb.-in.) torque wrench. _Note:_ The correct reading is the one that keeps the pinion rotating, not the one that starts it rotating.

(continued)

Manual Drive Train & Axles

_____ 8. Torque the drive pinion nut to the preload specifications, then rotate the companion flange a few turns by hand.

_____ 9. Re-check the torque reading once more to make sure the preload setting did not change when you turned the bearings.

_____ 10. Safely lower the vehicle to the ground.

Performance ✓ Checklist

Name _____ Date _____ Class _____

PERFORMANCE STANDARDS:
Level 4–Performs skill without supervision and adapts to problem situations.
Level 3–Performs skill satisfactorily without assistance or supervision.
Level 2–Performs skill satisfactorily, but requires assistance/supervision.
Level 1–Performs parts of skill satisfactorily, but requires considerable assistance/supervision.

Attempt (circle one): **1 2 3 4**

Comments:

_____

_____

_____

_____

PERFORMANCE LEVEL ACHIEVED: _____

_____ 1. Safety rules and practices were followed at all times regarding this job.
_____ 2. Tools and equipment were used properly and stored upon completion of this job.
_____ 3. This completed job met the standards set and was done within the allotted time.
_____ 4. No injury or damage to property occurred during this job.
_____ 5. Upon completion of this job, the work area was cleaned correctly.

Instructor's Signature _____ Date_____

MEASURING & ADJUSTING SIDE BEARING PRELOAD AND PINION BACKLASH

NATEF Standard(s) for Manual Drive Train & Axles:

E1-8 Measure and adjust side bearing preload and ring and pinion gear total backlash and backlash variation on a differential carrier assembly (threaded cup or shim types).

Safety First
- ☐ Wear safety glasses at all times.
- ☐ Follow all safety rules when using common hand tools.
- ☐ Follow all safety rules when using a lift.

Tools & Equipment:
- Dial indicator
- Drain pan
- Vehicle service information
- Bearing adjuster tool
- Lift
- Torque wrench
- Common hand tools

PROCEDURES Refer to the vehicle service information for specifications and special procedures. Then measure and adjust the side bearing preload and pinion backlash in the vehicle provided by your instructor.

_____ 1. Write up a repair order.

_____ 2. Make sure you follow all procedures in the vehicle service information.

_____ 3. Look up the manufacturer's recommended procedure and specifications for measuring and adjusting the side bearing preload and pinion backlash.

Source: _____ *Page:* _____

_____ 4. With the vehicle already safely on a lift and the rear axle lubricant and driveline removed, install the dial indicator on either the axle housing or the differential holding fixture. *Note:* Put the plunger of the indicator at a right angle to a tooth face.

_____ 5. Holding the companion flange still, measure the pinion gear backlash by rocking the ring gear until the teeth mesh. Then record the dial indicator reading. *Note:* Repeat the procedure four times as you move the ring gear a full turn, one-quarter turn at a time.

_____   _____

_____   _____

_____ 6. Check the recorded pinion gear backlash readings against the specifications. Adjust the side bearing preload and ring gear backlash if the readings are not within the manufacturer's recommendations. *Note:* If the backlash is different at different places around the ring gear, either the ring gear or differential has excess runout.

_____ 7. Adjust the pinion gear backlash and runout according to the manufacturer's recommendations.

(continued)

_____ 8. Once the ring gear runout and pinion gear backlash are measured and adjusted, rock the ring gear quickly to check for any looseness and rattling that indicates a low side bearing preload.

_____ 9. Remove the adjusting nut locks and then adjust the preload nuts with the appropriate adjusting tool. **Note:** Tighten the nuts to increase preload and loosen the nuts to decrease the preload.

_____ 10. Retest the side bearing preload and adjust it with the adjustment nuts until it is difficult to rock the ring gear.

_____ 11. Tighten the adjustment nuts.

_____ 12. Safely lower the vehicle to the ground.

Performance ✓ Checklist

Name _____ Date _____ Class _____

PERFORMANCE STANDARDS:
Level 4–Performs skill without supervision and adapts to problem situations.
Level 3–Performs skill satisfactorily without assistance or supervision.
Level 2–Performs skill satisfactorily, but requires assistance/supervision.
Level 1–Performs parts of skill satisfactorily, but requires considerable assistance/supervision.

Attempt (circle one): **1 2 3 4**

Comments:

_____

_____

_____

_____

PERFORMANCE LEVEL ACHIEVED: _____

_____ 1. Safety rules and practices were followed at all times regarding this job.

_____ 2. Tools and equipment were used properly and stored upon completion of this job.

_____ 3. This completed job met the standards set and was done within the allotted time.

_____ 4. No injury or damage to property occurred during this job.

_____ 5. Upon completion of this job, the work area was cleaned correctly.

Instructor's Signature _____ Date_____

CHECKING RING AND PINION TOOTH CONTACT PATTERNS

NATEF Standard(s) for Manual Drive Train & Axles:
E1-9 Check ring and pinion tooth contact patterns; perform necessary action.

Safety First
- ☐ Wear safety glasses at all times.
- ☐ Follow all safety rules when using common hand tools.
- ☐ Follow all safety rules when using a lift.
- ☐ Check Material Safety Data Sheets for chemical safety information.

Tools & Equipment:
- Torque wrench
- Lift
- Vehicle service information
- Drop light
- Prussian blue
- Drain pan
- Common hand tools

PROCEDURES Refer to the vehicle service information for specifications and special procedures. Then check the ring and pinion tooth contact patterns in the vehicle provided by your instructor.

_____ 1. Write up a repair order.

_____ 2. Make sure you follow all procedures in the vehicle service information.

_____ 3. Safely raise the vehicle on a lift.

_____ 4. Remove the driveshaft(s).

_____ 5. Drain the differential.

_____ 6. Use spray cleaner to clean lubricant off the gears, and then allow them to dry.

_____ 7. Visually inspect the gears for damage.

_____ 8. Apply a coat of Prussian blue to several teeth on the ring gear.

_____ 9. Rotate the pinion gear several turns.

_____ 10. Using a drop light, examine the tooth contact patterns.

_____ 11. Compare the patterns to what is in the vehicle service information, and make adjustments to correct improper tooth contact patterns.

_____ 12. Safely lower the vehicle to the ground.

Manual Drive Train & Axles

(continued)

Performance ✓ Checklist

Name _____ Date _____ Class _____

PERFORMANCE STANDARDS:
Level 4–Performs skill without supervision and adapts to problem situations.
Level 3–Performs skill satisfactorily without assistance or supervision.
Level 2–Performs skill satisfactorily, but requires assistance/supervision.
Level 1–Performs parts of skill satisfactorily, but requires considerable assistance/supervision.

Attempt (circle one): **1 2 3 4**

Comments:

_____

_____

_____

_____

PERFORMANCE LEVEL ACHIEVED: _____

_____ 1. Safety rules and practices were followed at all times regarding this job.

_____ 2. Tools and equipment were used properly and stored upon completion of this job.

_____ 3. This completed job met the standards set and was done within the allotted time.

_____ 4. No injury or damage to property occurred during this job.

_____ 5. Upon completion of this job, the work area was cleaned correctly.

Instructor's Signature _____ Date_____

INSPECTING & REPLACING RING GEARS AND DIFFERENTIAL CASES

NATEF Standard(s) for Manual Drive Train & Axles:

E1-4 Inspect ring gear and measure runout; determine necessary action.

Safety First
- ☐ Wear safety glasses at all times.
- ☐ Follow all safety rules when using common hand tools.
- ☐ Check Material Safety Data Sheets for chemical safety information.

Tools & Equipment:
- Dial indicator
- Whetstone
- Common hand tools
- Torque wrench
- Sharp awl
- Vehicle service information
- Plastic mallet
- Cleaning solvent

PROCEDURES Refer to the vehicle service information for specifications and special procedures. Then inspect and replace the ring gear and differential case in the vehicle provided by your instructor.

_____ 1. Write up a repair order.

_____ 2. Make sure you follow all procedures in the vehicle service information.

_____ 3. Look up the manufacturer's recommended procedure and specifications for inspecting and replacing ring gears and differential cases.

 Source: _____ _Page:_ _____

_____ 4. Measure and record the ring gear runout with the dial indicator. **Note:** Rotate the ring gear one full turn while taking the measurement.

 _____

_____ 5. If the runout is not within specification, remove the differential case and check the differential side bearings. **Note:** If these are replaced, check the runout again.

_____ 6. If the runout is still too much, the ring gear or the differential case must be replaced. **Note:** Check the differential case runout, after removing it, to see which part is bad.

_____ 7. Remove the differential case, with the ring gear in it, and place it securely in a vise.

_____ 8. Mark the differential case and ring gear for positioning if you are able to reuse them both.

_____ 9. Remove the ring gear bolts. **Note:** These may be left-handed bolts.

_____ 10. Remove the ring gear carefully. **CAUTION:** You may need to tap with a plastic mallet or chisel, first on one side and then on the other side several times. Do not pry the ring gear from the case or allow it to fall out.

(continued)

_____ 11. Throw away the ring gear bolts. *Note:* You should always reinstall new ones.

_____ 12. Clean the ring gear with an appropriate solvent.

_____ 13. Inspect the ring gear for scoring, chipping, breaking, and other signs of wear. **CAUTION:** Excessive backlash, the wrong kind of lubricant, or too little lubricant can cause scoring.

_____ 14. Smooth away any sharp edges with a whetstone and clean the gear again to remove any abrasive debris. *Note:* If there is significant wear or damage, replace the gear.

_____ 15. To reassemble, be sure the differential case is clean and undamaged, align the marks on the ring gear and the differential case, and fit the gear onto the case.

_____ 16. Insert new ring gear bolts and hand-tighten them.

_____ 17. Torque the ring gear bolts in a star pattern to make sure that the ring gear is tightened evenly against the differential case.

_____ 18. Re-check the ring gear runout to make sure that it is within specifications.

Performance ✓ Checklist

Name _____ Date _____ Class _____

PERFORMANCE STANDARDS:
Level 4–Performs skill without supervision and adapts to problem situations.
Level 3–Performs skill satisfactorily without assistance or supervision.
Level 2–Performs skill satisfactorily, but requires assistance/supervision.
Level 1–Performs parts of skill satisfactorily, but requires considerable assistance/supervision.

Attempt (circle one): **1 2 3 4**

Comments:

_____

_____

_____

_____

PERFORMANCE LEVEL ACHIEVED: _____

_____ 1. Safety rules and practices were followed at all times regarding this job.

_____ 2. Tools and equipment were used properly and stored upon completion of this job.

_____ 3. This completed job met the standards set and was done within the allotted time.

_____ 4. No injury or damage to property occurred during this job.

_____ 5. Upon completion of this job, the work area was cleaned correctly.

Instructor's Signature _____ Date_____

REMOVING & INSPECTING DRIVE PINION GEAR ASSEMBLIES

NATEF Standard(s) for Manual Drive Train & Axles:

E1-5 Remove, inspect, and reinstall drive pinion and ring gear, spacers, sleeves, and bearings.

Safety First
- ☐ Wear safety glasses at all times.
- ☐ Follow all safety rules when using common hand tools.
- ☐ Follow all safety rules when using a lift.
- ☐ Check Material Safety Data Sheets for chemical safety information.
- ☐ Follow all safety rules when using compressed air.

Tools & Equipment:
- Bearing puller
- Crocus (emery) cloth
- Cleaning solvent
- Air compressor
- Plastic mallet
- Drain pan for axle grease
- Common hand tools
- Shop towels
- Bristle/stiff brush
- Lift
- Vehicle service information

PROCEDURES Refer to the vehicle service information for specifications and special procedures. Then remove and inspect the drive pinion gear assembly in the vehicle provided by your instructor.

_____ 1. Write up a repair order.

_____ 2. Make sure you follow all procedures in the vehicle service information.

_____ 3. Look up the manufacturer's recommended procedure and specifications for removing and inspecting the drive pinion gear assembly.

Source: _____ *Page:* _____

_____ 4. With the vehicle safely on a lift and the rear lubricant, rear wheels, brake drums, rear axle shafts, driveline, and differential case or differential housing removed, reverse the pinion nut on the drive pinion and rethread it onto the shaft to protect the shaft threads. *Note:* The nut can be used to chase the threads if they are damaged.

_____ 5. Place two or three shop towels in the bottom of the differential housing to protect the drive pinion assembly in case it drops when you take it out. *Note:* See your instructor if there is an integral housing.

_____ 6. Drive the pinion gear from the differential housing by tapping it with the plastic mallet.

_____ 7. Remove the pinion assembly from the housing and discard the collapsible spacer.

(continued)

_____ 8. Remove the outer races from the front and rear bearings using a punch and hammer, clean the bearings, and inspect them. *Note:* Some differential housings have alignment slots for the punch.

_____ 9. Remove the rear bearing from the shaft with the bearing puller.

_____ 10. Clean the differential housing with the stiff brush and an approved solvent, inspect it for damage, and blow dry it with low-pressure compressed air. **CAUTION:** Carefully inspect the bore where the oil seal and pinion bearings are installed and gently remove any debris, nicks, burrs, and scratches.

_____ 11. Inspect the drive pinion gear teeth and shaft for wear and damage. **CAUTION:** Chipped gears indicate foreign matter in the axle housing. Be very particular when cleaning the axle housing.

_____ 12. Smooth away any damage to the pinion assembly components with the crocus cloth and reassemble the pinion assembly and differential housing.

_____ 13. Replace any components with damage that the crocus cloth doesn't remove.

_____ 14. Safely lower the vehicle to the ground.

Performance ✓ Checklist

Name _____ Date _____ Class _____

PERFORMANCE STANDARDS:
Level 4–Performs skill without supervision and adapts to problem situations.
Level 3–Performs skill satisfactorily without assistance or supervision.
Level 2–Performs skill satisfactorily, but requires assistance/supervision.
Level 1–Performs parts of skill satisfactorily, but requires considerable assistance/supervision.

Attempt (circle one): **1 2 3 4**

Comments:

_____

_____

_____

PERFORMANCE LEVEL ACHIEVED: _____

_____ 1. Safety rules and practices were followed at all times regarding this job.
_____ 2. Tools and equipment were used properly and stored upon completion of this job.
_____ 3. This completed job met the standards set and was done within the allotted time.
_____ 4. No injury or damage to property occurred during this job.
_____ 5. Upon completion of this job, the work area was cleaned correctly.

Instructor's Signature _____ Date_____

MEASURING & ADJUSTING DRIVE PINION DEPTH

NATEF Standard(s) for Manual Drive Train & Axles:
E1-6 Measure and adjust drive pinion depth.

Safety First

☐ Wear safety glasses at all times.
☐ Follow all safety rules when using common hand tools.
☐ Follow all safety rules when using a lift.

Tools & Equipment:
- Bearing removal and installation tools
- Lift
- Pinion depth gauge
- Feeler gauge or dial indicator
- Torque wrench
- Common hand tools
- Vehicle service information

PROCEDURES Refer to the vehicle service information for specifications and special procedures. Then measure and adjust the drive pinion depth in the vehicle provided by your instructor.

_____ 1. Write up a repair order.

_____ 2. Make sure you follow all procedures in the vehicle service information.

_____ 3. Safely raise the vehicle on a lift.

_____ 4. Drain the lubricant.

_____ 5. Disassemble the differential assembly.

_____ 6. Install the pinion depth gauge and related parts.

_____ 7. Using the manufacturer's procedure, determine the correct shim needed.

_____ 8. Using the correct shim, reassemble the differential assembly.

_____ 9. Safely lower the vehicle to the ground.

(continued)

Manual Drive Train & Axles

Performance ✓ Checklist

Name _____ Date _____ Class _____

PERFORMANCE STANDARDS:
Level 4–Performs skill without supervision and adapts to problem situations.
Level 3–Performs skill satisfactorily without assistance or supervision.
Level 2–Performs skill satisfactorily, but requires assistance/supervision.
Level 1–Performs parts of skill satisfactorily, but requires considerable assistance/supervision.

Attempt (circle one): **1 2 3 4**

Comments:

_____
_____
_____
_____

PERFORMANCE LEVEL ACHIEVED: _____

_____ **1.** Safety rules and practices were followed at all times regarding this job.
_____ **2.** Tools and equipment were used properly and stored upon completion of this job.
_____ **3.** This completed job met the standards set and was done within the allotted time.
_____ **4.** No injury or damage to property occurred during this job.
_____ **5.** Upon completion of this job, the work area was cleaned correctly.

Instructor's Signature _____ Date_____

INSPECTING & REPLACING COMPANION FLANGES AND PINION SEALS

NATEF Standard(s) for Manual Drive Train & Axles:

E1-3 Inspect and replace companion flange and pinion seal; measure companion flange runout.

Safety First

☐ Wear safety glasses at all times.
☐ Follow all safety rules when using common hand tools.
☐ Follow all safety rules when using a lift.
☐ Check Material Safety Data Sheets for chemical safety information.

Tools & Equipment:

- Dial indicator
- Crocus (emery) cloth
- Common hand tools
- Torque wrench
- Sharp awl
- Vehicle service information
- Plastic mallet
- Lift

PROCEDURES Refer to the vehicle service information for specifications and special procedures. Then inspect and replace the companion flange and pinion seal in the vehicle provided by your instructor.

_____ 1. Write up a repair order.

_____ 2. Make sure you follow all procedures in the vehicle service information.

_____ 3. Look up the manufacturer's recommended procedure and specifications for inspecting and replacing companion flanges and pinions seals.

Source: _____ *Page:* _____

_____ 4. With the vehicle safely on a lift, measure and record the pinion bearing preload by putting the torque wrench on the pinion nut and steadily rotating the drive pinion with the torque wrench. *Note:* This is the torque that keeps the drive pin rotating, not the torque that starts the drive pin rotating.

_____

_____ 5. Measure and record the companion flange runout with the dial indicator and replace the flange if there is too much runout. *Note:* See your instructor for directions.

_____

_____ 6. Put reinstallation matchmarks on the companion flange using the sharp awl.

_____

(continued)

Manual Drive Train & Axles

_____ **7.** Remove the pinion nut with the socket wrenches by holding the companion flange with the companion flange holding tool and removing the nut and washer. ***Note:*** Manufacturers recommend that old pinion nuts be replaced with new ones.

_____ **8.** Remove the companion flange with the puller and clean the area around the oil seal. **CAUTION:** Put the drain pan under the oil seal to capture any used lubricant.

_____ **9.** Remove and discard the old oil seal with the heavy screwdriver.

_____ **10.** Clean and inspect the pinion splines, housing bore, and companion flange for damage.

_____ **11.** Remove any burrs with the crocus cloth and put a thin coat of lubricant inside the axle housing bore and on the outside and sealing edges of the replacement seal.

_____ **12.** Position the pinion seal squarely in the housing bore with the seal installation tool to install the new seal.

_____ **13.** Put a thin coat of rear axle lubricant on the replacement companion flange using a new washer and a new nut. ***Note:*** If the washer is collapsible, be sure the round side is out.

_____ **14.** Install the companion flange according to the manufacturer's recommendations and the matchmarks. ***Note:*** If the old flange is damaged or worn, it must be replaced.

_____ **15.** Tighten the pinion nut gradually by turning the pinion several revolutions each time. ***Note:*** Check the preload with the torque wrench.

_____ **16.** Torque the pinion nut to the manufacturer's specifications.

_____ **17.** Safely lower the vehicle to the ground.

Performance ✓ Checklist

Name _____ Date _____ Class _____

PERFORMANCE STANDARDS:
Level 4–Performs skill without supervision and adapts to problem situations.
Level 3–Performs skill satisfactorily without assistance or supervision.
Level 2–Performs skill satisfactorily, but requires assistance/supervision.
Level 1–Performs parts of skill satisfactorily, but requires considerable assistance/supervision.

Attempt (circle one): **1   2   3   4**

Comments:

_____

_____

_____

_____

PERFORMANCE LEVEL ACHIEVED: _____

_____ **1.** Safety rules and practices were followed at all times regarding this job.
_____ **2.** Tools and equipment were used properly and stored upon completion of this job.
_____ **3.** This completed job met the standards set and was done within the allotted time.
_____ **4.** No injury or damage to property occurred during this job.
_____ **5.** Upon completion of this job, the work area was cleaned correctly.

Instructor's Signature _____ Date_____

INSPECTING & REPLACING DRIVE AXLE SHAFT WHEEL STUDS

NATEF Standard(s) for Manual Drive Train & Axles:

E3-2 Inspect and replace drive axle shaft wheel studs.

Safety First

☐ Wear safety glasses at all times.
☐ Follow all safety rules when using common hand tools.

Tools & Equipment:

• Arbor press/portable press • Common hand tools • Vehicle service information

PROCEDURES Refer to the vehicle service information for specifications and special procedures. Then inspect and replace the drive axle shaft wheel studs in the vehicle provided by your instructor.

_____ 1. Write up a repair order.

_____ 2. Make sure you follow all procedures in the vehicle service information.

_____ 3. Look up the manufacturer's recommended procedure for inspecting and replacing the drive axle shaft wheel studs.

   Source: _____ _Page:_ _____

_____ 4. Inspect every wheel stud for looseness, stripped threads, and other damage. Then mark the studs that need to be replaced.

_____ 5. If the axle is off the vehicle, put the axle flange in the arbor press—wheel studs up—and support the flange well.

_____ 6. If the axle is on the vehicle, put the portable press on the wheel stud with the screw assembly of the press on the outside of the axle flange.

_____ 7. Press the wheel studs out the backside of the axle shaft flange one at a time.

_____ 8. Put three or four washers on the wheel stud, and thread the stud nut—with its flat side against the washers—onto the stud.

_____ 9. Use a ratchet to tighten each nut until the new studs are seated flush against the inside surface of the axle shaft flange.

(continued)

Manual Drive Train & Axles

Performance ✓ Checklist

Name _____ Date _____ Class _____

PERFORMANCE STANDARDS:
Level 4–Performs skill without supervision and adapts to problem situations.
Level 3–Performs skill satisfactorily without assistance or supervision.
Level 2–Performs skill satisfactorily, but requires assistance/supervision.
Level 1–Performs parts of skill satisfactorily, but requires considerable assistance/supervision.

Attempt (circle one): **1 2 3 4**

Comments:

_____

_____

_____

_____

PERFORMANCE LEVEL ACHIEVED: _____

_____ **1.** Safety rules and practices were followed at all times regarding this job.

_____ **2.** Tools and equipment were used properly and stored upon completion of this job.

_____ **3.** This completed job met the standards set and was done within the allotted time.

_____ **4.** No injury or damage to property occurred during this job.

_____ **5.** Upon completion of this job, the work area was cleaned correctly.

Instructor's Signature _____ Date_____

MEASURING DRIVE AXLE RUNOUT AND SHAFT ENDPLAY

NATEF Standard(s) for Manual Drive Train & Axles:

E3-5 Measure drive axle flange runout and shaft endplay; determine necessary action.

Safety First

☐ Wear safety glasses at all times.
☐ Follow all safety rules when using common hand tools.
☐ Follow all safety rules when using a lift.

Tools & Equipment:
- Dial indicator
- V-blocks
- Common hand tools
- Emery (crocus) cloth
- Plastic mallet
- Vehicle service information
- Torque wrench
- Lift

PROCEDURES Refer to the vehicle service information for specifications and special procedures. Then measure the drive axle runout and the shaft endplay in the vehicle provided by your instructor.

_____ 1. Write up a repair order.

_____ 2. Make sure you follow all procedures in the vehicle service information.

_____ 3. Look up the manufacturer's recommended procedures and specifications for measuring axle runout and shaft endplay.

Source: _____ Page: _____

_____ 4. Safely raise the vehicle and remove the wheels and brake drums.

_____ 5. With the indicator plunger at a right angle to the shaft, measure and record the runout at the center of the axle shaft according to the manufacturer's recommendations.

_____

_____ 6. Measure and record the runout at both ends of the axle shaft.

_____

_____ 7. Measure and record the runout with the plunger against the pilot diameter of the axle flange.

_____

_____ 8. Clean the back of the axle flange with the emery cloth, and then measure and record the runout at the flange back just outside of the bolt circle. **Note:** Hold the center of the shaft with your thumb to reduce any endplay while you turn the shaft.

_____

(continued)

Manual Drive Train & Axles

_____ 9. If the runout is not within the manufacturer's specifications, the axle is bent and must be replaced.

_____ 10. Before measuring the axle shaft endplay, strike the axle shaft end squarely with the plastic mallet, and then mount the dial indicator on the brake backing plate with the plunger at a right angle to the axle flange face.

_____ 11. Measure and record the axle shaft endplay on both axles and compare the readings to the manufacturer's recommendations.

_____

_____ 12. If the axle shaft endplay is not within specification, install the manufacturer's recommended shims to adjust the axle shaft. Continue to measure the endplay and adjust the shims until the measurement is within the manufacturer's guidelines.

_____ 13. Safely lower the vehicle to the ground.

Performance ✓ Checklist

Name _____ Date _____ Class _____

PERFORMANCE STANDARDS:
Level 4–Performs skill without supervision and adapts to problem situations.
Level 3–Performs skill satisfactorily without assistance or supervision.
Level 2–Performs skill satisfactorily, but requires assistance/supervision.
Level 1–Performs parts of skill satisfactorily, but requires considerable assistance/supervision.

Attempt (circle one): **1 2 3 4**

Comments:

_____

_____

_____

_____

PERFORMANCE LEVEL ACHIEVED: _____

_____ 1. Safety rules and practices were followed at all times regarding this job.
_____ 2. Tools and equipment were used properly and stored upon completion of this job.
_____ 3. This completed job met the standards set and was done within the allotted time.
_____ 4. No injury or damage to property occurred during this job.
_____ 5. Upon completion of this job, the work area was cleaned correctly.

Instructor's Signature _____ Date_____

DIAGNOSING NOISE, VIBRATION, AND FLUID LEAKAGE CONCERNS

NATEF Standard(s) for Manual Drive Train & Axles:

E3-1 Diagnose drive axle shafts, bearings, and seals for noise, vibration, and fluid leakage concerns; determine necessary action.

Safety First

☐ Wear safety glasses at all times.
☐ Follow all safety rules when using common hand tools.
☐ Follow all safety rules when using a lift or jack and jack stands.
☐ Follow all safety rules when working around rotating parts.
☐ Avoid contact with hot exhaust components.

Tools & Equipment:
- Vehicle service information
- Automotive stethoscope
- Common hand tools
- Lift or jack and jack stands
- Pry bars

PROCEDURES Refer to the vehicle service information for specifications and special procedures. Then diagnose noise, vibration, and fluid leakage concerns in the vehicle provided by your instructor.

_____ 1. Write up a repair order.

_____ 2. Make sure you follow all procedures in the vehicle service information.

_____ 3. Axle bearing noise can be diagnosed by raising the vehicle on a lift, running the vehicle, and listening to and comparing the sound emitted from each bearing with an automotive stethoscope.

_____ 4. An axle bearing may be stopped from rotating by pulling on that wheel's parking brake cable to compare noise changes.

_____ 5. If you have identified a noisy axle bearing, compare the axial (up and down) play of that axle to the other axle. An excessively worn bearing may have excessive axial play.

_____ 6. If axle seals are leaking, the differential vent should be checked for blockage. A blocked vent will cause excessive internal pressure, which contributes to seal leakage. The fluid level of the differential should also be checked. A higher-than-specified fluid level also contributes to seal leakage.

_____ 7. If the vent is clear and the fluid level is normal, leakage is usually the result of worn seals.

_____ 8. When removing an axle to replace a bearing, the axle should be carefully inspected for damage. Some axle bearings have rollers that directly contact the axle. If the bearing is defective, it is likely that it has damaged the axle. If this is the case, the axle as well as the bearing and seal will require replacement.

(continued)

Performance ✓ Checklist

Name _____ Date _____ Class _____

PERFORMANCE STANDARDS:
Level 4–Performs skill without supervision and adapts to problem situations.
Level 3–Performs skill satisfactorily without assistance or supervision.
Level 2–Performs skill satisfactorily, but requires assistance/supervision.
Level 1–Performs parts of skill satisfactorily, but requires considerable assistance/supervision.

Attempt (circle one): **1 2 3 4**

Comments:

_____

_____

_____

_____

PERFORMANCE LEVEL ACHIEVED: _____

_____ 1. Safety rules and practices were followed at all times regarding this job.

_____ 2. Tools and equipment were used properly and stored upon completion of this job.

_____ 3. This completed job met the standards set and was done within the allotted time.

_____ 4. No injury or damage to property occurred during this job.

_____ 5. Upon completion of this job, the work area was cleaned correctly.

Instructor's Signature _____ Date _____

REMOVING & REPLACING DRIVE AXLE SHAFTS

NATEF Standard(s) for Manual Drive Train & Axles:
E3-3 Remove and replace drive axle shafts.

Safety First
- ☐ Wear safety glasses at all times.
- ☐ Follow all safety rules when using common hand tools.
- ☐ Follow all safety rules when using a lift.

Tools & Equipment:
- Shop light
- Brass drift punch
- Emery (crocus) cloth
- Lift
- Drain pan
- Socket set
- Torque wrench
- Common hand tools
- Hammer
- Slide hammer/attachments
- Plastic mallet
- Vehicle service information

PROCEDURES Refer to the vehicle service information for specifications and special procedures. Then remove and replace the drive axle shafts in the vehicle provided by your instructor.

_____ 1. Write up a repair order.

_____ 2. Make sure you follow all procedures in the vehicle service information.

_____ 3. Look up the manufacturer's recommended procedure for removing and replacing the drive axle shafts.

   Source: _____ *Page:* _____

_____ 4. Properly raise the vehicle on a lift and remove the wheels and brake drums on both sides of the shaft. Remove the differential cover and drain.

_____ 5. To remove axles secured with C-locks, begin by rotating the differential case and exposing the differential pinion shaft lock bolt and the C-lock.

_____ 6. Clean the pinion shaft and C-lock with a clean rag and remove the differential pinion shaft lock bolt. **CAUTION:** Shaft lock bolts break very easily.

_____ 7. Make sure you don't rotate the differential case with the pinion shaft removed. If you do this, the case components will fall out.

_____ 8. Press on the outside end of the axle shaft, slide the C-lock out of its groove, and repeat this procedure on the opposite side.

_____ 9. Pull out each axle shaft. **CAUTION:** Don't let the shaft hang from the housing–it can damage the seals and the bearings.

_____ 10. To remove axles secured by retainer plates, rotate the axle flange enough to reach the retainer plate nuts through the access hole.

(continued)

_____ 11. Pull the axle out of the housing, taking care not to damage the seals or bearings. *Note:* You might have to use the slide hammer and axle adapter.

_____ 12. Clean the axles with a shop towel and inspect them for rust, burrs, and rough spots.

_____ 13. Smooth away imperfections with the emery (crocus) cloth and lubricate the axles.

_____ 14. Slide the axle shaft into the axle housing. **CAUTION:** Don't let the splines damage the axle oil seal. For axles secured by retainer plates, see the manufacturer's recommendations.

_____ 15. Mesh the shaft splines into the side gear splines and push until the end of the axle shaft is inside the differential case.

_____ 16. Complete the drive axle shaft reassembly process for your vehicle according to the manufacturer's procedure.

_____ 17. Safely lower the vehicle to the ground.

Performance ✓ Checklist

Name _____ Date _____ Class _____

PERFORMANCE STANDARDS:
Level 4–Performs skill without supervision and adapts to problem situations.
Level 3–Performs skill satisfactorily without assistance or supervision.
Level 2–Performs skill satisfactorily, but requires assistance/supervision.
Level 1–Performs parts of skill satisfactorily, but requires considerable assistance/supervision.

Attempt (circle one): **1 2 3 4**

Comments:

_____

_____

_____

_____

PERFORMANCE LEVEL ACHIEVED: _____

_____ 1. Safety rules and practices were followed at all times regarding this job.
_____ 2. Tools and equipment were used properly and stored upon completion of this job.
_____ 3. This completed job met the standards set and was done within the allotted time.
_____ 4. No injury or damage to property occurred during this job.
_____ 5. Upon completion of this job, the work area was cleaned correctly.

Instructor's Signature _____ Date _____

INSPECTING & REPLACING DRIVE AXLE SHAFT SEALS, BEARINGS, AND RETAINERS

NATEF Standard(s) for Manual Drive Train & Axles:

E3-4 Inspect and replace drive axle shaft seals, bearings, and retainers.

Safety First

☐ Wear safety glasses at all times.
☐ Follow all safety rules when using common hand tools.
☐ Check Material Safety Data Sheets for chemical safety information.
☐ Follow all safety rules when using a lift.

Tools & Equipment:

- Shop light
- Seal and bearing drivers
- Slide hammer/attachments
- Common hand tools
- Lift
- Vehicle service information

PROCEDURES Refer to the vehicle service information for specifications and special procedures. Then inspect and replace the drive axle shaft seals, bearings, and retainers in the vehicle provided by your instructor.

_____ 1. Write up a repair order.

_____ 2. Make sure you follow all procedures in the vehicle service information.

_____ 3. Look up the manufacturer's recommended procedure for inspecting and replacing the drive axle shaft seals, bearings, and retainers.

Source: _____ *Page:* _____

_____ 4. With the vehicle properly raised and supported, remove the oil seal from the axle housing with the slide hammer fitted with an axle shaft bearing remover attachment.

_____ 5. Clean the axle with a rag and inspect it for rust, burrs, and rough spots.

_____ 6. Lightly lubricate the inside of the axle housing where the seal and bearing will be installed.

_____ 7. Install the bearing squarely on the axle with the bearing installation tool. **CAUTION:** The bearing is harder steel than the axle housing, so if the bearing is not exactly square, it can damage the housing.

_____ 8. Lubricate the seal with rear axle lubricant, and install it in its proper location on the axle with the axle seal installer.

_____ 9. Safely lower the vehicle to the ground.

(continued)

Copyright © Glencoe/McGraw-Hill,
a division of The McGraw-Hill Companies, Inc.

Manual Drive Train & Axles

Performance ✓ Checklist

Name _____ Date _____ Class _____

PERFORMANCE STANDARDS:
Level 4–Performs skill without supervision and adapts to problem situations.
Level 3–Performs skill satisfactorily without assistance or supervision.
Level 2–Performs skill satisfactorily, but requires assistance/supervision.
Level 1–Performs parts of skill satisfactorily, but requires considerable assistance/supervision.

Attempt (circle one): **1 2 3 4**

Comments:

_____

_____

_____

_____

PERFORMANCE LEVEL ACHIEVED: _____

_____ **1.** Safety rules and practices were followed at all times regarding this job.

_____ **2.** Tools and equipment were used properly and stored upon completion of this job.

_____ **3.** This completed job met the standards set and was done within the allotted time.

_____ **4.** No injury or damage to property occurred during this job.

_____ **5.** Upon completion of this job, the work area was cleaned correctly.

Instructor's Signature _____ Date_____

INSPECTING, TESTING & REPLACING SENSORS AND SWITCHES

NATEF Standard(s) for Manual Drive Train & Axles:
C16 Inspect, test, and replace transmission/transaxle sensors and switches.

Safety First
- ☐ Wear safety glasses at all times.
- ☐ Follow all safety rules when using common hand tools.

Tools & Equipment:
- Vehicle service information
- DVOM
- Common hand tools
- Scan tool

PROCEDURES Refer to the vehicle service information for specifications and special procedures. Then inspect, test, and replace sensors and switches on the vehicle provided by your instructor.

_____ 1. Write up a repair order.

_____ 2. Make sure you follow all procedures in the vehicle service information.

_____ 3. Inspect the sensor or switch for signs of physical deterioration, loose connection, or misalignment.

_____ 4. Test the switch or sensor according to the manufacturer's specifications using a DVOM or recommended tool.

_____ 5. Replace the switch or sensor if it does not meet specifications.

Manual Drive Train & Axles

(continued)

Performance ✓ Checklist

Name _____ Date _____ Class _____

PERFORMANCE STANDARDS:
Level 4–Performs skill without supervision and adapts to problem situations.
Level 3–Performs skill satisfactorily without assistance or supervision.
Level 2–Performs skill satisfactorily, but requires assistance/supervision.
Level 1–Performs parts of skill satisfactorily, but requires considerable assistance/supervision.

Attempt (circle one): **1 2 3 4**

Comments:

_____

_____

_____

_____

PERFORMANCE LEVEL ACHIEVED: _____

_____ 1. Safety rules and practices were followed at all times regarding this job.

_____ 2. Tools and equipment were used properly and stored upon completion of this job.

_____ 3. This completed job met the standards set and was done within the allotted time.

_____ 4. No injury or damage to property occurred during this job.

_____ 5. Upon completion of this job, the work area was cleaned correctly.

Instructor's Signature _____ Date _____

SERVICING TRANSAXLES AND DIFFERENTIALS

NATEF Standard(s) for Manual Drive Train & Axles:
C14 Remove, inspect, measure, adjust, and reinstall transaxle final drive pinion gears (spiders), shaft, side gears, side bearings, thrust washers, and case assembly.

Safety First

☐ Wear safety glasses at all times.
☐ Follow all safety rules when using common hand tools.
☐ Check Material Safety Data Sheets for chemical safety information.

Tools & Equipment:
- Specified measuring tools
- Vehicle service information
- Cleaning solvent
- Common hand tools

PROCEDURES Refer to the vehicle service information for specifications and special procedures. Then service the transaxle and differential assemblies in the vehicle provided by your instructor.

_____ 1. Write up a repair order.

_____ 2. Make sure you follow all procedures in the vehicle service information.

_____ 3. Look up the manufacturer's recommended procedure for servicing the transaxle and differential.

Source: _____ Page: _____

_____ 4. Follow all the manufacturer's safety recommendations.

_____ 5. Measure the ring gear runout, the side bearing preload, and the ring-to-pinion-gear backlash by following the manufacturer's recommended service procedure. *Note:* Ring gear backlash should always be checked whenever the transaxle and differential sections are removed from the vehicle.

_____ 6. Check the ring gear tooth contact patterns.

_____ 7. Remove, inspect, and clean the differential case assembly and housing. *Note:* The differential assembly should be cleaned and inspected whenever the transaxle is removed.

_____ 8. Remove and inspect the ring gear. Replace it only if it is damaged. *Note:* Always replace the output shaft drive gear when you replace the ring gear on a transaxle.

_____ 9. Reinstall the transaxle and differential assemblies according to the manufacturer's recommended procedure.

(continued)

Manual Drive Train & Axles

Performance ✓ Checklist

Name _____ Date _____ Class _____

PERFORMANCE STANDARDS:
Level 4–Performs skill without supervision and adapts to problem situations.
Level 3–Performs skill satisfactorily without assistance or supervision.
Level 2–Performs skill satisfactorily, but requires assistance/supervision.
Level 1–Performs parts of skill satisfactorily, but requires considerable assistance/supervision.

Attempt (circle one): 1 2 3 4

Comments:

_____

_____

_____

_____

PERFORMANCE LEVEL ACHIEVED: _____

_____ **1.** Safety rules and practices were followed at all times regarding this job.
_____ **2.** Tools and equipment were used properly and stored upon completion of this job.
_____ **3.** This completed job met the standards set and was done within the allotted time.
_____ **4.** No injury or damage to property occurred during this job.
_____ **5.** Upon completion of this job, the work area was cleaned correctly.

Instructor's Signature _____ Date_____

MEASURING & ADJUSTING TRANSAXLE ENDPLAY AND PRELOAD

NATEF Standard(s) for Manual Drive Train & Axles:
C10 Measure endplay or preload (shim or spacer selection procedure) on transmission/ transaxle shafts; perform necessary action.

Safety First
- ☐ Wear safety glasses at all times.
- ☐ Follow all safety rules when using common hand tools.
- ☐ Follow all safety rules when using a lift.

Tools & Equipment:
- Feeler gauges
- Lift
- Vehicle service information
- Dial indicator
- Bristle brush
- Depth gauge
- Common hand tools

PROCEDURES Refer to the vehicle service information for specifications and special procedures. Then measure and adjust the transaxle endplay and preload in the vehicle provided by your instructor.

_____ 1. Write up a repair order.

_____ 2. Make sure you follow all procedures in the vehicle service information.

_____ 3. Raise the vehicle and identify the transmission installed in the vehicle by the transmission identification tag or number. *Note:* You may have to clean it with the bristle brush to read it.

_____ 4. Look up the manufacturer's recommended procedure for measuring and adjusting the transaxle endplay and preload.

     Source: _____ *Page:* _____

_____ 5. Follow all the manufacturer's safety recommendations.

_____ 6. Measure and adjust the transaxle preload and endplay according to the specific manufacturer's procedure.

_____ 7. Safely lower the vehicle to the ground.

(continued)

Performance ✓ Checklist

Name _____ Date _____ Class _____

PERFORMANCE STANDARDS:
Level 4–Performs skill without supervision and adapts to problem situations.
Level 3–Performs skill satisfactorily without assistance or supervision.
Level 2–Performs skill satisfactorily, but requires assistance/supervision.
Level 1–Performs parts of skill satisfactorily, but requires considerable assistance/supervision.

Attempt (circle one): **1 2 3 4**

Comments:

_____

_____

_____

_____

PERFORMANCE LEVEL ACHIEVED: _____

_____ **1.** Safety rules and practices were followed at all times regarding this job.

_____ **2.** Tools and equipment were used properly and stored upon completion of this job.

_____ **3.** This completed job met the standards set and was done within the allotted time.

_____ **4.** No injury or damage to property occurred during this job.

_____ **5.** Upon completion of this job, the work area was cleaned correctly.

Instructor's Signature _____ Date_____

DIAGNOSING CONSTANT-VELOCITY JOINT NOISE

NATEF Standard(s) for Manual Drive Train & Axles:

D1 Diagnose constant-velocity (CV) joint noise and vibration concerns; determine necessary action.

Safety First
- ☐ Wear safety glasses at all times.
- ☐ Follow all safety rules when using common hand tools.
- ☐ Follow all safety rules when using an automotive lift or jack and jack stands.

Tools & Equipment:
- Vehicle service information
- Common hand tools
- Automotive lift or jack and jack stands

PROCEDURES Refer to the vehicle service information for specifications and special procedures. Then diagnose constant velocity joints in the vehicle provided by your instructor.

_____ 1. Write up a repair order.

_____ 2. Make sure you follow all procedures in the vehicle service information.

_____ 3. Always inspect for leaking or damaged CV joint boots. If the damaged boot is discovered before wear occurs, the boot can be replaced. If wear has occurred within the CV joint, the joint or complete axle assembly must be replaced.

_____ 4. Listen for popping or clicking noises when turning. This is the most common symptom of a worn outer Rzeppa joint. A worn outer joint usually makes no noise when driving straight. It may click when turning. The noise may be louder when backing or turning. The noise is due to excessive clearance between the balls, cage, tracks, and races. Wear of this type requires replacement of the joint or axle assembly.

_____ 5. Listen for a clunk when accelerating or decelerating. This noise is produced by a worn inner CV joint. This noise can also be caused by excess backlash in the transaxle differential gears. If the clunk is louder when backing up, it is probably due to a worn inner CV joint. Wear of this type requires replacement of the joint or axle assembly.

(continued)

Manual Drive Train & Axles

_____ 6. A bent axle shaft or worn CV joints can cause vehicle vibration. A bent axle can be observed in motion by lifting the vehicle on an automotive lift or with a jack and placed on jack stands under the lower control arms. Never run the vehicle in gear while it is on a frame contact automotive lift with the suspension hanging free. This will place the axles and CV joints at a severely abnormal angle and can cause vibration and damage. With the jack stands supporting the vehicle under the lower control arms and the wheels clear of the floor, the axles will be operating at the normal angle. Observe the axles while they're in motion for vibration or runout.

_____ 7. Other than replacement of a ruptured CV joint boot (when it is found before damage has occurred), the most cost-effective repair is to replace the entire axle assembly with a remanufactured unit.

Performance ✓ Checklist

Name _____ Date _____ Class _____

PERFORMANCE STANDARDS:
Level 4–Performs skill without supervision and adapts to problem situations.
Level 3–Performs skill satisfactorily without assistance or supervision.
Level 2–Performs skill satisfactorily, but requires assistance/supervision.
Level 1–Performs parts of skill satisfactorily, but requires considerable assistance/supervision.

Attempt (circle one): **1 2 3 4**

Comments:

_____

_____

_____

_____

PERFORMANCE LEVEL ACHIEVED: _____

_____ 1. Safety rules and practices were followed at all times regarding this job.
_____ 2. Tools and equipment were used properly and stored upon completion of this job.
_____ 3. This completed job met the standards set and was done within the allotted time.
_____ 4. No injury or damage to property occurred during this job.
_____ 5. Upon completion of this job, the work area was cleaned correctly.

Instructor's Signature _____ Date_____

INSPECTING & REPLACING MANUAL FRONT-DRIVE AXLES, JOINTS, AND BOOTS

NATEF Standard(s) for Manual Drive Train & Axles:
D4 Inspect, service, and replace shafts, yokes, boots, and CV joints.

Safety First

☐ Wear safety glasses at all times.
☐ Follow all safety rules when using common hand tools.
☐ Follow all safety rules when using a lift.
☐ Check Material Safety Data Sheets for chemical safety information.

Tools & Equipment:
- Bench vise
- Brass drift
- Lift
- Solvent
- Side cutters
- Snap-ring pliers
- Common hand tools
- Hammer
- Plastic mallet
- Vehicle service information

PROCEDURES Refer to the vehicle service information for specifications and special procedures. Then inspect and replace, if necessary, the manual front-drive axle, joint, and boot in the vehicle provided by your instructor.

_____ 1. Write up a repair order.

_____ 2. Make sure you follow all procedures in the vehicle service information.

_____ 3. Look up the manufacturer's recommended procedure and specifications for inspecting and replacing manual front-drive axle joints and boots.

Source: _____ Page:_____

_____ 4. Shift the transmission into NEUTRAL and safely raise the vehicle on the lift.

_____ 5. Examine the inner sidewalls, fenders, and suspension for grease that might have been thrown from a torn boot. Inspect all the boots for damage or wear and rotate the drive wheel so all sides of the boots are exposed for inspection.

_____ 6. If there is any sign of damage or deterioration, replace the boot.

_____ 7. Following the manufacturer's recommendations, remove the front-drive axle from the vehicle and, if possible, put it in a bench vise.

_____ 8. Cut off the clamp by using the side cutters at the small end of the boot. _Note:_ Mark the halfshaft with masking tape at the small end of the old boot for positioning a new boot on the shaft.

_____ 9. Cut the clamps off both ends of the boot. Split the boot horizontally and remove it. _Note:_ Using a clean rag, remove as much of the grease off the joint as you can. Flush the remaining grease with solvent. Examine the grease for debris and replace the joint if it is contaminated.

(continued)

_____ **10.** Place the new boot and clamp in the correct order on the halfshaft and move them out of the way.

_____ **11.** Put the joint on the shaft and pack it with grease according to the manufacturer's recommendations.

_____ **12.** Install the boot over the joint. *Note:* Make sure to position the shaft end of the boot in its grooves and the small end toward the masking tape.

_____ **13.** Install the clamp. *Note:* If the boot pleats are kinked or dimpled, gently pull them out.

_____ **14.** Look up the procedure for removing and replacing front-drive axles (either Rzeppa or Tripod) in the vehicle service information. *Note:* See your instructor for the detailed instructions to service your type of axle.

Source: _____ Page: _____

Performance ✓ Checklist

Name _____ Date _____ Class _____

PERFORMANCE STANDARDS:
Level 4–Performs skill without supervision and adapts to problem situations.
Level 3–Performs skill satisfactorily without assistance or supervision.
Level 2–Performs skill satisfactorily, but requires assistance/supervision.
Level 1–Performs parts of skill satisfactorily, but requires considerable assistance/supervision.

Attempt (circle one): **1 2 3 4**

Comments:

_____

_____

_____

_____

PERFORMANCE LEVEL ACHIEVED: _____

_____ **1.** Safety rules and practices were followed at all times regarding this job.
_____ **2.** Tools and equipment were used properly and stored upon completion of this job.
_____ **3.** This completed job met the standards set and was done within the allotted time.
_____ **4.** No injury or damage to property occurred during this job.
_____ **5.** Upon completion of this job, the work area was cleaned correctly.

Instructor's Signature _____ Date_____

REMOVING & REPLACING FRONT-WHEEL-DRIVE FRONT WHEEL BEARING

NATEF Standard(s) for Manual Drive Train & Axles:
D3 Remove and replace front wheel drive (FWD) front wheel bearing.

Safety First
- ☐ Wear safety glasses at all times.
- ☐ Follow all safety rules when using common hand tools.
- ☐ Follow all safety rules when using an automotive lift.

Tools & Equipment:
- Vehicle service information
- Axle/hub puller
- Common hand tools
- Torque wrench
- Air impact wrench
- Automotive lift

PROCEDURES Refer to the vehicle service information for specifications and special procedures. Then replace the front-wheel-drive front wheel bearing in the vehicle provided by your instructor.

_____ 1. Write up a repair order.

_____ 2. Make sure you follow all procedures in the vehicle service information.

_____ 3. Lift the vehicle and remove the wheel.

_____ 4. Remove and hang the disc brake caliper assembly and remove the brake rotor.

_____ 5. Pull the cotter pin and remove the large axle nut and washer.

_____ 6. Remove the bolts behind the axle hub flange that secure the hub/bearing assembly to the steering knuckle.

_____ 7. If the axle splines fit loosely in the hub, a hammer handle can be used to push the axle shaft from the hub. If the axle splines fit tightly in the hub, an axle/hub puller will be needed.

_____ 8. If necessary, attach an axle/hub puller to the lug studs of the hub. Turn the screw in the puller to push back on the axle while removing the hub/bearing assembly from the steering knuckle.

_____ 9. Attach a new hub/bearing assembly to the steering knuckle with the bolts previously removed and torque to specifications.

_____ 10. Install the large washer and a new axle nut to the axle and tighten the nut to pull the axle in place in the hub.

_____ 11. Re-install the brake rotor and caliper.

_____ 12. Re-install the wheel.

(continued)

Manual Drive Train & Axles

_____ **13.** Let the vehicle down, placing the wheel on the floor.

_____ **14.** Have someone firmly apply the brake pedal while you torque the axle nut to the proper specification and install a new cotter pin.

Note: Some vehicles require the removal of the steering knuckle and the use of a hydraulic press to press out the hub and bearing and to press in a new bearing.

Performance ✓ Checklist

Name _____ Date _____ Class _____

PERFORMANCE STANDARDS:
Level 4–Performs skill without supervision and adapts to problem situations.
Level 3–Performs skill satisfactorily without assistance or supervision.
Level 2–Performs skill satisfactorily, but requires assistance/supervision.
Level 1–Performs parts of skill satisfactorily, but requires considerable assistance/supervision.

Attempt (circle one): **1 2 3 4**

Comments:

_____

_____

_____

_____

PERFORMANCE LEVEL ACHIEVED: _____

_____ **1.** Safety rules and practices were followed at all times regarding this job.

_____ **2.** Tools and equipment were used properly and stored upon completion of this job.

_____ **3.** This completed job met the standards set and was done within the allotted time.

_____ **4.** No injury or damage to property occurred during this job.

_____ **5.** Upon completion of this job, the work area was cleaned correctly.

Instructor's Signature _____ Date_____

INSPECTING, ADJUSTING & REPAIRING SHIFTING CONTROLS

NATEF Standard(s) for Manual Drive Train & Axles:

F2 Inspect, adjust, and repair shifting controls (mechanical, electrical, and vacuum), bushings, mounts, levers, and brackets.

Safety First

- ☐ Wear safety glasses at all times.
- ☐ Follow all safety rules when using common hand tools.
- ☐ Follow all safety rules when using a lift.

Tools & Equipment:
- Locking pin (if required)
- Lift
- Common hand tools
- Vehicle service information

PROCEDURES Refer to the vehicle service information for specifications and special procedures. Then inspect, adjust, and repair the shifting controls in the vehicle provided by your instructor.

_____ 1. Write up a repair order.

_____ 2. Make sure you follow all procedures in the vehicle service information.

_____ 3. Safely raise the vehicle on a lift.

_____ 4. Visually check cables, linkages, and bellcranks for wear, looseness, and damage. Replace if necessary.

_____ 5. Make linkage adjustments as required. *Note:* To do this, the transmission may have to be in a particular gear or be locked using a special tool.

_____ 6. Safely lower the vehicle to the ground.

(continued)

Performance ✓ Checklist

Name _____ Date _____ Class _____

PERFORMANCE STANDARDS:
Level 4–Performs skill without supervision and adapts to problem situations.
Level 3–Performs skill satisfactorily without assistance or supervision.
Level 2–Performs skill satisfactorily, but requires assistance/supervision.
Level 1–Performs parts of skill satisfactorily, but requires considerable assistance/supervision.

Attempt (circle one): **1 2 3 4**

Comments:

_____

_____

_____

_____

PERFORMANCE LEVEL ACHIEVED: _____

_____ 1. Safety rules and practices were followed at all times regarding this job.
_____ 2. Tools and equipment were used properly and stored upon completion of this job.
_____ 3. This completed job met the standards set and was done within the allotted time.
_____ 4. No injury or damage to property occurred during this job.
_____ 5. Upon completion of this job, the work area was cleaned correctly.

Instructor's Signature _____ Date_____

REMOVING & INSTALLING TRANSFER CASES

NATEF Standard(s) for Manual Drive Train & Axles:

F3 Remove and reinstall transfer case.

Safety First

☐ Wear safety glasses at all times.

☐ Follow all safety rules when using common hand tools.

☐ Follow all safety rules when using a lift.

☐ Follow all safety rules when using a transmission jack.

☐ Check Material Safety Data Sheets for chemical safety information.

☐ Disconnect the battery to prevent shock.

Tools & Equipment:
- Lift
- Vehicle service information
- Transmission jack
- Common hand tools

PROCEDURES Refer to the vehicle service information for specifications and special procedures. Then remove and reinstall the transfer case in the vehicle provided by your instructor.

_____ 1. Write up a repair order.

_____ 2. Make sure you follow all procedures in the vehicle service information.

_____ 3. Look up the manufacturer's recommended procedure and specifications for removing and installing transfer cases.

Source: _____ _Page:_ _____

_____ 4. With the vehicle safely on a lift, remove the skid plate, if there is one.

_____ 5. Mark and remove any vacuum and electrical components that will be in the way when you remove the transfer case.

_____ 6. Remove the speedometer cable.

_____ 7. Mark the output shafts and the front and rear drivelines. _Note:_ You will use these marks later to reinstall these parts in their correct locations.

_____ 8. Disconnect the drivelines.

_____ 9. Disconnect the transfer case shift linkages from the case housing.

_____ 10. If necessary, remove exhaust components.

_____ 11. Support the transfer case on a transmission jack.

_____ 12. Remove the crossmember support, and then remove the bolts that hold the transfer case to the transmission.

(_continued_)

Manual Drive Train & Axles

_____ **13.** If there is a transmission adapter, remove the stud nuts or bolts that are holding the transfer case to the adapter.

_____ **14.** Remove the transfer case by pulling it rearward and downward from the vehicle.

_____ **15.** Place the transfer case on a bench to service.

_____ **16.** Reinstall the transfer case according to the manufacturer's recommendations after it has been serviced.

_____ **17.** Safely lower the vehicle to the ground.

Performance ✓ Checklist

Name _____ Date _____ Class _____

PERFORMANCE STANDARDS:
Level 4–Performs skill without supervision and adapts to problem situations.
Level 3–Performs skill satisfactorily without assistance or supervision.
Level 2–Performs skill satisfactorily, but requires assistance/supervision.
Level 1–Performs parts of skill satisfactorily, but requires considerable assistance/supervision.

Attempt (circle one): **1 2 3 4**

Comments:

_____

_____

_____

_____

PERFORMANCE LEVEL ACHIEVED: _____

_____ **1.** Safety rules and practices were followed at all times regarding this job.
_____ **2.** Tools and equipment were used properly and stored upon completion of this job.
_____ **3.** This completed job met the standards set and was done within the allotted time.
_____ **4.** No injury or damage to property occurred during this job.
_____ **5.** Upon completion of this job, the work area was cleaned correctly.

Instructor's Signature _____ Date _____

DISASSEMBLING & SERVICING TRANSFER CASES

NATEF Standard(s) for Manual Drive Train & Axles:

F4 Disassemble, service, and reassemble transfer case and components.

Safety First

☐ Wear safety glasses at all times.
☐ Follow all safety rules when using common hand tools.
☐ Check Material Safety Data Sheets for chemical safety information.

Tools & Equipment:

• Petroleum jelly   • Common hand tools   • Vehicle service information

---

PROCEDURES Refer to the vehicle service information for specifications and special procedures. Then disassemble, service, and reassemble the transfer case and components in the vehicle provided by your instructor.

_____ 1. Write up a repair order.

_____ 2. Make sure you follow all procedures in the vehicle service information.

_____ 3. Look up the manufacturer's recommended procedures for disassembling, inspecting, and servicing the transfer case.

   Source: _____ _Page:_ _____

_____ 4. Clean the transfer case exterior.

_____ 5. Drain the lubricant and properly dispose of it.

_____ 6. Remove all external shift controls, vacuum and electrical switches, and sensors. **Note:** If necessary, mark their position.

_____ 7. Remove all transfer case nuts and split the case halves using a screwdriver. **CAUTION:** Most transfer cases are made of aluminum. Be careful not to damage the case during disassembly.

_____ 8. Remove the drive chain, tensioner, oil slinger, shift forks, rails, and gear set. **Note:** Make sure you lay out all parts in the order of disassembly.

_____ 9. Following the vehicle service information's specifications, inspect all parts for wear or damage.

(continued)

Performance ✓ Checklist

Name _____ Date _____ Class _____

PERFORMANCE STANDARDS:
Level 4–Performs skill without supervision and adapts to problem situations.
Level 3–Performs skill satisfactorily without assistance or supervision.
Level 2–Performs skill satisfactorily, but requires assistance/supervision.
Level 1–Performs parts of skill satisfactorily, but requires considerable assistance/supervision.

Attempt (circle one): **1 2 3 4**

Comments:

_____

_____

_____

_____

PERFORMANCE LEVEL ACHIEVED: _____

_____ 1. Safety rules and practices were followed at all times regarding this job.

_____ 2. Tools and equipment were used properly and stored upon completion of this job.

_____ 3. This completed job met the standards set and was done within the allotted time.

_____ 4. No injury or damage to property occurred during this job.

_____ 5. Upon completion of this job, the work area was cleaned correctly.

Instructor's Signature _____ Date _____

INSPECTING FRONT-WHEEL BEARINGS AND MANUAL LOCKING HUBS

NATEF Standard(s) for Manual Drive Train & Axles:
F5 Inspect front-wheel bearings and locking hubs; perform necessary action.

Safety First
- ☐ Wear safety glasses at all times.
- ☐ Follow all safety rules when using common hand tools.
- ☐ Follow all safety rules when using a lift.
- ☐ Check Material Safety Data Sheets for chemical safety information.

Tools & Equipment:
- Technician's wire
- Lift
- Vehicle service information
- Axle nut wrenches
- Snap-ring pliers
- Solvent
- Bearing driver
- Common hand tools

PROCEDURES Refer to the vehicle service information for specifications and special procedures. Then inspect the front-wheel bearings and manual locking hubs in the vehicle provided by your instructor.

_____ 1. Write up a repair order.

_____ 2. Make sure you follow all procedures in the vehicle service information.

_____ 3. Safely raise the vehicle on a lift.

_____ 4. Remove the front wheels.

_____ 5. Remove the brake calipers and wire them out of the way to prevent damage to the brake hoses.

_____ 6. Remove the brake rotors.

_____ 7. Disassemble the locking hubs. *Note:* Locking hubs contain snap rings, springs, spacers, and other small components. Lay all parts out in order of disassembly.

_____ 8. Clean the bearings with solvent and inspect for wear, heat discoloration, or impact damage. If necessary, replace the bearings.

_____ 9. Pack the bearings with fresh grease and reassemble using vehicle service information procedures.

Source: _____ *Page:* _____

_____ 10. Check the hubs for proper operation. Make sure that they lock and unlock.

_____ 11. Replace the front wheels.

_____ 12. Safely lower the vehicle to the ground.

(continued)

Manual Drive Train & Axles

Performance ✓ Checklist

Name _____ Date _____ Class _____

PERFORMANCE STANDARDS:
Level 4–Performs skill without supervision and adapts to problem situations.
Level 3–Performs skill satisfactorily without assistance or supervision.
Level 2–Performs skill satisfactorily, but requires assistance/supervision.
Level 1–Performs parts of skill satisfactorily, but requires considerable assistance/supervision.

Attempt (circle one): **1 2 3 4**

Comments:

_____

_____

_____

_____

PERFORMANCE LEVEL ACHIEVED: _____

_____ 1. Safety rules and practices were followed at all times regarding this job.
_____ 2. Tools and equipment were used properly and stored upon completion of this job.
_____ 3. This completed job met the standards set and was done within the allotted time.
_____ 4. No injury or damage to property occurred during this job.
_____ 5. Upon completion of this job, the work area was cleaned correctly.

Instructor's Signature _____ Date _____